Exam Ref 70-398 Planning for and Managing Devices in the Enterprise

Brian Svidergol
Robert Clements
Charles Pluta

PUBLISHED BY
Microsoft Press
A division of Microsoft Corporation
One Microsoft Way
Redmond, Washington 98052-6399

Library of Congress Control Number: 2015956896
ISBN: 978-1-5093-0221-5

Printed and bound in the United States of America.

First Printing

Microsoft Press books are available through booksellers and distributors worldwide. If you need support related to this book, email Microsoft Press Support at mspinput@microsoft.com. Please tell us what you think of this book at http://aka.ms/tellpress.

This book is provided "as-is" and expresses the author's views and opinions. The views, opinions and information expressed in this book, including URL and other Internet website references, may change without notice.

Some examples depicted herein are provided for illustration only and are fictitious. No real association or connection is intended or should be inferred.

Acquisitions Editor: Karen Szall
Developmental Editor: Karen Szall
Editorial Production: Troy Mott, Ellie Volckhausen
Technical Reviewer: Charlie Russel; Technical Review services provided by Content Master, a member of CM Group, Ltd.
Copyeditor: Tara McGoldrick-Walsh
Indexer: Julie Grady
Cover: Twist Creative • Seattle

Contents at a glance

Contents

What do you think of this book? We want to hear from you!

Microsoft is interested in hearing your feedback so we can continually improve our
books and learning resources for you. To participate in a brief online survey, please visit:

www.microsoft.com/learning/booksurvey/

Introduction

For this book, we focused on two primary objectives: write about exam skills with some real-world information throughout and approach each exam skill from an exam item writer mentality. We asked ourselves, if we were writing exam questions for this topic, which questions would we write? How would we test somebody's knowledge of a topic? We thought about how each of the concepts applied in the real-world, in your day-to-day job tasks. Then, we incorporated that information in the book to make it easily consumable. We think we ended up with a good balance of information and exam preparation material. Good luck on the exam!

This book covers every skill identified as measured on the exam on the exam web page, but it does not cover every exam question. Only the Microsoft exam team has access to the exam questions themselves and Microsoft regularly adds new questions to the exam, making it impossible to cover specific questions. You should consider this book a supplement to your relevant real-world experience and other study materials. If you encounter a topic in this book that you do not feel completely comfortable with, use the links you'll find in text to find more information and take the time to research and study the topic. Great information is available on MSDN, TechNet, and in blogs and forums.

Microsoft certifications

Microsoft certifications distinguish you by proving your command of a broad set of skills and experience with current Microsoft products and technologies. The exams and corresponding certifications are developed to validate your mastery of critical competencies as you design and develop, or implement and support, solutions with Microsoft products and technologies both on-premises and in the cloud. Certification brings a variety of benefits to the individual and to employers and organizations.

> **MORE INFO ALL MICROSOFT CERTIFICATIONS**
>
> For information about Microsoft certifications, including a full list of available certifications, go to *http://www.microsoft.com/learning*.

Acknowledgments

We would like to thank Karen Szall and the rest of the team at Microsoft Press for working with us on this project. Their stellar support throughout the process improved the quality of the book and helped ensure a smooth running project! We'd also like to thank Charlie Russel for combing through the book as the technical reviewer. He helped track down hard to find issues and helped bring additional consistency to the book.

Brian Svidergol would like to thank his wife Lindsay, his son Jack, and his daughter Leah for making life such a joy! He would also like to thank Bob Clements and Charles Pluta for joining him on this project. Bob and Charles brought expertise and extensive real-world experience to help create a well-rounded book.

Bob Clements would like to thank his wife Diane for her continued support and frequent words of encouragement. He would like to thank his daughter Abigail for her contagious laughter and joyful spirit. He would like to thank his magnificent son Samuel, who arrived just a few months prior to the start of this project. Finally, he would like to say a big thank you to Brian and Charles for their professionalism, technical knowledge, and commitment to this project.

Charles Pluta would like to thank his wife Jennifer for her amazing support every day. He would also like to thank Brian and Bob for their hard work and dedication to making this book a success.

Free ebooks from Microsoft Press

From technical overviews to in-depth information on special topics, the free ebooks from Microsoft Press cover a wide range of topics. These ebooks are available in PDF, EPUB, and Mobi for Kindle formats, ready for you to download at:

http://aka.ms/mspressfree

Check back often to see what is new!

Microsoft Virtual Academy

Build your knowledge of Microsoft technologies with free expert-led online training from Microsoft Virtual Academy (MVA). MVA offers a comprehensive library of videos, live events, and more to help you learn the latest technologies and prepare for certification exams. You'll find what you need here:

http://www.microsoftvirtualacademy.com

Errata, updates, & book support

We've made every effort to ensure the accuracy of this book and its companion content. You can access updates to this book—in the form of a list of submitted errata and their related corrections—at:

http://aka.ms/ER398/errata

If you discover an error that is not already listed, please submit it to us at the same page.

If you need additional support, email Microsoft Press Book Support at *mspinput@microsoft.com*.

Please note that product support for Microsoft software and hardware is not offered through the previous addresses. For help with Microsoft software or hardware, go to *http://support.microsoft.com*.

We want to hear from you

At Microsoft Press, your satisfaction is our top priority, and your feedback our most valuable asset. Please tell us what you think of this book at:

http://aka.ms/tellpress

The survey is short, and we read every one of your comments and ideas. Thanks in advance for your input!

Stay in touch

Let's keep the conversation going! We're on Twitter: *http://twitter.com/MicrosoftPress*.

Important: How to use this book to study for the exam

Certification exams validate your on-the-job experience and product knowledge. To gauge your readiness to take an exam, use this Exam Ref to help you check your understanding of the skills tested by the exam. Determine the topics you know well and the areas in which you need more experience. To help you refresh your skills in specific areas, we have also provided "Need more review?" pointers, which direct you to more in-depth information outside the book.

The Exam Ref is not a substitute for hands-on experience. This book is not designed to teach you new skills.

We recommend that you round out your exam preparation by using a combination of available study materials and courses. Learn more about available classroom training at *http://www.microsoft.com/learning*. Microsoft Official Practice Tests are available for many exams at *http://aka.ms/practicetests*. You can also find free online courses and live events from Microsoft Virtual Academy at *http://www.microsoftvirtualacademy.com*.

This book is organized by the "Skills measured" list published for the exam. The "Skills measured" list for each exam is available on the Microsoft Learning website: *http://aka.ms/examlist*.

Note that this Exam Ref is based on this publicly available information and the author's experience. To safeguard the integrity of the exam, authors do not have access to the exam questions.

CHAPTER 1

Design for cloud/hybrid identity

The public cloud is growing. More cloud providers are offering services. New services are being offered. And the capabilities of cloud services are ever expanding. As an administrator, you need to be familiar with the different cloud offerings and know how to integrate them with your on-premises environment to create a seamless hybrid environment. In a hybrid environment, users might have an identity in your on-premises environment, an identity in your public cloud environment, or a single identity that enables them to authenticate across your on-premises environment and your public cloud environment.

> **IMPORTANT**
>
> ### Have you read page xv?
>
> It contains valuable information regarding the skills you need to pass the exam.

The 70-398 exam focuses on planning and managing devices in an enterprise environment. But, today, most enterprise environments are using the cloud in some capacities and often are integrated closely with the cloud. Thus, you need to be comfortable in a cloud environment, especially a hybrid cloud environment. This chapter covers many Microsoft Azure Active Directory (Azure AD) features and tools. Azure's ongoing development includes portal enhancements, new features, and updated features. Thus, some of the technologies described in this chapter are currently in preview, which is an Azure mode that enables users to test-drive features before they become generally available and supported in a production environment.

Skills covered in this chapter:

- Plan for Azure Active Directory identities
- Design for Active Directory synchronization with Azure AD Connect

Skill 1.1: Plan for Azure Active Directory identities

Many administrators have experience working with identities, especially on-premises identities, through technologies such as Active Directory Domain Services (AD DS). But relying on just AD DS for identity and access management becomes difficult when your organization begins using the public cloud to host applications and services. To address the limitations of on-premises AD DS for public cloud and hybrid cloud environments, Azure AD was introduced. With Azure AD, your identities live in the cloud, alongside your cloud applications and services, either independently or as part of an integration with an on-premises AD DS environment. Along with the identities, premium features expand the capabilities of Azure AD.

> **This section covers how to design for:**
> - Azure AD identities
> - Active Directory integration
> - Azure Multi-Factor Authentication (Azure MFA)
> - User self-service from the Azure Access Panel
> - Azure AD reporting
> - Company branding
> - Azure AD Premium features

Design Azure AD identities

Microsoft Azure AD has three defined identities you can use. The use of *identity* in this way refers to a user account that you can use to authenticate to applications and services. These identity types are often referred to as integration scenarios in Microsoft documentation. The three identities are:

- **Cloud identity**   A cloud identity is any user account that you have created in Azure AD (often for a cloud-based application). For the purposes of this exam, a cloud identity is an Azure AD user account. You create and maintain these accounts in the cloud. Cloud identities are best suited for small organizations because when you manually create and maintain user accounts your administrative overhead increases. Large enterprise environments have a large number of users and should choose one of the other options to reduce the administrative overhead of managing identities.

- **Synced identity**   A synced identity is a user account in an on-premises AD DS environment that is synced to Azure AD. The sync process requires a directory synchronization application such as Azure Active Directory Connect (Azure AD Connect). It is most common to sync users and password hashes. This allows Azure AD to authenticate users for cloud-based applications while AD DS authenticates users for on-premises applications.

- **Federated identity** A federated identity is an identity that is linked to an on-premises AD DS user account. Federated identities require a federation trust, which is an agreement between two organizations (such as your company and Microsoft) that establishes a federation where on-premises AD DS/Active Directory Federation Services (AD FS) establishes a federation trust with Microsoft Azure. When a user authenticates to an Azure-based resource, the authentication is routed back to the on-premises AD DS environment for handling.

From a design perspective, you need to consider several factors before you choose which identity is best suited for your environment. Compare the three identities across the key considerations in Table 1-1, which shows how the three identities rate from low to high in terms of administrative overhead, security, and complexity, and also rates overall user experience.

TABLE 1-1 Cloud identities and their distinguishing characteristics

Identity	Administrative overhead	Security	Complexity	User experience
Cloud identity	High	Medium	Low	Poor
Synced identity	Low	Medium-High	Medium	Good
Federated identity	Medium	High	High	Best

✔ **Quick check**
- You need to choose an identity that maximizes the user experience while minimizing administrative overhead and reducing complexity. Which identity should you choose?

Quick check answer
- You should choose a synced identity. It offers a good user experience, low administrative overhead, and medium complexity. When compared to a cloud identity, a synced identity rates favorably in two of the three categories in Table 1-1. When compared to a federated identity, a synced identity rates favorably in two of the three categories.

The rest of this section walks through the identities and includes diagrams to show which components are used in each identity scenario and how authentication works.

Cloud identity

You can use the cloud identity in a couple of different ways. First, you can use a cloud identity as the only identity a user has, as shown in Figure 1-1. This is often best suited for companies that solely use the cloud for all of their application and service needs. Every time users access an application or a service in the cloud, they enter their cloud identity username and password and authentication is performed in Azure AD. Figure 1-1 is a diagram of a simple cloud identity environment.

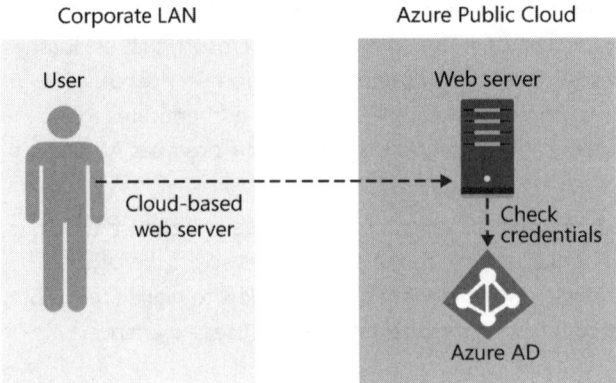

FIGURE 1-1 A simple cloud identity environment, where users always authenticate to Azure AD

For larger companies and those that have an internal AD DS environment, you can use a cloud identity in addition to the on-premises identity. For such cases, users who access on-premises applications can authenticate with their AD DS credentials. When they access cloud-based applications, they authenticate with their Azure AD credentials. Figure 1-2 is a diagram of a cloud identity environment with AD DS being used on-premises.

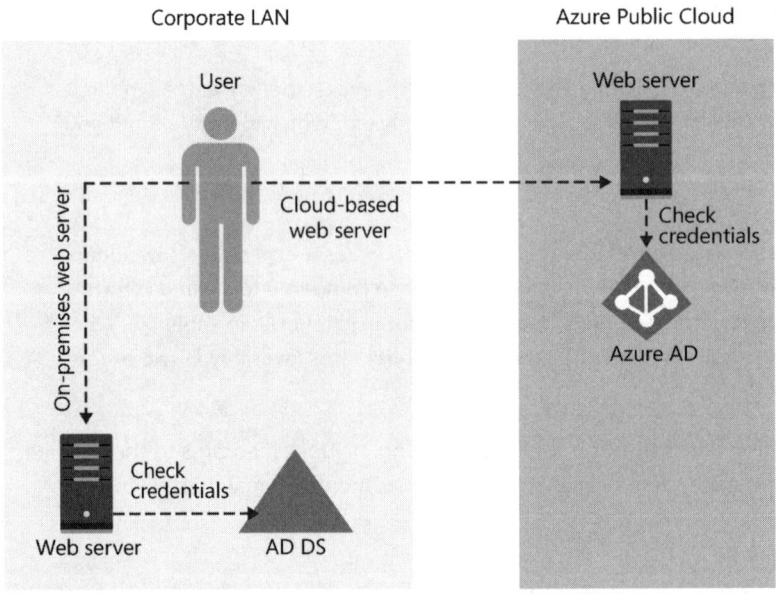

FIGURE 1-2 A cloud identity environment, where a user has an AD DS user account and a separate Azure AD user account

The primary downside to using a cloud identity when you already have AD DS on-premises is that users have two user accounts. Each user account has its own password. Password policies might be different. Users have to deal with having two separate accounts (adhere to the password policies, update passwords based on maximum password age, and remember which account to use for specific scenarios). Often, the thought of having multiple accounts is enough to cause you to look at a synced identity instead, which you can consider when you review it in the next section.

Synced identity

A synced identity environment provides users with the benefit of using a single username and password to authenticate to applications, whether those applications are on-premises or in the cloud. You can accomplish this by installing and configuring Microsoft Azure Active Directory Connect (Azure AD Connect). Azure AD Connect synchronizes from an on-premises AD DS environment to Azure AD. In a synced identity environment, password hash syncing is the optional setting that is the key because it provides users with the benefit of using only one username and password.

Figure 1-3 shows a synced identity environment.

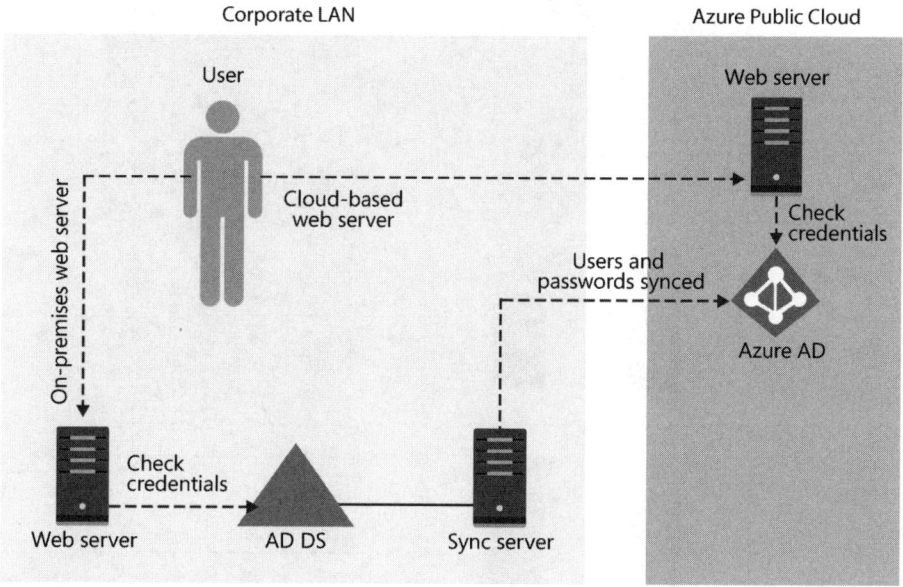

FIGURE 1-3 A synced identity environment, where a user uses a single username and password across on-premises and cloud-based applications

For companies that have AD DS on-premises, a synced identity environment provides a better user experience compared to a cloud identity environment. The downside is that authentication takes place in two different places. That complicates the auditing process and makes troubleshooting a bit more complex. Additionally, if a user is terminated and is disabled in the on-premises AD DS environment, it might take up to three hours before the user account is disabled in Azure AD. The three-hour delay is based on the default sync time of every three hours. You can run a sync on demand or configure your sync to occur more often to reduce the risk of disabled users still having some access through Azure AD after disablement.

Federated identity

A federated identity environment is an environment where all authentication is handled by the on-premises AD DS environment. It is the most complex environment because it requires the most components and requires the most planning. Figure 1-4 is a diagram of a federated identity environment.

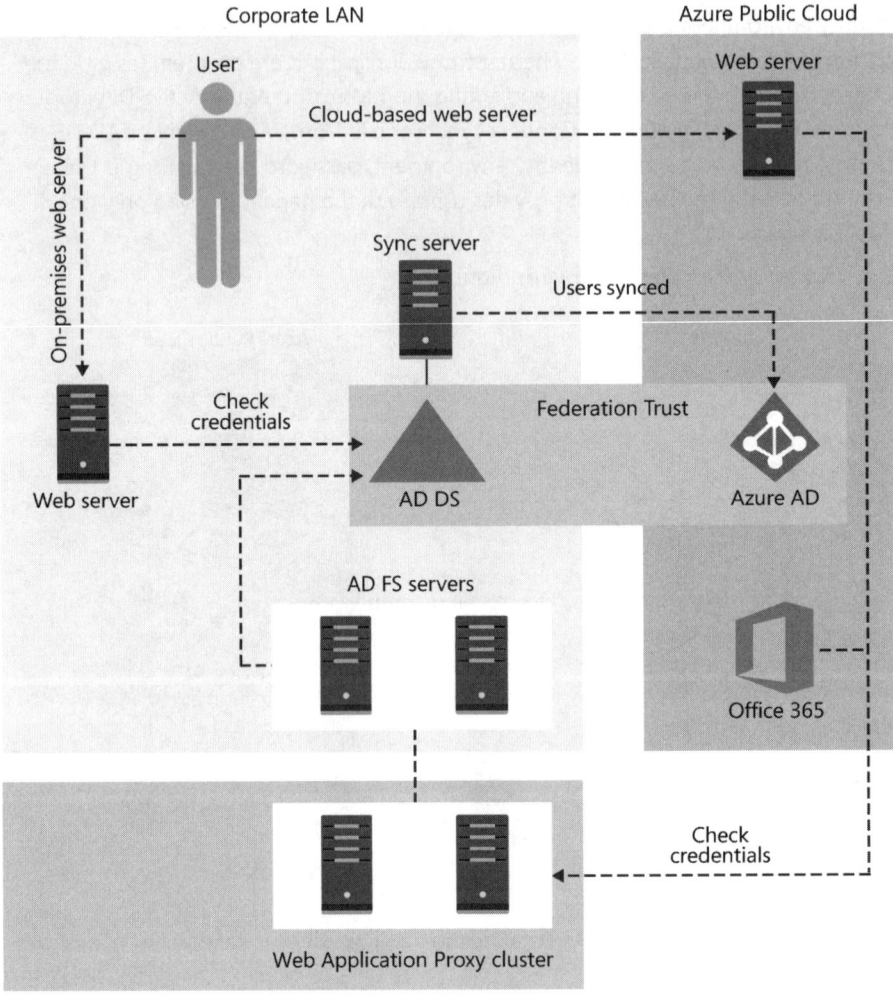

FIGURE 1-4 A federated identity environment, where the on-premises AD DS environment handles all authentication

In the federated identity environment shown in Figure 1-4, a company has an internal environment with AD DS, AD FS, Web Application Proxy servers, a web server, and a sync server. In Microsoft Azure, the company subscribes to Office 365 and has a web server in an Azure VM. Users are synced from AD DS to Azure AD but passwords are not synced. When a user authenticates to a resource in Azure, credentials are validated by the internal AD DS environment (first through the Web Application Proxy cluster, then through the AD FS servers, then to AD DS). For companies that have a security or compliance policies that mandate that all authentication is handled by their on-premises AD DS environment, you should consider a federated identity environment. Hybrid environments, where a company has some resources on-premises and some resources in the cloud, are also good candidates for a federated identity environment. One key security benefit of using federated identities is that you can immediately disable a user and all access is immediately revoked. The primary downside of using federated identities is that the overall environment is complex and requires more cost and administrative overhead.

From an exam perspective, you need to be able to distinguish the characteristics between the identities. You might be presented with a scenario and a single fact in the scenario might disqualify one design option. For example, a scenario might be presented where a company does not have a perimeter network and they do not allow direct Internet communication between computers on the local area network (LAN) and computers on the Internet. In that case, identities that require syncing or federation would be disqualified, because those solutions require a sync through the Internet. Administrative overhead is a common trait that you need to distinguish between the identity solutions. In a scenario where a company's IT department is overworked and understaffed, introducing a federated identity might not be optimal without considering the impact on the IT department resources.

Take a look at one more scenario. A company is planning to subscribe to several cloud-based applications. The company wants to maximize the user experience. The scenario might end there and you might be asked to choose one of the listed solutions. Providing single sign-on is probably the answer. But the solution might not present that information or say that directly. It might say something about using federated identities or something similar. Or, the scenario could offer some constraints such as reducing the complexity while also maximizing the user experience (seemingly conflicting goals). The answer might be to use synced identities. By knowing the pros and cons of the identity solution, you are prepared to answer scenario questions such as these on the exam.

Each of the identity solutions has some prerequisites. While some of that information was mentioned at a high level earlier in this section, the following list documents all of the high-level prerequisites for the solutions.

- **Cloud identity** The cloud identity has the least amount of prerequisites. Thus, it is often the quickest and easiest to use, especially for initial testing and proof of concept (PoC) work. The only prerequisite is having a Microsoft Azure subscription, Azure AD, and an administrative account to create and manage identities.

- **Synced identity**   You need to meet the following prerequisites to use synced identities:

  - On-premises AD DS environment. This is the source of the sync.

  - Synchronization. Azure AD Connect is the free software that Microsoft makes to handle syncing an on-premises AD DS environment with Azure AD. You need to have it running and syncing with Azure AD.

  - Microsoft Azure subscription, Azure AD, and administrative account. While this can often be assumed, watch for scenarios that don't call it out and answer choices that call out purchasing a subscription.

- **Federated identity**   You need to have the following prerequisites to use federated identities:

  - On-premises AD DS environment. This is the service that authenticates users.

  - Synchronization. Azure AD Connect is the free software that Microsoft makes to handle syncing an on-premises AD DS environment with Azure AD. You need to have it running and syncing with Azure AD.

  - AD FS servers on the LAN. These are the servers that help handle the authentication. The proxy servers communicate with these servers. By having proxy servers in the perimeter network, you avoid having to expose your AD FS servers to the Internet.

  - Web Application Proxy in the perimeter network (optional but a good practice). Beware of answers that call out AD FS proxy servers in the perimeter network, especially if one answer calls out AD FS proxy server and another calls out Web Application Proxy. The Web Application Proxy is the new name for an AD FS proxy. The AD FS proxy technology is still valid, but for older server operating systems such as Windows 2012 and older. The Web Application Proxy was first introduced with Windows Server 2012 R2.

  - Microsoft Azure subscription, Azure AD, and administrative account.

  - Federation trust with Microsoft Federation Gateway (MFG).

Active Directory integration

Integrating your on-premises Active Directory environment with Azure enables you to use synced identities and federated identities. In addition, you can synchronize users and enable several other features such as password synchronization, password writeback, and more.

Review the reasons why you would integrate your on-premises AD DS environment with Microsoft Azure:

- **You want to provide single sign-on (SSO) for users.**   Full SSO is achieved through the use of on-premises AD DS, AD FS, and federation. On the exam, pay close attention to the details so you know whether a question is referring to federation or synced identities.

- **You want users to use the same username and password for on-premises and cloud applications.** This is another form of SSO. While users use the same username and password, the user accounts are actually separate – one in AD DS and one in Azure AD. They are synced. Users sometimes can't distinguish between this and federation-based SSO because they use the same username and password.

- **You want all cloud-based authentication to be handled by your on-premises AD DS.** Because you reviewed the federated identity earlier in this chapter, you already know how on-premises AD DS handles the authentication. For the exam, an answer could be Integrate Active Directory with Azure AD and never mention the type of identity being used. Thus, you need to know the benefits of integration, outside of just the identity solution.

- **You want to simplify the provisioning process.** When new employees start at a company, the process to get all of the IT systems ready for them is often called "onboarding". When employees leave a company, the process to remove them from the IT systems is often called "off boarding". As companies begin to use cloud-based applications, onboarding and off boarding becomes more complex. Automating the processes becomes more challenging. And this is often the point at which administrators begin realizing that they need a simpler way to handle it. Integrating AD DS with Azure AD is the start of that simplification because you can reduce the number of identities for a user, reduce the amount of applications that you have to configure, and reduce the administrative time required to maintain all of the systems and processes.

Besides knowing about the benefits of integration, you should also understand the differences between the versions of Azure AD and the features that you can use for the Azure AD you have selected. Azure AD offers three versions: Free, Basic, and Premium.

EXAM TIP

For the exam, you need to be able to distinguish between Azure AD Free, Azure AD Basic, and Azure AD Premium along with the features that each offers.

For this section, we'll focus on the integration features. We'll also call out differences between the versions in other sections, where applicable.

In the following section, the three Azure AD versions are shown along with the features that are supported for each.

- **Features supported by all three versions** With all three versions, you get directory as a service, user and group management using the UI or Windows PowerShell cmdlets, device registration, Access Panel portal for SSO-based user access to SaaS and custom applications, self-service password change for cloud users, Azure AD Connect, standard security reports, B2B collaboration (cross-organization collaboration).

- **Features supported by Basic and Premium only**   With Basic or Premium, you get customization of company logo and colors to the Sign In and Access Panel pages, self-service password reset for cloud users, Application Proxy: Secure Remote Access and SSO to on-premises web applications, group-based application access management and provisioning, high availability SLA uptime (99.9 percent).

- **Features supported only by Premium**   With Premium, you get advanced application usage reporting, self-service group management for cloud users, self-service password reset with on-premises writeback, Microsoft Identity Manager (MIM) user licenses–for on-premises identity and access management, advanced anomaly security reports machine learning-based, cloud app discovery, Multi-Factor Authentication service for cloud users, Multi-Factor Authentication server for on-premises users, Azure Active Directory Connect Health to monitor the health of on-premises Active Directory infrastructure and getting usage statistics.

Office 365 licenses enable some features too, but that is beyond the scope of this exam and book.

> *NEED MORE REVIEW?*   **THE COMPLETE LIST OF FEATURES IN VERSIONS OF AZURE AD**
>
> Microsoft maintains a complete list of Azure AD features, which also details the supported features for each Azure AD version. Additionally, Microsoft documents which Office 365 features are available under Office 365 licensing. See *https://azure.microsoft.com/ documentation/articles/active-directory-editions/.*

Azure Multi-Factor Authentication

On October 4, 2012, Microsoft announced that it had purchased a company named Phone-Factor, which specializes in two-factor authentication. Ultimately, the purchase enabled Microsoft to offer Azure MFA, a service that enables you to require multiple authentication factors when users authenticate to an application or service. Figure 1-5 shows the high-level process flow for Azure MFA.

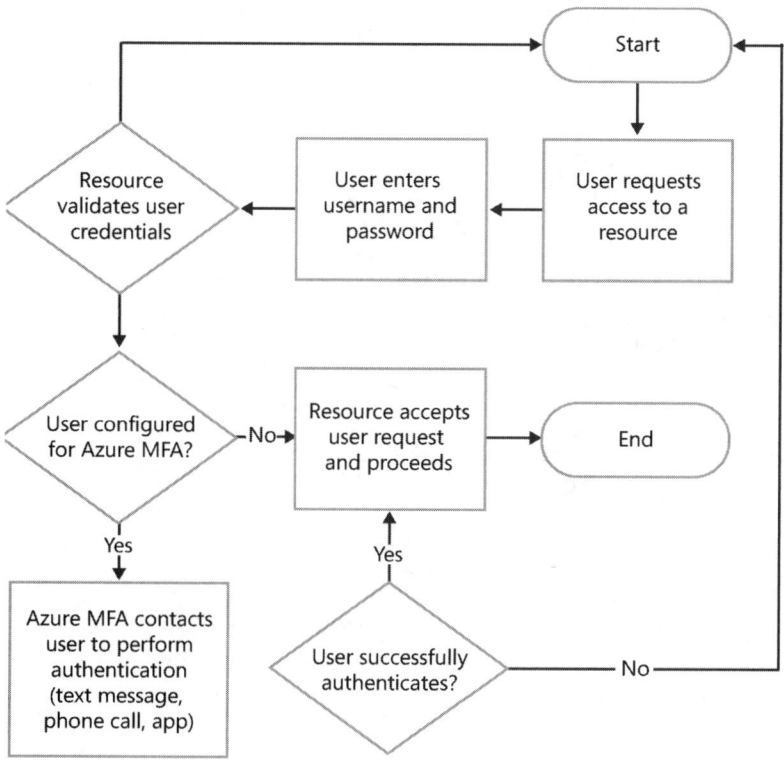

FIGURE 1-5 A flowchart showing the high-level Azure MFA process

You need to know a couple of key points about the process flow:

- The multi-factor authentication process starts after a user successfully authenticates (most often by username/password).

- The multi-factor authentication is only initiated for users that are configured for Azure MFA.

Azure MFA is a cloud service run by Microsoft in the Azure public cloud. In the simplest configuration, you just enable users to use Azure MFA. In a more complex configuration, you configure on-premises Azure MFA servers and provide multi-factor authentication for your Azure resources and your on-premises resources. Figure 1-6 shows how the service works in a common scenario.

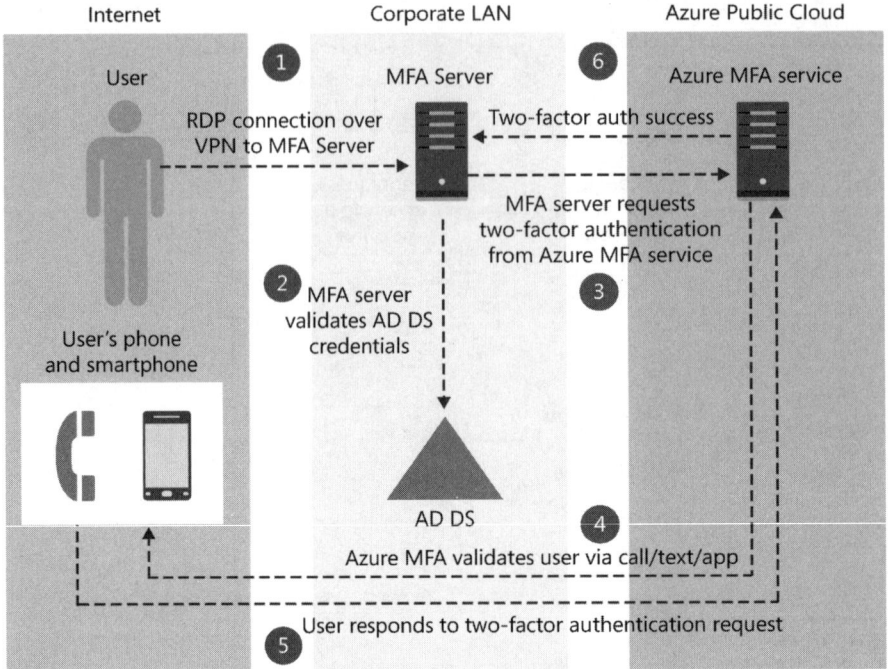

FIGURE 1-6 A user on the Internet connecting to the corporate LAN and being authenticated by Azure MFA

The following steps describe the steps shown in Figure 1-6.

1. In Step 1, a user is connected to the corporate network by using a VPN. The user initiates a connection to the MFA server by using Remote Desktop Connection. The MFA server is configured to perform multi-factor authentication on terminal services connections.

2. In Step 2, the MFA server validates the user's AD DS credentials with the on-premises AD DS domain. If successful, then the multi-factor authentication tasks begin. If unsuccessful, the user might authenticate again until successful or until the connection is ended.

3. In Step 3, the MFA server communicates with the Azure MFA service over TCP port 443 and requests that the service perform the two-factor authentication.

4. In Step 4, the Azure MFA service performs the two-factor authentication process based on the user's configuration. For example, if the user is configured for a phone call, then Azure MFA calls the user's configured phone number.

5. In Step 5, the user receives notification (phone call, text message, or app notification) that they need to confirm the authentication request. The user must respond to the notification, which varies based on the configuration. For example, by default a phone call from the Azure MFA service requires that the user press the pound (#) key to confirm the authentication.

6. In Step 6, after a successful response from the user in Step 5, the Azure MFA service notifies the on-premises MFA server that the authentication was successful. The MFA server confirms the authentication and the Remote Desktop Connection session begins.

An authentication factor is a form of authentication. One factor is a username and password. Another factor is an app on a smartphone. Azure MFA supports multiple authentication factors, as shown in a typical user's configuration page in Figure 1-7.

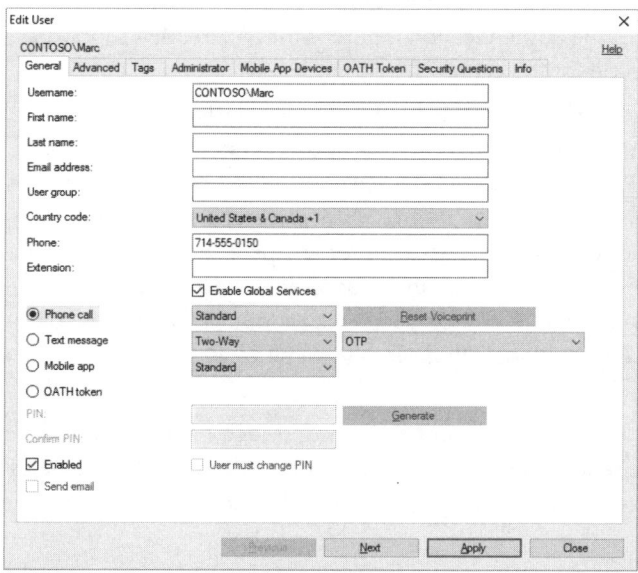

FIGURE 1-7 On a user's configuration page, you can specify the second authentication factor–a phone call, text message, mobile app, or OATH token

A username and password aren't shown as one of the factors when configuring Azure MFA because the username and password are authenticated prior to Azure MFA's second factor authentication. The factors that Azure MFA supports are:

- **Phone call** You can have Azure MFA call any phone number, including a land line and smartphone. After you answer the call, you press the pound (#) key to complete the authentication. Optionally, you can configure Azure MFA to require a PIN instead of the pound key.

- **Text message** You can have a text message sent to a smartphone. Upon receiving the message, you have to send a text message back with the code provided in the received message. That completes the authentication.

- **Smartphone app** Microsoft provides a free smartphone app named Azure Authenticator. Upon authentication, the app notifies you that an authentication request needs to be verified. You tap the Verify button in the app and that completes the multi-factor authentication process.

- **OATH token**   An OATH token is an access token issued by a third-party that provides another authentication factor in a multi-factor authentication scenario. The term OATH and OAUTH mean the same thing. While OAUTH is the most popular term for this open authentication solution, Azure's MFA uses the term OATH in spots in the application. To use OATH tokens with Azure MFA, you need to add the tokens to the MFA server or import them to the MFA server. Then, each token is associated with a user. To use them, each user must be configured to use OATH.

Now that you have an overview of the authentication methods that Azure MFA supports, take a look at some import information for the exam:

- **App passwords may be required**   Some versions of Outlook, Lync, and some other non-browser based apps that require authentication do not support native multi-factor authentication. For those apps, you must generate and use app passwords, which are passwords generated for each app that take the place of a second factor authentication method. You can generate a password for an app or for a device. When you generate a password for a device, it can be used for all apps on that device that do not support Azure MFA. As an administrator, you can clear out all app passwords for a user, which is useful when troubleshooting. Note the limit of 40 app passwords per user.

- **Self-service setup**   Users can choose their preferred second-factor authentication method–phone call, text message, or smartphone app. And, they can perform some of their own account management from the Azure Access Panel, which is a topic covered in the next section.

- **Azure MFA requires Azure AD Premium or the Enterprise Mobility Suite**   You must have Azure AD Premium or the Azure Mobility Suite to take advantage of Azure MFA for cloud users or on-premises users. The only exception is users with an Office 365 subscription. Office 365 comes with Azure MFA but only for Office 365 services (whereas you can use Azure MFA, when it comes with Azure AD Premium, across a wide variety of services in the cloud and on-premises).

- **You can choose between two usage models**   Azure MFA offers two usage models. It is important to know about these because generally, you choose a model in the architecture and planning phase of a project. The two usage models are:
  - **Per authentication**   In this model, you are charged for every ten authentications that occur over Azure MFA. This model makes sense if your users are authenticating occasionally and you have a large number of users. The breakeven point, in the current pricing model, is 10 authentications per month. If your users authenticate with Azure MFA 10 or less times per month, then the per authentication usage model is the most cost effective.
  - **Per enabled user**   In this model, you are charged a flat rate for every user enabled for Azure MFA, no matter how many authentications they have over Azure MFA. This model makes sense if your users perform authentications often and the total number of users is not large. The breakeven point, in the current pricing model, is

10 authentications per month. If users authenticate more than that, then you can save money by choosing the per-enabled usage model.

- **A subset of Azure MFA is available just for administrators, for free**   If you haven't provisioned Azure MFA for directory users, a subset of Azure MFA is available for free for Azure Global Administrators.

> ✔ **Quick check**
>
> - You just implemented Azure MFA. Users are reporting that they cannot get into the native email on their Apple iPhones due to an authentication issue. What should you do?
>
> **Quick check answer**
>
> - Some non-browser applications do not support Azure MFA directly. To fix the issue, you need to advise the users to generate an app password and use the app password when they run configure Apple Mail. App passwords can be generated per app or per device (where one password can be used for all apps that do not support Azure MFA directly). Users can generate an app password from the Office 365 portal (for use with Office 365 related apps), from the Access Panel, or from the Azure portal.

User self-service from the Azure Access Panel

Users can go to the Azure Access Panel at *https://myapps.microsoft.com* to access applications and perform self-service tasks such as resetting their password and updating their contact information. Before users can do that, however, you need to enable the option to allow users to reset their own password and choose the methods that they can use. Figure 1-8 shows the portion of the Azure AD Configure tab with the applicable settings.

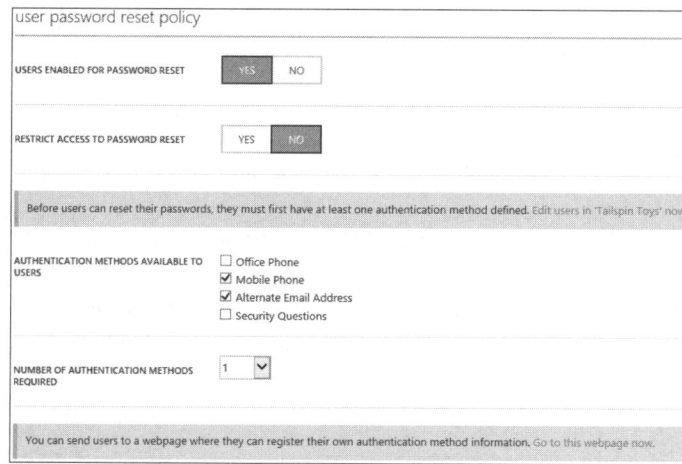

FIGURE 1-8 The self-service user password reset settings in Azure AD dictate if users can reset their own password

After users are configured for multi-factor authentication, you will occasionally perform additional management tasks. One common task is to force users to provide their updated contact methods for multi-factor authentication. This is handy when users have updated phone numbers because they can perform the updates themselves, without having to involve IT. When you view your Azure AD from the Azure portal, there is a link to manage multi-factor auth. A separate web page opens where you can manage users (enable/disable) and manage user settings (one of which requires users to verify their contact information the next time that they sign in to an Azure portal). The setting to require users to verify their contact information is shown in Figure 1-9.

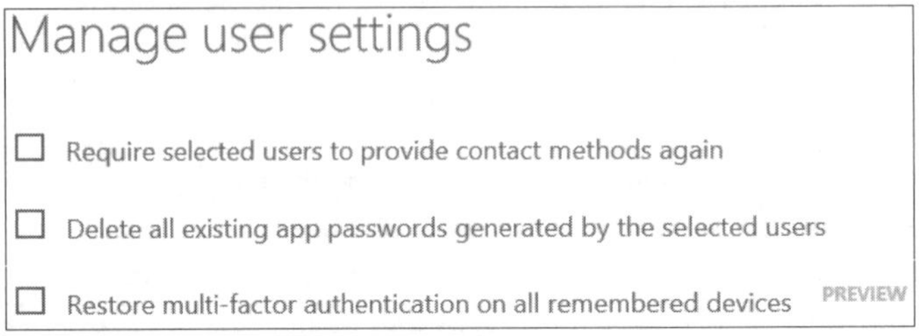

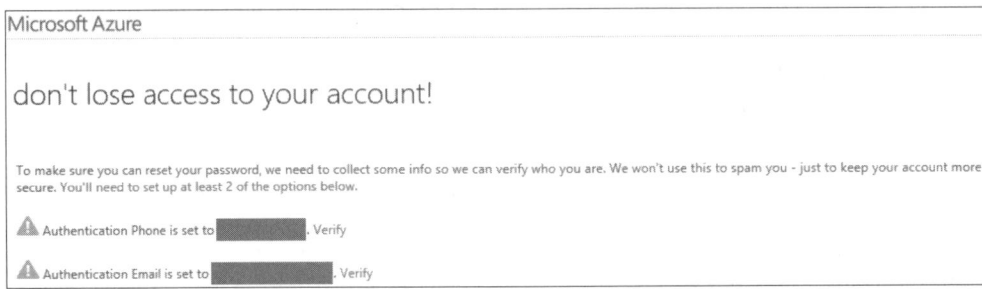

FIGURE 1-9 You can enable users to update their contact information by selecting the Require Selected Users To Provide Contact Methods Again check box

When a user is configured to have to verify their contact information, they see a prompt during their next sign-in. If they choose to perform the verification during that sign-in, they'll see a web page similar to the page in Figure 1-10.

FIGURE 1-10 Users can perform self-service to update and verify their contact information

From an exam perspective, know that self-service reduces IT administrative overhead (and thus overall costs). It improves the user experience by empowering users and allowing them to take care of some of their own account management tasks without resorting to calling the Helpdesk or putting in a trouble ticket. In environments where the IT department is understaffed, self-service would be helpful. You can review more about self-service, especially around group management, later in this chapter.

Azure AD reporting

Azure AD has a number of built-in reports. Table 1-2 shows the available anomalous activity reports and which version of Azure AD you need to have to access the reports.

TABLE 1-2 Azure anomalous activity reports

Report name	Azure Description	Supported Azure AD versions
Sign ins from unknown sources	Might indicate an attempt to sign in without being traced	Free/basic/premium
Signs ins after multiple failures	Might indicate a successful brute force attack	Free/basic/premium
Sign ins from multiple geographies	Might indicate that multiple users are signing in with the same account	Free/basic/premium
Sign ins from IP addresses with suspicious activity	Might indicate a successful sign in after a sustained intrusion attempt	Premium only
Sign ins from possibly infected devices	Might indicate an attempt to sign in from possibly infected devices	Premium only
Irregular sign in activity	Might indicate event anomalous to users' sign in patterns	Premium only
Users with leaked credentials	Users with leaked credentials.	Premium only
Users with threatened credentials	Users with threatened credentials	Premium only

Table 1-3, shows the available activity log reports and which version of Azure AD you need to have to access the reports.

TABLE 1-3 Azure activity log reports

Report name	Azure Description	Supported Azure AD versions
Audit report	Audited events in your directory	Free/basic/premium
Password reset activity	Provides a detailed view of password resets that occur in your organization	Premium only
Password reset registration activity	Provides a detailed view of password reset registrations that occur in your organization	Premium only
Self service groups activity	Provides an activity log to all group self-service activity in your directory	Premium only

Table 1-4 shows the available integrated applications reports and which version of Azure AD you need to have to access the reports.

TABLE 1-4 Azure integrated applications reports

Report name	Azure Description	Supported Azure AD versions
Application usage	Provides a usage summary for all SaaS applications integrated with your directory	Premium only
Account provisioning activity	Provides a history of attempts to provision accounts to external applications	Free/basic/premium
Password rollover status	Provides a detailed overview of automatic password rollover status of SaaS applications	Premium only
Account provisioning errors	Indicates an impact to users' access to external applications	Free/basic/premium

You can run reports on demand and save reports to .CSV format. From a management perspective, there isn't much to it. From a design consideration perspective, the key item to focus on is the Azure AD version. As an administrator, you need to know your organization's requirements, especially around security, to figure out whether Azure AD Premium is necessary to meet the requirements.

Company branding

One of the initial downsides to moving to the cloud is the loss of your company identity in some of the tools and portals. The familiar designs, logos, and associated branding material are not part of a default cloud configuration for most applications. Luckily, Azure offers some customizations. You can customize the following elements for your Sign-In page and Access Panel page:

- **Banner logo**   The size should be a maximum of 60x300. To maximize performance, the graphics file should be between 5 KB and 10 KB in size. This logo can be reduced in size for smaller displays, such as those of a smartphone.

- **Square logo**   The square logo can have dimensions up to 240x240. It is used to represent user accounts in various parts of the portals. It should be between 5 KB and 10 KB in size to maximize performance.

- **Square logo, dark theme**   The dark theme square logo takes the place of the square logo in certain situations such as when a Windows 10 device is joined to Azure AD. If your logo already looks good in a dark colored theme, you do not have to use a dark themed square logo. Instead, use your square logo.

- **User ID placeholder**   On the Sign-In page, the default text in the email address text box shows the email address of *someone@example.com*. Instead, you can opt to use your domain's email syntax such as *first.last@tailspintoys.com*. This can be helpful to users, especially when users are still new to Azure.

- **Sign-In page text heading**   You can customize the text that is displayed above your customized Sign-In page text.

- **Sign-In page text**   This is the text shown near the bottom of the sign-in page. It can be up to 500 characters long and is mostly used to give users instructions on who to contact or how to obtain help.

- **Sign-In page illustration**   This is the graphic that is displayed on the left side of the sign-in page. The recommended resolution is 1420 by 1200. The maximum size of the file is 500 KB. The graphic can be reduced in size for smaller displays.

- **Sign-In page background color**   If users have a high-latency connection, the sign-in page illustration might not load. In such cases, use the specified background color. You can match the color with your company color, if desired. You must use an RGB color code to specify the color.

- **Hide KMSI**   This option enables you to hide the Keep Me Signed In check box on the Sign-In page. By default, KMSI is displayed to users.

- **Post logout link label**   After users log out of an Azure AD web application, you can display a label (along with a corresponding URL).

- **Post logout link URL**   After users log out of an Azure AD web application, you can display a URL (along with a corresponding label).

Beyond customizing these elements, you can also have a customized Sign-In page based on the language. For example, for Spanish language users, you can set your customized text to Spanish and use a different sign-in page illustration.

In Figure 1-11, a customized Sign-In page for Tailspin Toys is shown. The figure shows where the Sign-In page illustration is, where the banner logo is, where the user ID placeholder is, and where the Sign-In page text is.

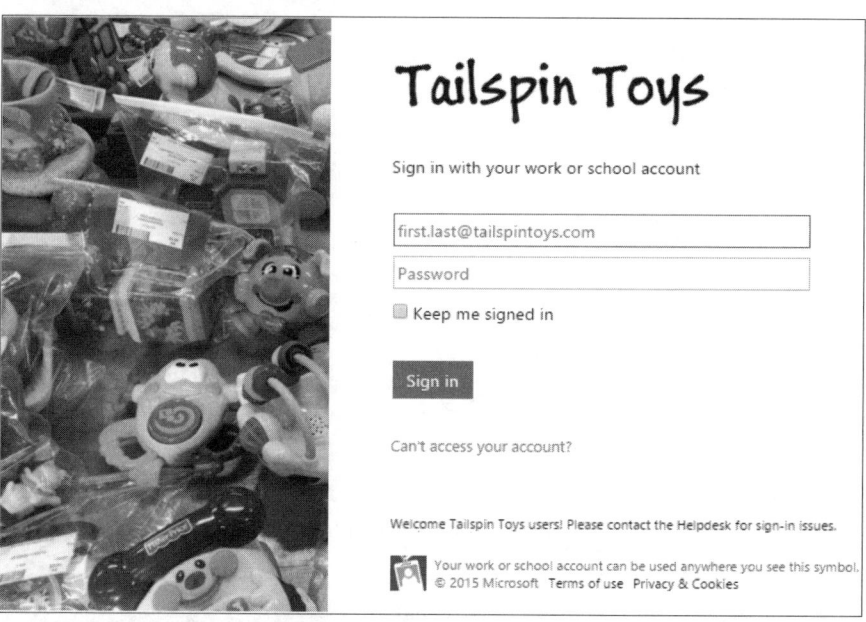

FIGURE 1-11 A screen shot showing a customized Sign-In page

To customize your Sign-In and Access Panel pages, perform the following high-level steps.

1. Sign in to the Azure portal. Navigate to Active Directory. Click the Configure tab in the top pane. Then click the Customizing Branding button, as shown in Figure 1-12. If you don't see the button, it might be because you do not have Azure Active Directory Premium.

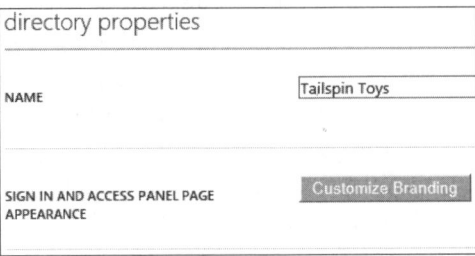

FIGURE 1-12 A properties page with the Customize Branding button

2. On the Customize Branding page, click the right arrow button in the lower-right corner.

3. On the Customize Default Branding page (page 2), browse for logo files and update the desired text. In Figure 1-13, an administrator at Tailspin Toys is using tt-logo4.jpg as the banner logo and has updated the text. When finished, click the right arrow.

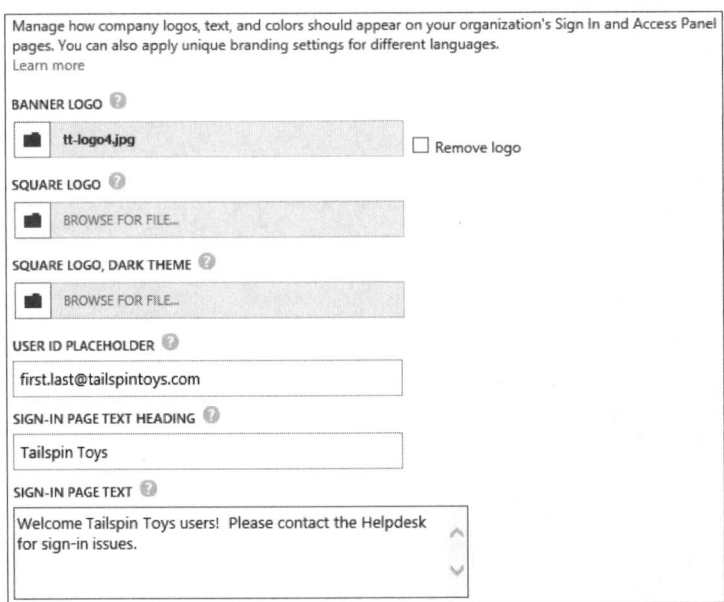

FIGURE 1-13 The customization options available for the Sign-In and Access Panel pages

4. On the Customize Default Branding page (page 3), browse for the Sign-In page illustration file, type an RGB value for a background color, set the KMSI behavior, and add logout URLs, if desired. Figure 1-14 shows the page. An administrator at Tailspin Toys is planning to use Toys.jpg as the Sign-In page illustration. When finished, click the check mark to save the changes.

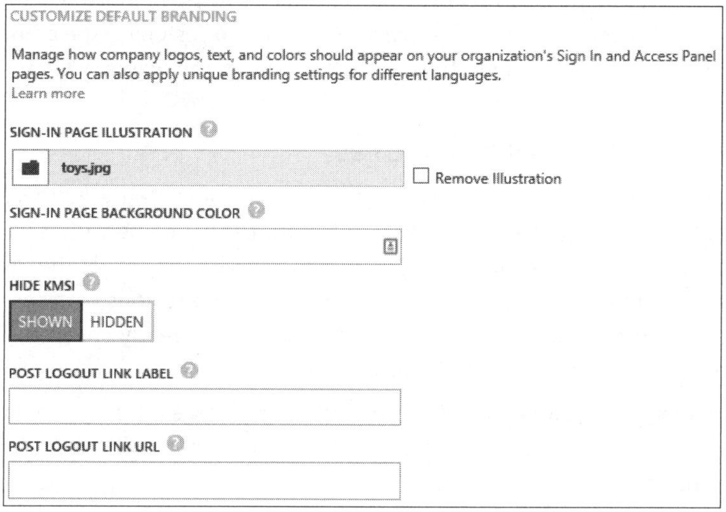

FIGURE 1-14 The customization options available for the Sign-In and Access Panel pages

After you customize your Sign-In and Access Panel page, it can take up to 1 hour for the changes to be visible on the pages. Thereafter, you can go back anytime and perform additional customizations.

Look at some important information about customization for the exam:

- **Customization requires Azure AD Basic or Azure AD Premium** You cannot customize your Sign-In or Access Panel pages with the Free version of Azure AD. An exception is that Office 365 also enables customization.

- **Customization improves the user experience** In scenarios where the user experience is an important design consideration, think about customization because it helps users become comfortable with a solution more quickly, which often leads to more successful migration projects.

- **To maximize the customizations across devices, you must test** Your customized pages might be viewed on high-resolution monitors, outdated smartphones with small screens, and tablet computers. To ensure that your customizations work well across all of those scenarios, you need to test the customizations across a variety of devices, operating systems, and platforms. Based on the testing, you can make minor adjustments to the graphics files and text to enhance the look and feel across all of the devices.

Design Azure AD Premium features

Some of the most advanced features of Azure AD are only available with the Azure AD Premium. For the exam, you should be familiar with the features that come with Azure AD Premium. While you have reviewed Azure MFA, you need to review the rest of the Premium features as well. Also note that the customizations discussed in the previous section are a feature of the Basic and Premium versions of Azure AD (and thus already covered). The Premium features you have yet to review in detail yet are:

- **Advanced application usage reporting**   The advanced usage reporting feature brings new advanced reports to customers. The list of the reports is shown in Table 1-2, Table 1-3, and Table 1-4.

- **Self-service group management for cloud users**   This feature gives cloud users (users with a cloud identity) a way to create and manage cloud-based security groups. The users use a web portal to create and manage groups and the groups can be used for application access.

- **Self-service password reset with on-premises writeback**   Self-service password reset is one feature and it comes with the Basic and Premium versions of Azure AD. The on-premises writeback feature of self-service password reset comes only with the Premium version of Azure AD. It enables users to reset their on-premises password and their cloud identity password, or both, from the Azure portal.

- **Microsoft Identity Manager (MIM) user licenses**   These MIM user licenses can be used with on-premises identity and access management solutions. MIM offers a self-service portal for users, group management, and some automated provisioning and deprovisioning.

- **Advanced anomaly security reports (machine learning-based)**   The advanced reports are focused on security, such as authentications from possibly infected hosts. See Table 1-2, Table 1-3, and Table 1-4 for a list of the reports.

- **Cloud app discovery**   This feature uses client computer agents to look for cloud apps in use. It works for cloud apps that are communicated with only over HTTP and HTTPS. You can run reports to find out things such as which apps are the most widely used, how many apps are being used, and whether or not the apps are integrated with Azure AD.

- **Azure Active Directory Connect Health**   This feature monitors the health of your directory environment including Active Directory Federation Services (AD FS), AD DS (on-premises), and Azure Active Directory Connect (the sync). You can quickly ascertain if the environment is healthy. The next section of this chapter offers details about this feature.

Advanced application usage reporting

As outlined in Table 1-2, Table 1-3, and Table 1-4, the Premium version of Azure AD offers some advanced application usage reporting. You can run reports from the Azure portal. Figure 1-15 displays the application usage report.

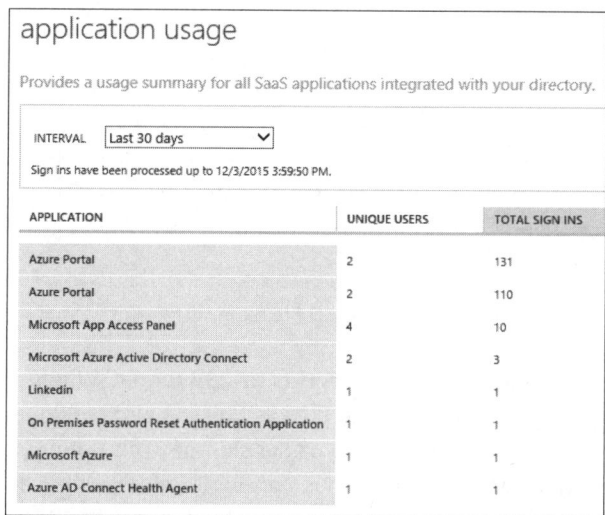

FIGURE 1-15 The application usage report for Azure AD Premium

Figure 1-16 shows the output of the irregular sign in activity report. For this report, one user is shown as having irregular sign in activity.

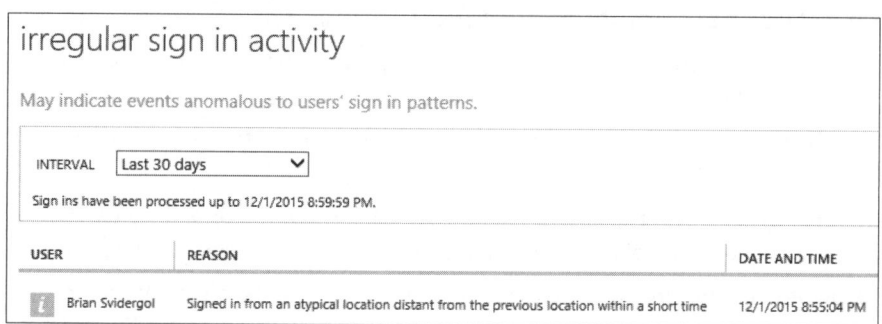

FIGURE 1-16 The irregular sign in activity report for Azure AD Premium

In Figure 1-16, note the small text that indicates that sign-ins up until a certain date and time have been processed. Most of these reports are not in real time. They are often up to 12 hours behind real time. If you run a report and expect to see an entry but don't, check the date and time that the report indicates to see if the activity hasn't been processed yet.

Another premium report is the password reset registration activity report. This report, shown in Figure 1-17, displays activity related to users registering for self-service password reset such as registering a phone number.

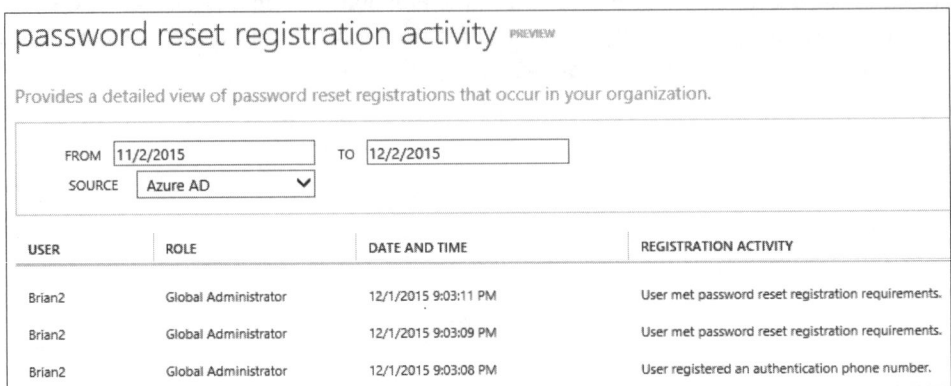

FIGURE 1-17 The password reset registration activity report for Azure AD Premium

Besides just viewing reports, you can download individual reports to .CSV format. When viewing a report, the bottom of the web page has a Download button. You can also configure Azure AD to send email (or not send email) notifications of anomalous sign-ins. This option emails global administrators when there have been 10 anomalous sign-ins (or more) within a 30day window. The option to email the anomalous sign-in activity is enabled by default. At the time of this writing, you cannot configure which email addresses are notified (currently, the global administrators' primary and secondary email address are notified) and you cannot email notifications for any other reports.

Self-service group management for cloud users

Self-service group management enables users to create, manage, and delete groups. Owners of a group can perform the following tasks:

- **Add members** The owner of a group can add group members to the group. Members can be Azure users or synced users.
- **Edit the group** By editing the group, the owner can change the display name, the description, and the approval (whether joining a group requires owner approval or not).
- **Set the owners of the group** The owner of a group has rights to manage the group. You can set one owner or multiple owners. Only groups that you own are displayed in the Azure Access Panel.

- **Leave the group** If the group is configured to allow people to join it without approval, you can opt to leave a group.

- **Delete the group** Deleting a group renders the group unusable and all permissions are revoked at the deletion time.

For the exam, it is important to know what you can do with self-service group management. But it is even more important to understand the design considerations for self-service group management. First, self-service group management is not enabled by default. You can enable and configure self-service group management on the Configure tab of Azure AD. The configuration options are shown in Figure 1-18.

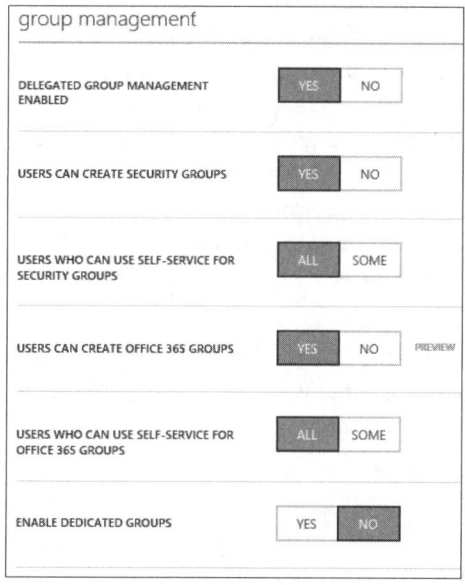

FIGURE 1-18 Azure AD self-service group management options

EXAM TIP

You cannot create security groups by using Azure AD self-service group management.

Look at some of the design considerations for self-service group management.

- **Group management becomes decentralized** In traditional environments without self-service group management, the IT team creates and maintains groups. Often, a naming standard is used for groups so that users can identify the purpose of a group. Additionally, groups are often stored together in an OU for easy delegation to other administrators or managers. With self-service, group management becomes decentralized. In such cases, naming conventions are difficult to enforce. Sometimes, multiple users decide to create a group to do the same thing without realizing that they are creating duplicate groups. Eventually, organizations sometimes end up with a very large number of groups.

This is sometimes referred to as "group sprawl." To minimize the risk of these things, you can limit some of the self-service to subsets of users instead of all users.

- **IT administrative overhead is reduced**   Reducing IT administrative overhead is an important consideration. Organizations are focusing on having IT work on value-added activities such as deploying solutions that impact sales and profitability. Managing groups is a repetitive task that often keeps IT resources away from the value-added activities. Self-service, in general, is a good way to reduce administrative overhead and get more value from IT. Self-service groups is a good way to start.

- **There is added complexity in a hybrid environment**   In organizations that have an on-premises AD DS environment and Azure AD, you manage groups on-premises and in Azure. While self-service helps with Azure AD groups, you have to use a separate self-service solution on-premises or rely on IT to perform all on-premises group management. If an organization also has Office 365, it increases complexity. Users might require extra education about managing groups and the IT team might have to spend more time on group management than they would in a less complex environment (such as one that only existed in Azure). To minimize the chances of a complex group management environment, you can move all your groups to Azure AD and use self-service management. You can also disable self-service group management until you move all the groups to Azure AD.

Once you enable self-service group management, you can use the Access Panel to create new groups, view groups that you own, and manage groups that you own. Figure 1-19 shows the view in the Access Panel. In this example, you own three groups.

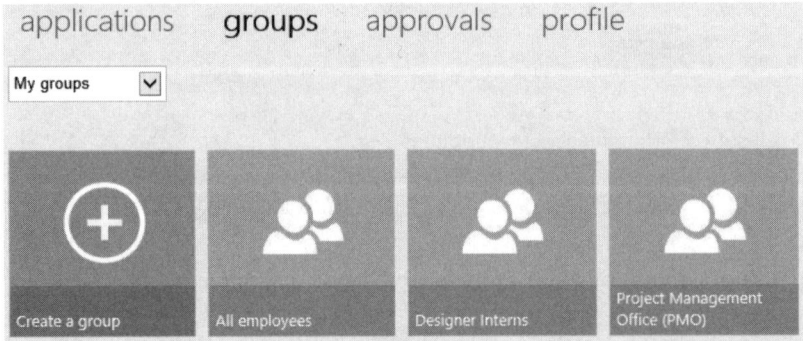

FIGURE 1-19 A screen shot of the Azure Access Panel shows the list of groups a user owns

You can get a high-level overview of a group by clicking the group, as shown in Figure 1-20.

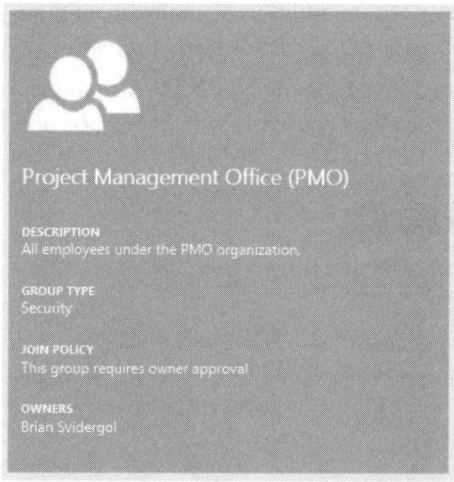

FIGURE 1-20 A screen shot of the Azure Access Panel shows the high-level overview of a group

Besides merely viewing the high-level information about the group, the user also has buttons to perform the management actions discussed earlier in the section (add members, edit, set owners, leave group, delete group), as shown in Figure 1-21.

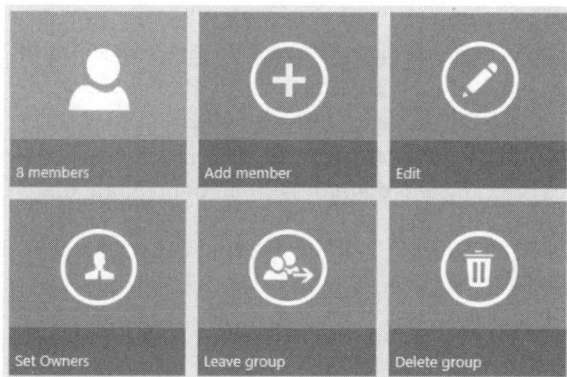

FIGURE 1-21 The Azure Access Panel shows the group management tasks for a group

Now that we have covered the self-service group management feature, take a look at another self-service feature, but this time for password resets.

Self-service password reset with on-premises writeback

Earlier in this chapter, you reviewed the self-service password reset feature. In this section, it is time to specifically discuss the on-premises writeback enhancement. The Password Writeback feature syncs password resets performed in the cloud with your on-premises AD DS, if the user account is a synced account. Based on that description, you can probably infer that you need an on-premises AD DS environment syncing to Azure AD in order to take advantage of this feature. And you would be right! In addition, you must configure your Azure AD Connect for password writeback and your Azure AD for password writeback.

First, take a look at the Azure AD Connect password writeback configuration. In Figure 1-22, the Azure AD Connect configuration is shown. The Password Hash Synchronization and the Password Writeback features are enabled.

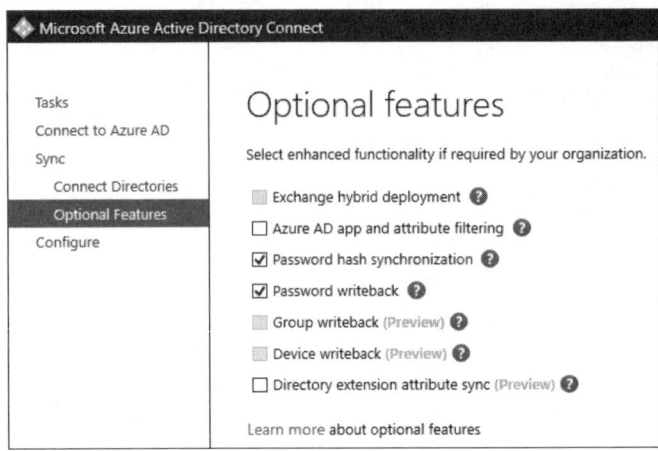

FIGURE 1-22 The Azure AD Connect configuration options

Next, look at the Azure portal configuration for password writeback. Figure 1-23 shows the option to enable writeback. Notice that the Password Writeback Service Status shows as Configured. This is required to have writeback functionality and is a good indication that everything is configured correctly and functional for the on-premises configuration and the Azure configuration. If the writeback status is set to something other than Configured, there is likely a configuration issue in the environment.

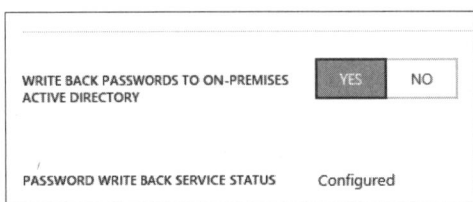

FIGURE 1-23 The Azure portal password writeback option

Microsoft Identity Manager (MIM) user licenses

MIM is the Microsoft flagship identity management product that provides identity management and self-service features for your on-premises environment. One of the benefits of Azure AD Premium is that you get user licenses for MIM. For hybrid environments, this allows you to use self-service in Azure and MIM for self-service for on-premises. For the exam, you need to be aware of scenarios that are presented where MIM could be a key solution to a requirement or problem, such as the following scenarios:

- You have multiple directories on-premises. MIM can synchronize your internal directories. This enables you to sync a single forest to Azure AD. Or, if your users are located in a directory other than AD DS, MIM can sync users and groups to AD DS, which enables you to sync users to Azure AD.

- You want to provide privileged identity management to employees. MIM offers privileged identity management, which enables you to provide temporary administrative access to resources and control administrative access with granularity.

- You want to provide users with self-service group management in the cloud and on-premises. While Azure AD offers self-service group management as a feature, it is for groups in Azure AD. You can opt to use group writeback to bring those groups back to AD DS. Or, you can use MIM to provide self-service group management for your on-premises AD DS environment.

EXAM TIP

There are other scenarios where MIM could be valuable. The exam does not identity MIM as a skill for the exam other than the benefit of obtaining MIM licenses with Azure AD Premium. However, it might come up in scenario-based questions, case studies, or as an answer choice on some questions.

NEED MORE REVIEW? MICROSOFT IDENTITY MANAGER

To review further details on MIM, see *https://technet.microsoft.com/library/mt150253.aspx.*

Advanced anomaly security reports (machine learning-based)

Earlier in this chapter, you reviewed the lists of all of the reports, including the advanced anomaly security reports. You also looked at the actions you can perform as an administrator such as running reports and saving reports to .CSV. From the exam perspective, the most important thing to know is the additional reports that you get with Azure AD Premium:

- Sign ins from IP addresses with suspicious activity
- Sign ins from possibly infected devices
- Irregular sign-in activity
- Users with leaked credentials
- Users with threatened credentials

Cloud app discovery

The cloud app discovery technology finds apps that users are using on the Internet and tracks the total number of users and the total amount of activity (amount of data transferred, files upload, files downloaded, and a few other data points). The information is made available in report form in the Azure portal under Cloud App Discovery. The following two components make up this technology:

- **Azure Cloud App Discovery** You can add the app for free as part of Azure AD Premium. Once added, you can view application usage reports and set up apps with Azure AD integration.

- **Cloud App Discovery endpoint agent** To discover apps, you need to need the Cloud App Discovery endpoint agent on your users' computers. The app listens for HTTP and HTTPS connections (whether in a browser or in an app) and logs the details of the connections for report data. The information captured includes the IP address, the destination URL, the username, and information about the connectivity (data transferred and similar).

The promise of cloud app discovery is the centralization of user identities. Today, many organizations' users have multiple identities—one or more on-premises identities, an Azure cloud identity, and many cloud-based app identities. Managing those identities is a challenge, especially when IT manages some of them and the user manages the rest of them. To add to the challenge, users are often using their own devices for work-related activities. So work identities and personal identities are comingling and the lines between them are becoming blurred. To reduce the associated challenges, you can centralize identities to Azure AD. It starts with syncing and/or federating with Azure. But to take full advantage of the benefits, you need deeper integration, of which the use of cloud app discovery is one part. Integrating as many of the apps into Azure AD as possible reduces the overhead of identity management and strengthens the overall security of your environment by reducing the reliance on so many identities.

After you begin using cloud app discovery and gather some data from endpoint agents, you can view activity in the Azure portal. Figure 1-24 shows the apps discovered by agents. Once apps are integrated with Azure AD, they show a status of Managed. Apps that were discovered but have not been integrated with Azure AD show a status of Unmanaged.

APP	CATEGORY	STATUS
Office 365 Exchange Online	Mail	Unmanaged
Windows Azure	Developer Services	Managed
LastPass	Security	Unmanaged
Dropbox for Business	Collaboration	Managed
Linkedin	Social	Managed
Visual Website Optimizer	Marketing	Unmanaged
Active Directory Access Panel	Data Services	Managed
DemandBase Ad Dashboard	CRM	Unmanaged

FIGURE 1-24 Apps discovered by the cloud app discovery endpoint agent

Now that you have reviewed what cloud app discovery does, walk through the process of integrating an application with Azure AD. First, you have to add the app. From the Azure AD Applications tab in the Azure portal, click the Add button at the bottom of the page. That opens a window asking you the type of app that you want to add, as shown in Figure 1-25.

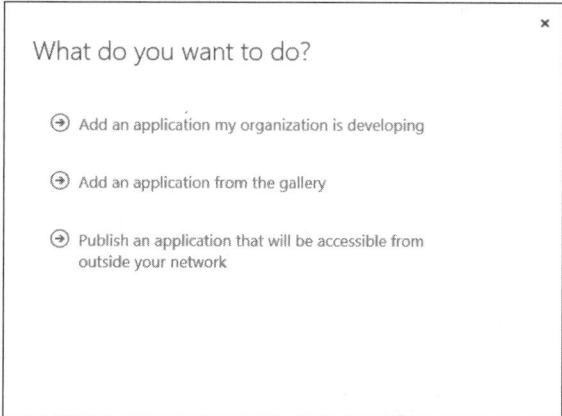

FIGURE 1-25 Types of applications you can add when adding an application to Azure AD for management

For this walk-through, you need to choose an app from the Azure gallery. At the time of this writing, the gallery contains 2,527 apps. In the gallery, click Microsoft Developer Network (MSDN), as shown in Figure 1-26. After selecting MSDN, click a check mark to complete the process (the check mark is not shown in Figure 1-26).

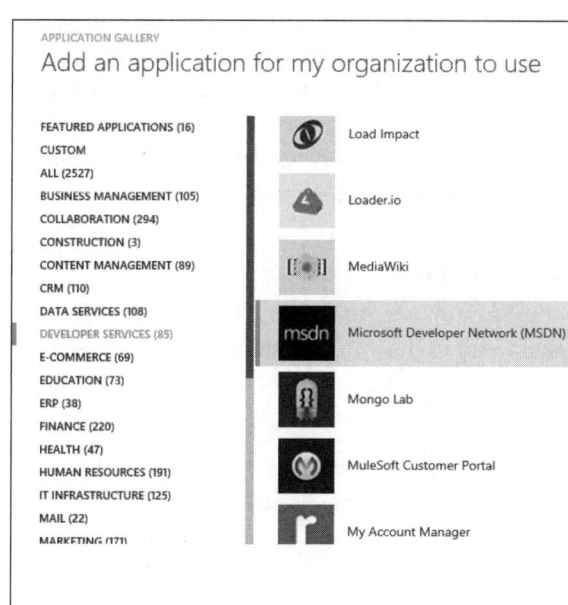

FIGURE 1-26 In the Azure gallery you choose to add applications from the thousands of apps currently available

After an app is added, it is displayed in the apps list and is ready to be configured, as shown in Figure 1-27.

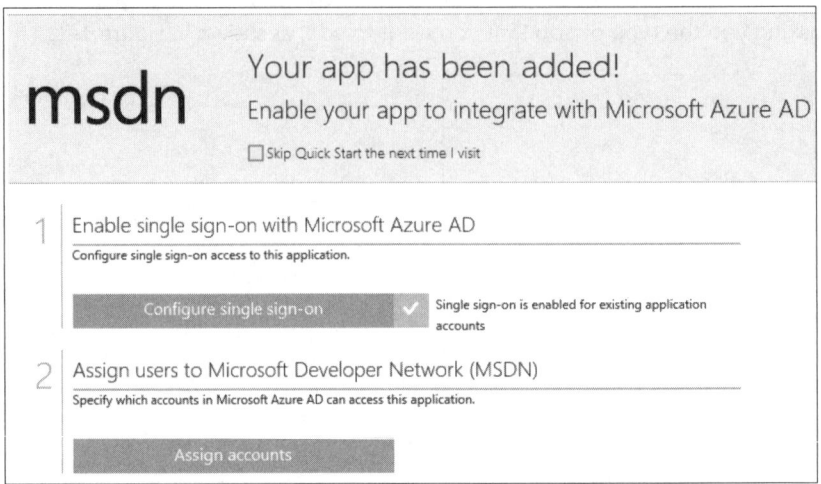

FIGURE 1-27 A notification from Azure that your app has been added for management

To configure the app for single sign-on, click the Configure Single Sign-On Button. A window displays showing the following three options:

- **Federated Single Sign-On**   This option is only displayed if you have federation with Azure already configured. If you don't have federation, this option is not displayed.

- **Password Single Sign-On**   This option enables users to store account credentials in Azure. For example, they would input their MSDN username and password into the Access Panel app and the credentials would be saved. An administrator can also enter the credentials on a user's behalf although this isn't as common of a method unless you are working with shared accounts or do not want users to know the password to an account. Password single sign-on requires a browser plug-in/add-on.

- **Existing Single Sign-On**   This option requires an existing SSO solution such as AD FS. If you have such a solution, then you should consider this option.

After you choose the SSO method, you must assign users to the application. If you don't assign users, they cannot use the Access Panel to go to the app. The steps to assign the application to a user are:

1. Click the Assign Accounts button.

2. At the top of the page, click the dropdown menu next to Show, click All Users, and then click the check mark on the right. This displays all of your user accounts.

3. Click the user that you want to assign the app to and click the Assign button at the bottom of the page.

4. In the Assign Users window, click the check mark. Optionally, you can click the option to enter credentials and then enter the credentials.

Summary

- You can use three identities with Azure: a *cloud identity* is an identity that is created and maintained in Azure AD. It is best suited for small organizations or organizations that have all of their services in the cloud. A synced identity is an identity that is synchronized from an on-premises AD DS environment to Azure AD. Password hashes are synced and a user with a *synced identity* can be authenticated by AD DS for on-premises resources or by Azure AD for cloud-based resources. A *federated identity* is an identity that is synchronized from an on-premises AD DS environment to Azure AD. Password hashes are not synced and all user authentication is handled by the on-premises AD DS environment.

- With Azure AD Premium, you get additional features including advanced application usage reporting, self-service group management for cloud users, self-service password reset with on-premises writeback, Microsoft Identity Manager (MIM) user licenses–for on-premises identity and access management, advanced anomaly security reports (machine learning-based), cloud app discovery, Multi-Factor Authentication service for cloud users, Multi-Factor Authentication server for on-premises users, Azure Active Directory Connect Health to monitor the health of on-premises Active Directory infrastructure and getting usage statistics.

- Azure MFA is a cloud-based multi-factor authentication service that enables you to require multiple authentication factors when authenticating to some cloud-based resources or some on-premises resources. Azure MFA supports a phone call, a TXT message, a smartphone app verification, or an OATH token as a second factor of authentication. You need to run an on-premises Azure MFA server to protect on-premises resources with multi-factor authentication.

- Users can go to the Azure Access Panel at *myapps.microsoft.com* to access applications and perform self-service tasks such as resetting their password, going to apps by using SSO, and updating their contact information.

- Azure AD offers built-in reporting to report on directory related activities such as sign-ins, password resets, and use of integrated applications. Some reports are only available with Azure AD Premium such as the password reset activity report and the self-service groups activity report. You should know which reports are only offered as part of Azure AD Premium.

- If you have Azure AD Premium, you can customize the look of the Azure Sign-In page (full-featured customization) and Access Panel page (limited customization). You can customize the banner, logo, user ID placeholder, text heading, help text, the graphic on the Sign-In page, and the logout text and URL.

- You can integrate SaaS applications with your Azure AD to enable SSO for users. Users subsequently add integrated applications to their Access Panel pages, which provide users with quick links to go to the applications. When you integrate applications with Azure AD, you can have your SaaS username and password saved and automatically entered when you visit the application page, or you can rely on federation or a third-party identity provider such as OKTA.

Skill 1.2: Design for Active Directory synchronization with Azure AD Connect

Now that you have reviewed many of the technologies around Azure AD identities, you should next look at synchronizing an on-premises AD DS environment with Azure AD. Synchronization is required for some features, such as single sign-on. It also enhances users' experiences when they access applications. Few users enjoy working with multiple usernames and passwords. You also need to review the design factors for single sign-on and work through some real-world integration scenarios. Then, look at the available tools, and discuss synchronization services and a feature named Connect Health, which monitors your environment for health when you are using Active Directory Federation Services (AD FS) as part of your environment.

> **This section covers how to design for:**
>
> - Single sign-on, Active Directory integration scenarios, Active Directory synchronization tools
> - Azure AD Synchronization Services
> - Connect Health

Design single sign-on, Active Directory Integration scenarios, and Active Directory synchronization tools

In the previous section, you focused on identities at a high level. While you looked at SSO for applications, integration, and tools, you have yet to get into the details of the integration (especially the syncing). In this section, you are going to look at the details of SSO, integration scenarios that you can use for each of the cloud identities, and the synchronization tools that you can use to integrate AD DS and Azure AD. By the end of this section, you should be comfortable with all of the design considerations and high-level implementation steps required to integrate AD DS and Azure AD.

Single sign-on

SSO enables users to use a single username and password to access resources. It also enables users to sign into applications with their existing credentials. For example, if a user signs into a device with their AD DS credentials, they can gain access to a company SharePoint site without having to enter their name and password again. With Azure AD, there are a couple of ways to achieve SSO, as follows:

- **Directory synchronization and federation with AD FS** This option is the most complex, as noted earlier in the chapter when discussing federated identities.

- **Directory synchronization and password synchronization** This option is easy to set up and configure and provides a good user experience.

Take a look at some design considerations and requirements to see when SSO is needed and when to use each SSO method.

- **An organization is moving everything to the cloud** In this scenario, a company is opting to put everything in the cloud and not have anything beyond computing devices on-premises. In this scenario, users have only a cloud identity. By default, users have SSO throughout the Microsoft. You can integrate apps with Azure AD to expand SSO to SaaS apps, too.

- **An organization is going to use Office 365** In this scenario, the company has on-premises technologies such as AD DS, file and print servers, and web servers. The company is going to use Office 365 for email, SharePoint, and Skype for Business. You can use cloud identities for Office 365 (thus, no SSO). However, the user experience is degraded. Users have to remember another password and user account management requires more IT administrative overhead. In this scenario, you can achieve a better user experience by using synced identities or federated identities. The user experience improves and you reduce IT administrative overhead in some areas.

- **An organization has a security requirement mandate that AD DS password hashes not be transmitted outside of the company network** In this scenario, you only have one way to achieve SSO -through federation, where all authentication is validated by the internal AD DS environment.

- **An organization wants SSO but with minimal IT administrative overhead** In this scenario, you should look at the options for SSO and compare the administrative overhead of each. For synced identities, the administrative overhead is lower than it is for federated identities. Thus, in this scenario, you should use synced identities.

Active Directory integration scenarios

You've looked at integration scenarios so far in this chapter. These scenarios are the cloud identity, synced identity, and federated identity. You also reviewed these in detail earlier in this chapter while reviewing identities.

> **NEED MORE REVIEW? DESIGNING A HYBRID IDENTITY ENVIRONMENT**
>
> To review further details on designing a hybrid identity environment, see the Azure Hybrid Identity Design Considerations Guide at *https://gallery.technet.microsoft.com/Azure-Hybrid-Identity-b06c8288*.

Active Directory synchronization tools

Not too long ago, you could use several tools to sync AD DS to Azure AD. The supported tools included DirSync, Azure AD Connect, Azure AD Sync, and Forefront Identity Manager (FIM). Microsoft has opted to build up the functionality in Azure AD Connect and make it the single tool for all directory synchronization needs. It is recommended that Azure AD Connect be the only tool to perform AD DS to Azure AD Synchronizations as it is the only one being actively developed and enhanced for AD DS to Azure AD Syncs. All of the other tools are still supported, but DirSync is going to be deprecated soon and eventually unsupported. It is likely that Azure AD Sync and FIM are going to be deprecated and unsupported eventually as well.

> **IMPORTANT A GOOD PRACTICE**
>
> While Azure AD Connect can be installed on a domain controller, it is good practice to install it on a dedicated server or a server that is not a domain controller. This enables administrators to manage and troubleshoot Azure AD Connect without having to connect to a domain controller through a console connection or remote desktop connection. Additionally, it is a good practice to minimize the installation of software on domain controllers to reduce conflict and maximize performance.

For the exam, you need to be intimately familiar with Azure AD Connect. You need to understand what its capabilities are, when you should use it (which was covered previously in this chapter), and how to install and configure Azure AD Connect with syncing. In addition to reading this book, you should spend a little time installing Azure AD Connect and configuring an AD DS to Azure AD Sync. It will help you on the exam.

CAPABILITIES

Take a look at some of the capabilities of Azure AD Connect:

- **Synchronize single forest and multi-forest AD DS environments to Azure AD**
  As previously discussed, Azure AD Connect's primary job is to sync AD DS to Azure AD.

- **Synchronize specified app attributes (shown as "Azure AD app and attribute filtering" in Azure AD Connect)**   By default, Azure AD Connect synchronizes a specific set of attributes that are required for functionality of apps and SSO. But, if needed, you can filter out some of these. For example, if you have a security policy that mandates that email be stored on-premises, you could filter out Exchange Online Attributes from the sync.

- **Writeback of devices**   This feature enables you to sync devices registered in Azure AD with your on-premises AD DS. The purpose of the device registration in AD DS is to use the devices for conditional access (access based on credentials and other factors, such as the device being used to gain access). There are two key things to know about this feature. One, you must prepare AD DS for device writeback. Two, you must configure the sync to support device writeback. On the Azure AD side, if you are not currently configured to enable users to join devices to Azure AD, you need to configure it.

- **Writeback of attributes**   This feature, shown as Exchange Hybrid Deployment in Azure AD Connect, enables you to write attribute updates from Azure AD to AD DS. You need to use this option for some Exchange hybrid scenarios.

- **Writeback of groups**   This feature enables you to sync Office 365 groups from Azure AD to your on-premises AD DS. This is used for hybrid Exchange environments. The groups synced from Azure AD to AD DS become distribution groups for on-premises use.

- **Writeback of users**   This feature was in Azure AD Connect but was temporarily removed in the August 2015 update. This feature was in early preview (one stage before the Preview stage). It will likely be coming back soon. The feature enables you to sync Azure AD users to an on-premises AD DS environment. The key limitation of this feature is that the AD DS environment has to be unused at the time and Azure AD must be the source for all user objects.

- **Writeback of passwords**   This feature enables password changes that originate in Azure AD to be written back to your on-premises AD DS environment. This enables users to change their password in their on-premises environment (by using AD DS) or in the cloud environment (using the Azure Access Panel and password writeback feature).

- **Synchronize specified attributes (shown as Directory extension attribute sync in Azure AD Connect)**   If you need to sync additional attributes from AD DS to Azure AD, this feature is what you use. For example, if you store employee nicknames in an attribute that is not synced, you can configure the attribute to be synced.

- **Synchronize password hashes for single or multi-forest AD DS environments to Azure AD**   When thinking about synchronizing password hashes, it is important to remember that passwords are not synced. Only password hashes are synced. And they are synced securely.

- **Provide SSO. One of the primary features of Azure AD Connect, besides synchronization, is SSO (a byproduct of synchronization)** SSO is often the reason why companies implement a sync to Azure AD.

INSTALLING AND CONFIGURING AZURE AD CONNECT

Now, review the initial installation and configuration process because there are some important key points for you to remember for the exam. In the following walk-through, you are going to install Azure AD Connect and configure it to sync the alpineskihouse.com domain. Alpine Ski House is going to use synced identities and wants password hashes to be synced from AD DS to Azure AD and from Azure AD to AD DS. Alpine Ski House has two AD DS forests in their on-premises environment. This walk-through explains the options that you select for this sync as well as other options that you don't use for the sync.

1. To begin, download Azure AD Connect from *https://www.microsoft.com/download/ details.aspx?id=47594*.

2. While signed in as a local Administrator on the server where you perform the installation, double-click AzureADConnect.msi.

3. On the Welcome To Azure AD Connect page, as shown in Figure 1-28, read the license terms and privacy notice that are linked to. Then, if you agree to the terms and privacy notice, select the I Agree To The License Terms And Privacy Notice check box. Then, click Continue.

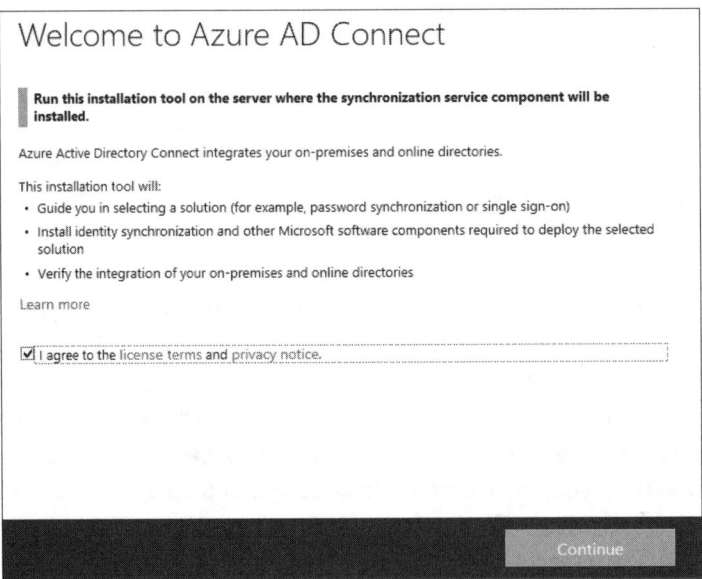

FIGURE 1-28 The Welcome page during the Azure AD Connect installation and configuration

4. On the Express Settings page, as shown in Figure 1-29, you can opt to use the express settings (sync current AD forest, sync password hashes, start syncing, and sync all attributes). Or, you can opt to customize the sync configuration. For this example, you customize the sync configuration because you need to enable the password writeback feature, which isn't part of the express settings. Click Customize to continue.

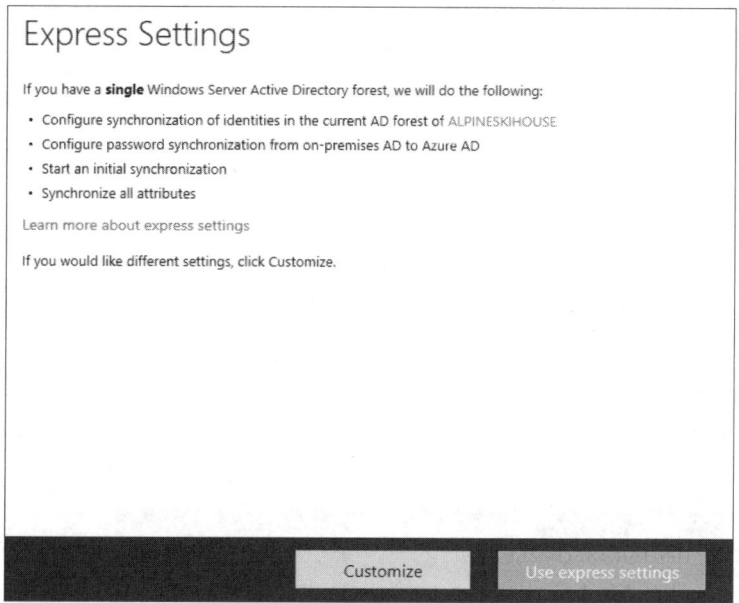

FIGURE 1-29 The Express Settings page where you can opt to customize the settings or continue with the default settings

5. On the Install Required Components page, shown in Figure 1-30, you can choose to use the optional configuration options. For your configuration, you do not opt for any of these features. Click Install to continue.

- **Specify a custom installation location** This option enables you to install Azure AD Connect in a different location than the default location (C:\Program Files\Microsoft Azure Active Directory Connect).

- **Use an existing SQL server** This option enables you to use an existing SQL server in your environment. If you do not opt to use this option, Azure AD Connect installs SQL Server 2012 Express locally.

- **Use an existing service account** By default, Azure AD Connect will create an AD DS user account to use to run the Microsoft Azure AD Sync service. However, you can opt to create your own service account instead. If you do, you must use this option to specify the account. If your company has service account naming conventions or other security policies that cannot be met by having Azure AD Connect create the service account, then you should use this option.

- **Specify custom sync groups**   At the time of this writing, this feature functions only if Azure AD Connect is installed on a domain controller. The feature enables you to choose specific AD DS groups that will be included in the sync. Azure AD Connect will support AD DS groups from member servers in the near future.

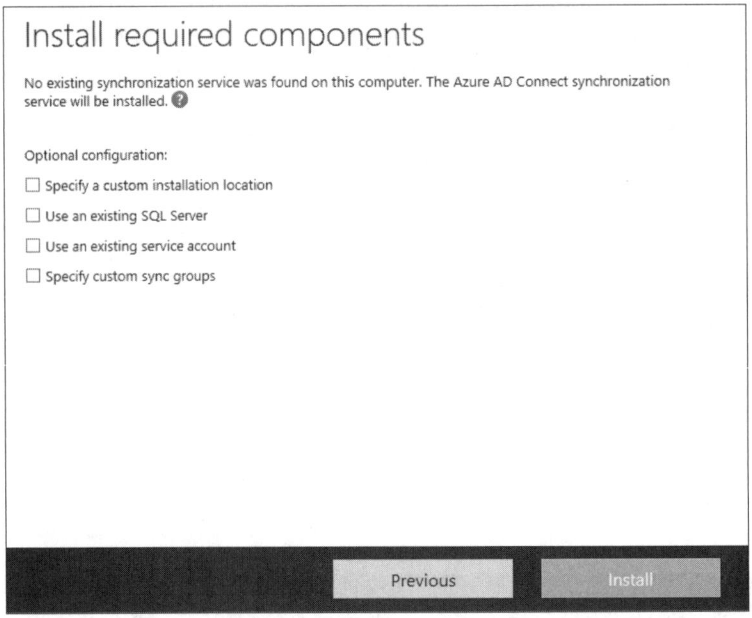

Install required components

No existing synchronization service was found on this computer. The Azure AD Connect synchronization service will be installed. ❓

Optional configuration:

☐ Specify a custom installation location

☐ Use an existing SQL Server

☐ Use an existing service account

☐ Specify custom sync groups

[Previous]   [Install]

FIGURE 1-30 The Install Required Components page enables you to customize Azure AD Connect

6. On the User Sign-In page, shown in Figure 1-31, you can choose to use password synchronization or federation with AD FS. Choose password synchronization if you plan to use synced identities. If you plan to use federation, choose the federation with AD FS option. Optionally, if you have a third-party SSO solution, such as OKTA, you can opt to not configure SSO as part of Azure AD Connect. For this walk-through, you should use the default option of password synchronization. Click Next to continue.

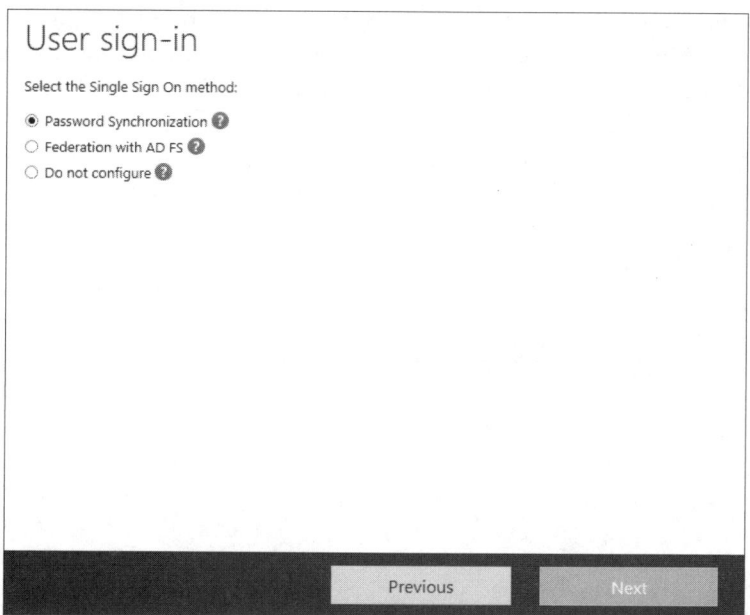

FIGURE 1-31 The User Sign-In page displays the SSO options that you can use for Azure AD Connect

7. On the Connect To Azure AD page, shown in Figure 1-32, specify the Azure AD credentials to configure the sync. You must specify an Azure global administrator account. Click Next to continue.

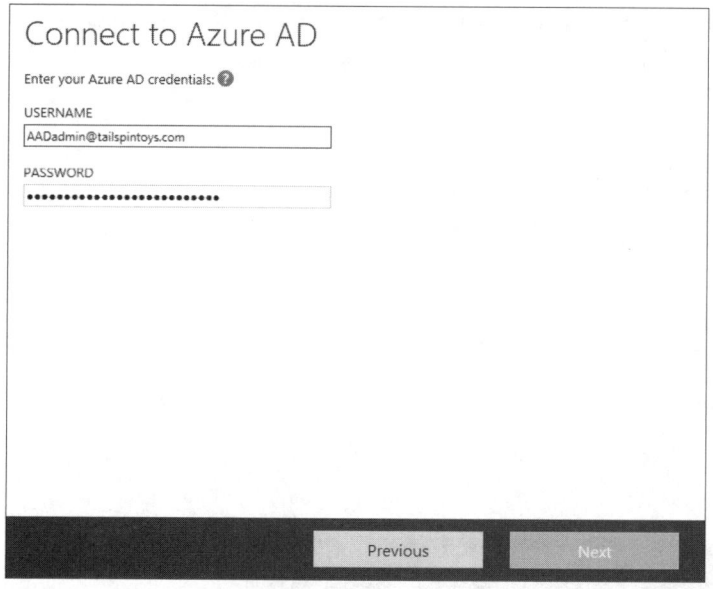

FIGURE 1-32 The Connect To Azure AD page prompts you for credentials to connect to Azure AD

8. On the Connect Your Directories page (first page of 2 pages), shown in Figure 1-33, you need to specify a user account that is a member of the Enterprise Admins group. Then, click Add Directory to add the forest to the sync.

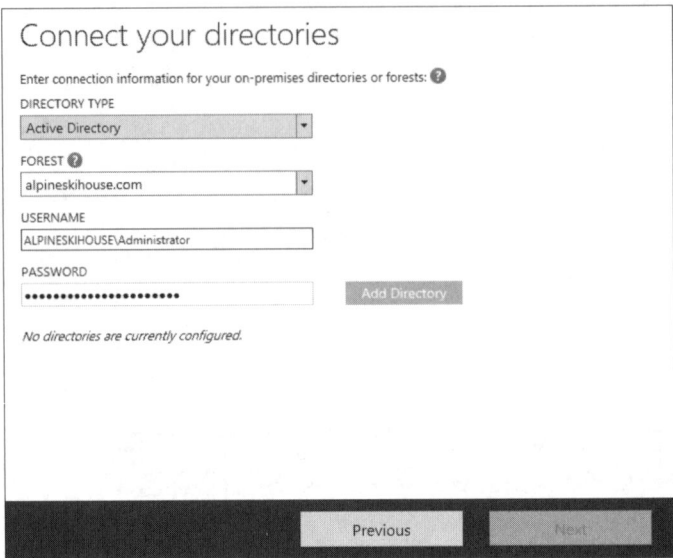

FIGURE 1-33 The Connect Your Directories page prompts you for credentials to your on-premises forest

9. On the Connect Your Directories page (second page of 2 pages), shown in Figure 1-34, the alpineskihouse.com forest is shown as configured. Click Next to continue.

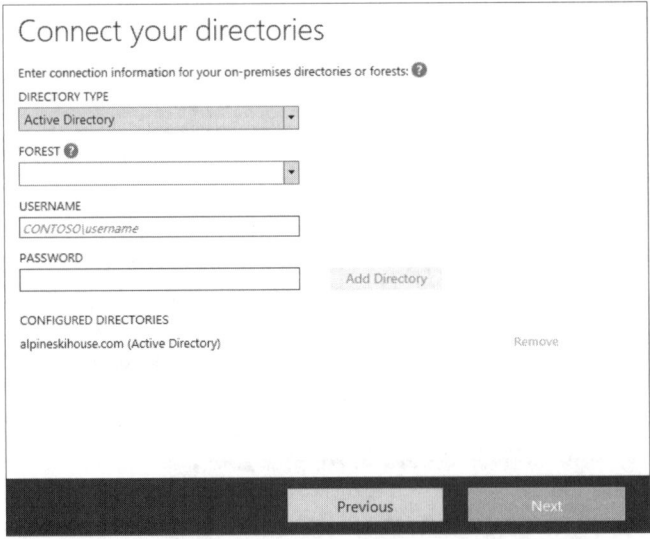

FIGURE 1-34 The Connect Your Directories page prompts you for credentials to your on-premises forest

10. On the Uniquely Identifying Your Users page, shown in Figure 1-35, you can select from a couple of different options. For our walk-through, click the User Identities Exist Across Multiple Directories option, click the SAMAaccountNAME And MailNickName Attributes option, and then click Next. This is because our example organization has multiple forests.

- **Users are represented only once across all directories** In a singled-domain forest, or in a multi-domain forest where all users only have a user account in one forest, you would use this option. Azure AD Connect doesn't have to take any custom actions to account for users that have accounts in multiple forests.

- **User identities exist across multiple directories** In a multi-forest environment, where users might have accounts in more than one forest, you need to use this option to ensure that users are only represented in Azure AD once. For example, if a user named Bob Kelly has an account in three forests, he will have only a single account in Azure AD. You need to select the method to match users so that Azure AD Connect can match up the user's accounts across the forests. For this walk-through, you are syncing alpineskihouse.com and already syncing tailspintoys.com. Thus, you are going to match users using the SAMAccountName and MailNickName attributes.

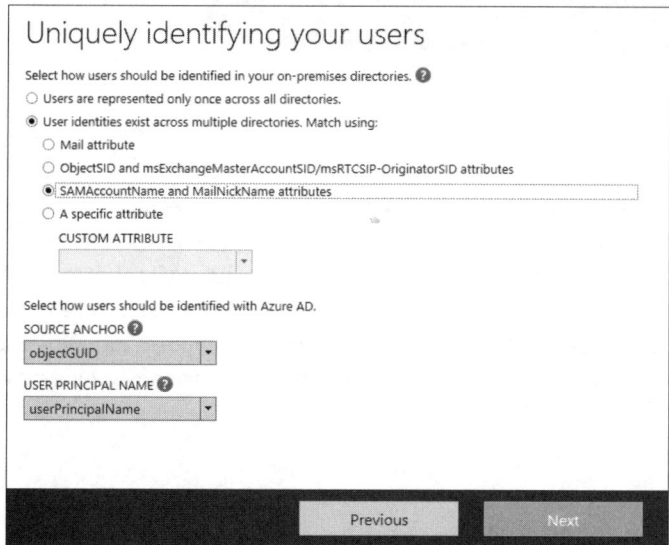

FIGURE 1-35 The Uniquely Identifying Your Users page enables you to account for users that have accounts in multiple synced forests, if applicable

11. On the Filter Users And Devices page, shown in Figure 1-36, you can choose to synchronize all users and devices or only those that are members of a specified

AD DS group. If you are testing, then you should synchronize a subset of users to reduce risk and complexity. In some cases, you might also want to synchronize a subset of users to easily control which accounts get synchronized. For this walk-through, maintain the default to synchronize all users and click Next.

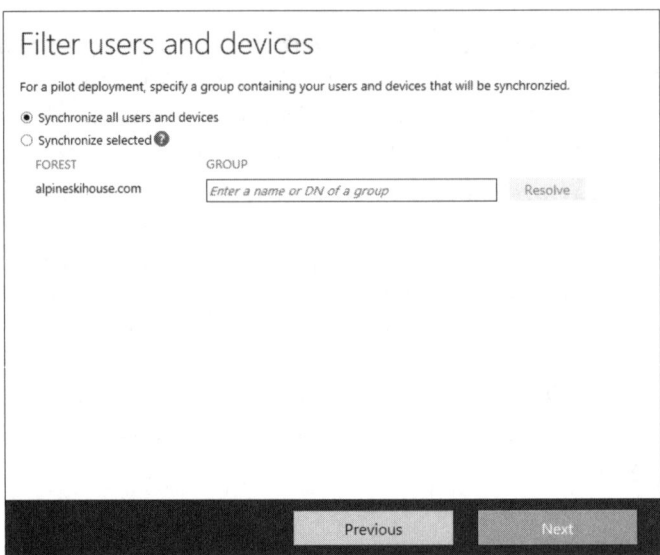

FIGURE 1-36 The Filter Users And Devices page enables you to synchronize all users and devices or a defined subset of them

12. On the Optional Features page, shown in Figure 1-37, you can choose to enable optional features. By default, only the Password Hash Synchronization is selected. For our example, Alpine Ski House wants to sync password changes in both directions (password change from AD DS syncs to Azure AD and password change from Azure AD Syncs to AD DS). Select the Password Writeback check box and then click Next. Note that some of the optional features require configuration before configuring the sync option. When your environment is not configured for optional features, they appear dimmed (unavailable).

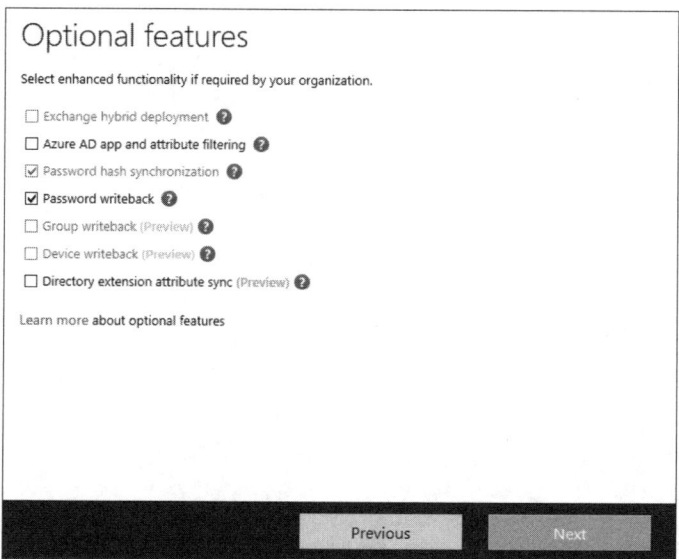

FIGURE 1-37 The Optional Features page enables you to choose optional features that enable additional functionality for your sync

13. On the Ready To Configure page, shown in Figure 1-38, you have two options. For this walk-through, you start the sync after the configuration completes. Click Install.

- **Start the synchronization process as soon as the configuration completes (selected by default)** You normally should use this option so that the sync begins as soon as possible. This way you know if the sync is working as expected, or if you need to change the configuration.

- **Enable Staging mode: When selected, synchronization does not export any data to AD or Azure AD** Thinking of Staging mode as a read-only mode. This is a key option and it might come up on the exam. Staging mode is a way for you to have a second sync server with the same configuration as your primary server, which you could use in the event of a failure. For example, if your primary sync server is in datacenter #1, you could create a staging sync server in datacenter #2 and the staging server could be made the active server if datacenter #1 has a catastrophic failure.

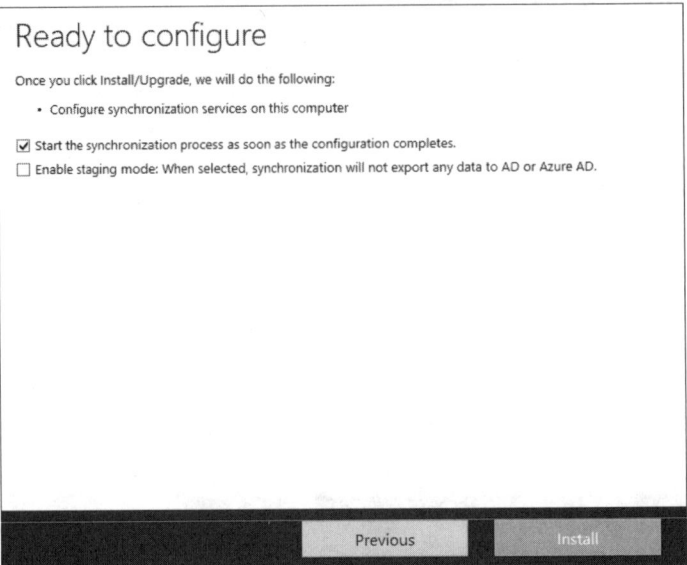

FIGURE 1-38 The Ready To Configure page enables you to set the staging mode for the server, if desired, and begins the installation

14. On the Configuration Complete page, shown in Figure 1-39, a completion message is displayed. Click Exit to finish.

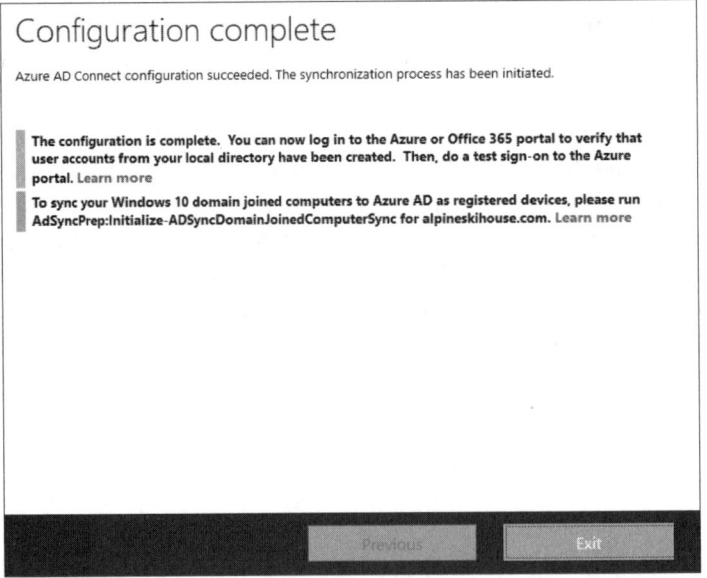

FIGURE 1-39 The Configuration Complete page displays the installation status

Once Azure AD Connect is installed and configured, the syncing process runs as a scheduled task (using a dedicated user account created during the installation). You should check Azure AD to verify that user and group objects are syncing. Optionally, you can manipulate the scheduled task to have the sync run more often or less often, depending on your company's requirements.

If you need to make changes to your sync configuration, you can run Azure AD Connect and customize the sync options.

> **NEED MORE REVIEW?** **REQUIRED PERMISSIONS**
>
> To review further details on the required permissions for setup and configuration, especially for additional scenarios such as those involving AD FS, see *https://azure.microsoft.com/documentation/articles/active-directory-Azure ADconnect-accounts-permissions/*.

Plan for Azure AD Synchronization services

Before Microsoft standardized on Azure AD Connect, the primary sync tool was Azure Active Directory Sync (Azure AD Sync). Azure AD Sync replaced DirSync. Now Azure AD Connect has replaced Azure AD Sync. Both tools share the same core functionality. In fact, the look and feel of the tools is very similar and they support the same integration scenarios. However, Azure AD Sync doesn't support all of the newest optional features that Azure AD Connect supports. Figure 1-40 shows the optional features that are supported by Azure AD Sync.

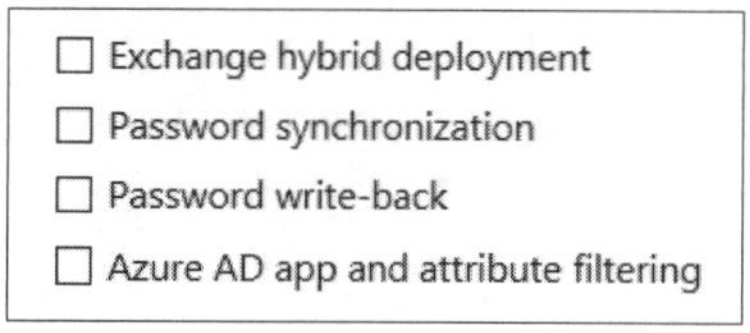

☐ Exchange hybrid deployment
☐ Password synchronization
☐ Password write-back
☐ Azure AD app and attribute filtering

FIGURE 1-40 Azure AD Sync's optional features

This ER 70-398 exam measures specific skills that are listed at *https://www.microsoft.com/learning/exam-70-398.aspx*. The skills measured are defined in advance of the exam item development. Thus, sometimes a technology or term is listed as a measured skill but by the time the exam items are developed, a new name is being used or a new technology has taken over as the primary product. This is especially true with fast developing technologies such as Azure. For the exam, the skills you need to know are covered in the previous section, which discusses Azure AD Connect.

Design for Connect Health

Azure Active Directory Connect Health (Azure AD Connect Health) is a new feature in Azure that enables you to view the health and operations of your hybrid directory environment. It provides the following health information about your environment:

- **AD FS health status** If you have AD FS in your hybrid identity environment, you can quickly see if the environment is healthy or unhealthy.

- **Azure AD Connect health status** The health of your Azure AD Connect environment is displayed as health or unhealthy.

- **AD DS health status** This displays the health status of your on-premises AD DS environment. Note: The feature is not available at the time of this writing.

Look at some of the health information in the Azure portal. In Figure 1-41, the overview of two Azure AD Connect servers is shown, along with the operational alerts, and the status of the last export to Azure AD.

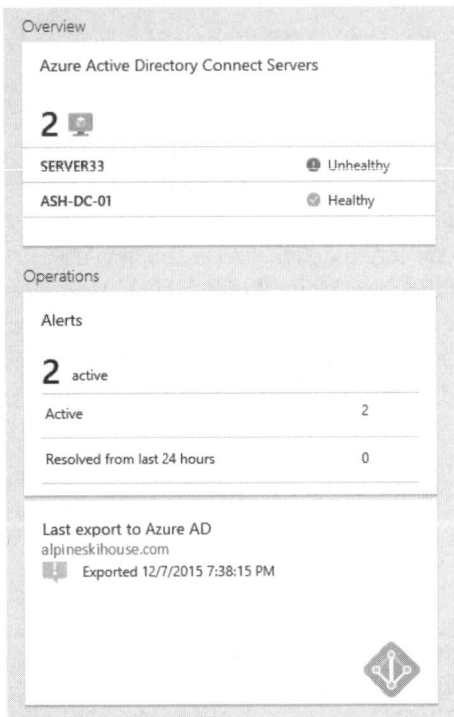

FIGURE 1-41 The overview, operations, and last export date of two servers in the Azure portal

In Figure 1-41, you can see two sync servers. One server, ASH-DC-01, is healthy and is shown with a green check mark. The other server, SERVER33, is unhealthy and is shown with a red exclamation point. You can click SERVER33 to find out more information about the problem. By doing so, the operational alerts for SERVER33 are displayed. You can then click the active alerts to display the alerts, as shown in Figure 1-42.

NAME	TYPE	SCOPE	RAISED	LAST DETECTED
ACTIVE ALERTS				
Health service data is not up to date.	⚠ Warning	SERVER33	12/3/2015 23:24:24	12/7/2015 19:30:30
Password Synchronization heartbeat was skipped in last 120 minutes.	ⓘ Error	SERVER33	12/3/2015 22:24:36	12/7/2015 20:41:38
RESOLVED ALERTS				
No items for this.				

FIGURE 1-42 The Azure AD Connect Health alerts for an unhealthy server

By clicking the individual alerts, you can display more information about the alerts and view information about fixing any issues. In Figure 1-43, the information about the password synchronization issue is displayed.

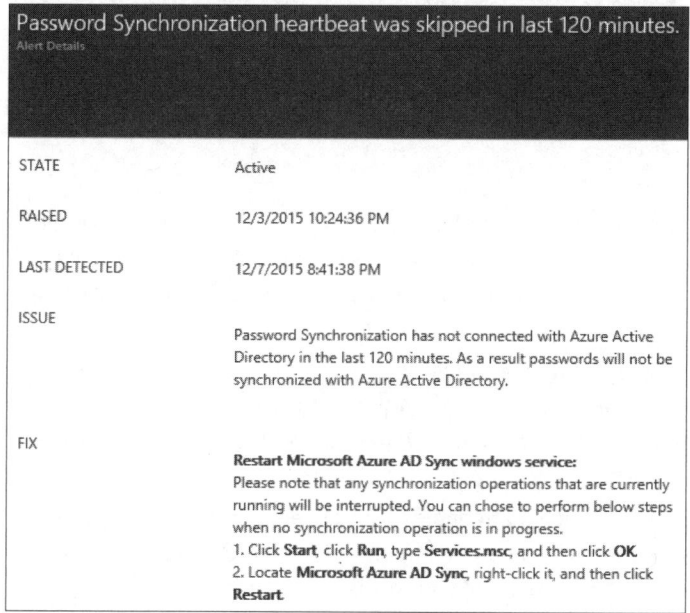

FIGURE 1-43 Azure AD Connect Health displays potential fixes to health issues

Whether you are working with AD FS, Azure AD Connect, or your internal AD DS environment, or even all of them, you can perform the same tasks in Azure AD Connect Health to view health status, view operational alerts, and view detailed issue resolution steps.

Review the following additional information on this subject, which you need to know for the exam:

- **Installing Azure AD Connect and configuring it to sync automatically provides health information to Azure AD Connect Health**   Included with Azure AD Connect is the Azure AD Connect Health agent.

- **You need to download Azure AD Connect Health agents for monitoring AD FS**   This is applicable if you have AD FS and you want to add it to the monitoring. Without the agents, Azure AD Connect Health cannot monitor your AD FS environment.

- **You can enable or disable automatic updating of your Azure AD Connect Health agent in the portal**   By default, automatic updating is enabled.

- **You can enable Microsoft to access your health information**   This is used for troubleshooting, especially when on a support call with Microsoft.

- **You need to open TCP port 80, 443, and 5671 in your firewalls**   The agent requires those ports to communicate with the Azure AD Connect Health service.

> *NEED MORE REVIEW?* **AZURE AD CONNECT HEALTH**
>
> To review further details on Azure AD Connect Health, see *https://azure.microsoft.com/documentation/articles/active-directory-aadconnect-health/.*

Summary

Azure AD Connect is the synchronization tool that you use to sync your on-premises AD DS environment with Azure AD. It offers several optional features to enhance the functionality of the sync including writeback of groups, writeback of passwords, writeback of attributes, and writeback of devices.

- While there have been other synchronization tools such as Azure AD Sync and DirSync, Microsoft is moving forward with a single sync solution with Azure AD Connect. The other tools are deprecated and no longer being updated. Eventually, support for the older sync tools will also expire.

- Azure AD Connect Health is a service that enables you to view the health and operations of your directory services environment including the health of AD FS, the health of your Azure AD Connect, and the status of your on-premises AD DS environment. Know what the system prerequisites are for Microsoft Intune, like assigning Intune as your MDM authority.

Thought experiment

In this thought experiment, demonstrate your skills and knowledge of the topics covered in this chapter. You can find answer to this thought experiment in the next section.

You are a system administrator for Alpine Ski House, a luxury mountain sports provider of mountain lodging, recreational activities, and special events facilities and services. The current IT environment consists of two small datacenters: a primary and a secondary. The company uses AD DS for all authentication and authorization, IIS for public and internal web applications, and Windows Server 2012 R2 and Windows Server 2016 for a variety of other services, including file services and database services. The company relies heavily on social media to market its products and services. The following pain points have been identified:

1. Phishing incidents have enabled attackers to gain access to some employees' user credentials.

2. Employees have reported difficulty in working with a large number of credentials to access corporate resources and cloud-based resources and this often results in routine account lockouts and password resets.

3. The small datacenters that the company maintains are at capacity.

4. When the company opened its second location, the time it took to build out the datacenter and prepare the location for IT services delayed the grand opening.

The company has recently decided to expand its services to other geographies and you have been asked to come up with a solution to provide IT services to the new locations. The following requirements have been identified:

- Minimize the use of on-premises servers.

- Improve the time to market for new locations.

- Reduce the impact of phishing incidents.

- Improve the user experience, especially for users with multiple credentials.

You need to design a solution to fix the existing issues and meet the company requirements. Answer the following questions based on the scenario:

1. Which type of identity should you use for the organization? Why?

2. How can you minimize the impact of usernames and passwords being compromised, such as from phishing attacks?

3. Which solution should you use to reduce the administrative overhead of managing user identities?

4. How would you fix the user experience issues regarding management of multiple credentials?

Thought experiment answer

This section contains the solution to the thought experiment. Each answer explains why the answer choice is correct.

1. One of the company's requirements is to reduce the use of on-premises servers. A federated identity would go against that requirement because you would need to add Web Application Proxy servers in the perimeter network and AD FS servers on the internal network. Instead, you should use synced identities. You can install Azure AD Connect on an existing server. Or, you could deploy a single new server for Azure AD Connect, which meets the requirement of minimizing the use of on-premises servers.

2. You should introduce Azure MFA to reduce the impacts felt by compromised usernames and passwords. With Azure MFA, even if usernames and passwords become compromised, attackers wouldn't be able to sign into applications that are configured with multi-factor authentication. And at first attempt, users would be notified to authenticate by Azure MFA, which works as a monitor and alarm when something is amiss.

3. You should implement the Azure Access Panel along with SSO. With Azure Access Panel, you can enable self-service features such as password reset, profile management, and group management. By offloading some user management tasks to users, you reduce the administrative overhead for your IT teams.

4. To improve the user experience regarding managing multiple credentials, you should integrate applications with Azure AD and configure SSO. Then, you should have users add the applications to their Access Panel page. Thereafter, users can go to applications from their Access Panel page without having to authenticate or remember different usernames and passwords. Their company AD DS credentials would effectively give them access to all of the integrated applications.

CHAPTER 2

Design for device access and protection

The device management lifecycle covers multiple stages. Enrollment, user access, and data protection make up the fundamentals of a basic mobile device management solution. For this exam we review a few key areas you need to be familiar with when dealing with device access and protection. These areas include: how device enrollment works with Microsoft Intune, what steps to take to provide access to corporate services using tools like the Company Portal, and how to protect data using features such as device encryption and selective wipe.

Skills covered in this chapter:

- Plan for device enrollment
- Plan for the Company Portal
- Plan protection for data on devices

Skill 2.1: Plan for device enrollment

The first stage in the device management lifecycle, and arguably the most important, is device enrollment. The enrollment process is commonly a user-facing activity through services like self-enrollment. Because the user bears most of the heavy lifting in the enrollment experience, it is essential to develop a unified and simplistic process to ensure an effective deployment. If self enrollment is confusing or overwhelming, user adoption will suffer and your helpdesk will likely see an influx in traffic. When a user needs to connect a device to your organization, the instructions should be as simple as (Go to this URL and sign in with your Active Directory credentials.) If you have created a multi-page, how-to document, you will have to pare down the document to more concise instructions. For this skill review, you will understand how to plan a device inventory and how to read reports, review the requirements for enrolling a device, and identify ways to customize the enrollment experience.

Plan device inventory

As a system administrator it is important that you not only have the tools to do your job, but know which one is right for each task. For example, your environment may have devices connected through Exchange ActiveSync (EAS) and enrolled in Microsoft Intune. If you are asked by management to provide a report on all of the devices running Android KitKat 4.4, which service will you choose? Microsoft Intune would be your best choice in this situation because all of your enrolled devices report operating system version information that you can quickly retrieve in a report. Refer to Table 2-1 for the available options when it comes to collecting device inventory.

TABLE 2-1 Device inventory and reporting capabilities

Management Platform	Capabilities	Reporting Methods
MDM for Office 365	Basic hardware and operating system inventory for supported mobile devices	Office 365 admin console
Exchange ActiveSync	Basic hardware and operating system inventory for supported mobile devices	Microsoft Exchange Management Console, PowerShell
Microsoft Intune	Detailed hardware and software inventory for supported mobile devices and client computers	Microsoft Intune admin console
System Center Configuration Manager 2012 R2	Detailed hardware and software inventory for supported client computers, with the ability to add additional inventory classes	System Center Configuration Manager 2012 R2 console, SQL Server Reporting Services
System Center Configuration Manager 2012 R2 with Microsoft Intune integration	Detailed hardware and software inventory for supported mobile devices and client computers, with the ability to add additional inventory classes	System Center Configuration Manager 2012 R2 console, SQL Server Reporting Services

Without a dedicated management solution in place, your visibility to connected devices is limited. Mail delivery services like Exchange ActiveSync provide a high-level view of connected devices, reporting basic hardware and software inventory like operating system versions and hardware models. However, Exchange ActiveSync was not designed to manage mobile devices; therefore, gathering device inventory into a report can be a time-consuming process, and the results difficult to interpret.

On the other end of the spectrum you have dedicated solutions like System Center Configuration Manager. While Configuration Manager is designed to gather device inventory at a much more granular level, providing extensive data and reporting options, as a standalone solution it is limited on what devices it can manage. For example, you can manage desktop-level operating systems like Microsoft Windows 10 with Configuration Manager, but mobile operating systems like Windows Phone are not compatible. With Microsoft Intune you gain the ability to manage both types of devices. You also have the option of integrating your Intune subscription with Configuration Manager for a full coverage solution, making Intune the recommended technology to address the shortcomings of alternative solutions.

Now take a closer look at how inventory works with a dedicated Microsoft Intune subscription. Once a device is enrolled in Microsoft Intune, it begins collecting hardware and software inventory based on a set of predefined inventory classes. These classes include common attributes such as operating system, manufacturer, model, and much more. Device inventory offers deeper visibility into your environment, providing live data on current needs and the ability to plan for upcoming refresh cycles or required infrastructure growth.

✔ **Quick check**

- You are a system administrator for Tailspin Toys. The company has 2,500 employees and a wide range of devices to support. These include an assortment of client computers running Windows 10 and a mixture of Windows Phones, iPhones, and Android mobile devices. Your manager has asked you to identify an MDM solution that will support these device types and provide detailed inventory for each platform, including the ability to expand inventory classes if needed. Which technology provides the best solution for this environment?

Quick check answer

- A Microsoft Intune subscription that is integrated with an on-premises Confiagation Manager site is the best solution.

NEED MORE REVIEW?   **UNDERSTAND YOUR DEVICE INVENTORY**

For more information about the various inventory classes supported by Microsoft Intune, visit *https://technet.microsoft.com/library/jj733634.aspx*.

You can review, export, and print the inventory data that Microsoft Intune has collected through the Reports workspace in the Intune admin console. The following list contains basic breakdowns of the reports Intune provides:

- **Update Reports**   This report displays the current software update status for enrolled computers. Status updates include those that are needed, pending, successful, and failed.

- **Detected Software Reports**   This report displays the software installed on enrolled computers. Attributes include product name, publisher, and version number.

- **Computer Inventory Reports**   This report displays machine-specific inventory for enrolled computers. Attributes include, but are not limited to, name, model, manufacturer, operating system, and last user to sign in.

- **Mobile Device Inventory Reports**   This report displays device-specific inventory for enrolled mobile devices. Attributes include, but are not limited to, name, model, manufacturer, operating system, compliance, and jailbroken status.

- **License Purchase Reports**   This report displays all licensed software based on the agreements that have been added in the Licenses workspace.

- **License Installation Reports**   This report displays the licensing results after comparing installed software with any existing license agreements that have been added to Intune.

- **Terms and Conditions Reports**   This report displays the status of acceptance for existing terms and conditions policies on a per-user basis.

- **Noncompliant Apps Reports**   This report displays users and their corresponding mobile devices based on application compliance.

- **Certificate Compliance Reports**   This report displays the certificates that have been issued to users and devices. Attributes include issued, expired, and revoked.

- **Device History Reports**   This report displays an audit log for the following device actions: retire, wipe, and delete.

- **Mac OS X Hardware Report**   This report displays machine-specific inventory for enrolled Macs.

- **Mac OS X Software Report**   This report displays the software installed on enrolled Macs. Attributes include product name and version.

NEED MORE REVIEW?   **UNDERSTAND INTUNE OPERATIONS BY USING REPORTS**

For more information about the built-in reports included in the Microsoft Intune admin console, visit *https://technet.microsoft.com/library/dn646977.aspx*.

Next, consider this real-world scenario, where inventory and reports play a key role. As a system administrator you might be given the task to deploy new software within your organization. In the following example, you are the system administrator at Tailspin Toys and must prepare to deploy Microsoft Office 2016 to your Windows 10 clients. Before proceeding with the deployment, you need to review existing device inventory data to ensure that all client computers meet the minimum system requirements. The following steps show how to create a computer inventory report, which includes attributes for available disk space and physical memory, as shown in Figure 2-1.

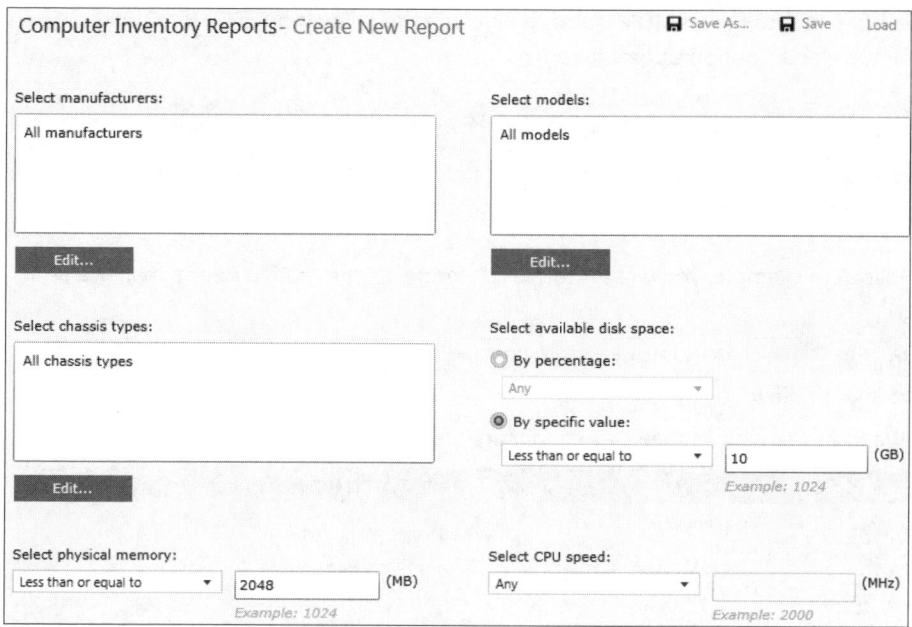

FIGURE 2-1 The Computer Inventory Reports page

1. Sign in to the Microsoft Intune admin console at *manage.microsoft.com/.*

2. Click the Reports workspace.

3. In the tasks list, click Computer Inventory Reports.

4. On the Computer Inventory Reports page, under Select Physical Memory, select Less Than Or Equal To from the dropdown, and enter **2048** for the value.

5. Under Select Available Disk Space, click By Specific Value, select Less Than Or Equal To from the dropdown, and enter **10** for the value.

6. Optionally, you can choose Save As at the top of the Computer Inventory Reports page to save these settings for later use.

7. Click View Report to run this report and view the results.

Assign mobile device management authority

Prior to enrolling your mobile devices, you need to define a mobile device management (MDM) authority within the Intune admin console. The MDM authority defines which system has permissions to manage your enrolled devices. For example, in an environment that has a hybrid solution running Configuration Manager with an integrated Intune subscription, the MDM authority will be set to Configuration Mannager and all enrolled devices will require the Configuration Manager client. If Microsoft Intune is the only management solution, it would be set as the MDM authority and all devices would require the Intune agent.

The available options for an MDM authority include the following: MDM for Office 365, Microsoft Intune, and Configuration Manager.

> **IMPORTANT   CHOOSING YOUR MOBILE DEVICE MANAGEMENT AUTHORITY**
>
> It is important to remember that once an MDM authority is defined in the Intune admin console it cannot be changed.

In the following example, you will set Microsoft Intune as the mobile device management authority.

1. Sign in to the Microsoft Intune admin console at *manage.microsoft.com/.*

2. Click the Admin workspace.

3. In the tasks list, click Mobile Device Management, as shown in Figure 2-2.

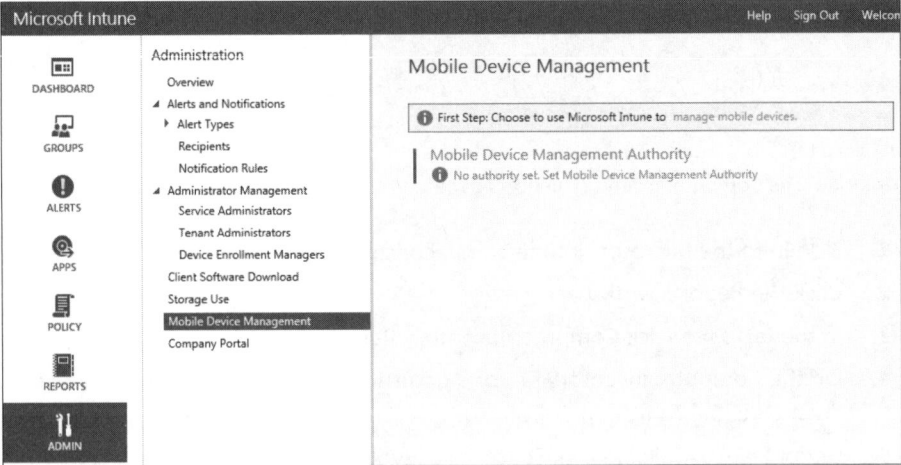

FIGURE 2-2 The Mobile Device Management page

4. On the Mobile Device Management page, click Set Mobile Device Management Authority.

The next time you select Mobile Device Management from the tasks list, you will see the assigned MDM authority, along with a number of new platform management options, as shown in Figure 2-3.

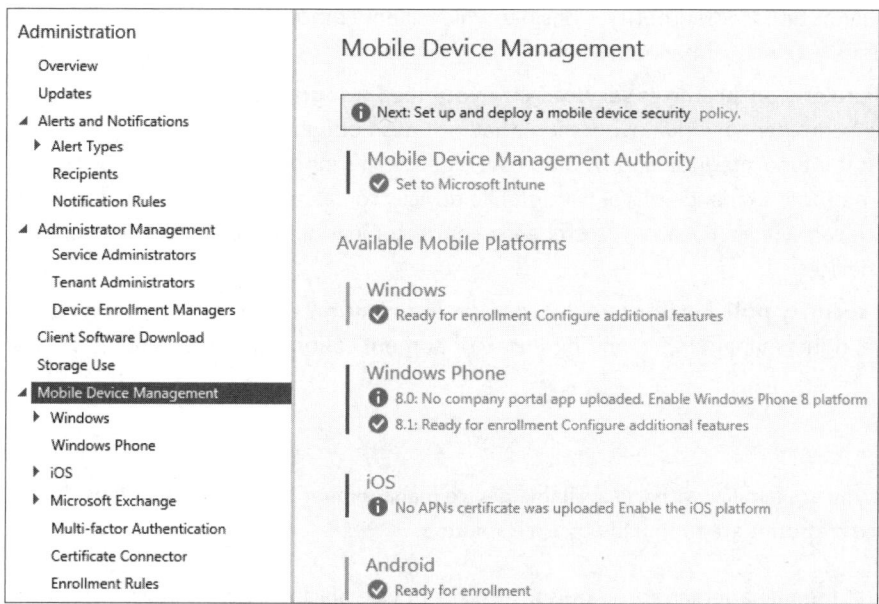

FIGURE 2-3 The Mobile Device Management page with Microsoft Intune set as the MDM authority

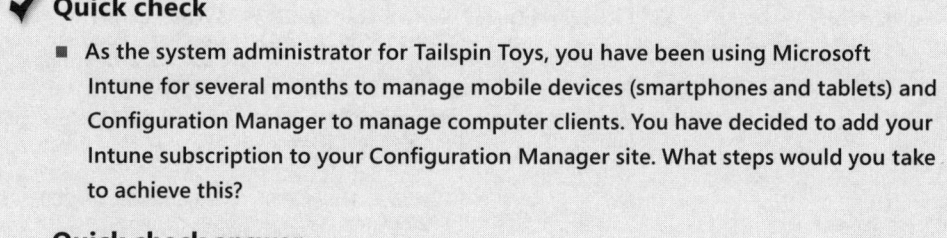

Quick check

- As the system administrator for Tailspin Toys, you have been using Microsoft Intune for several months to manage mobile devices (smartphones and tablets) and Configuration Manager to manage computer clients. You have decided to add your Intune subscription to your Configuration Manager site. What steps would you take to achieve this?

Quick check answer

- It cannot be added. Once the MDM authority has been set to Microsoft Intune, it cannot be undone.

Meet device management prerequisites

Before you can begin managing mobile devices with Microsoft Intune, you must review and plan for certain prerequisites. To achieve a successful deployment you need to understand the device management requirements for your organization. The following questions will help you assess the needs of your environment and prepare you for implementing Microsoft Intune:

1. **Which types of devices need to be managed in your organization?** Consider devices running Microsoft Windows Server or UNIX, which require Configuration

Manager as an MDM authority. Consider which client computers and mobile device platforms you will be managing.

2. **Are there any on-premises services that you need integrated with Microsoft Intune?** Consider a hybrid environment that includes Configuration Manager with Microsoft Intune integration, providing coverage for all supported platforms. Consider an on-premises Exchange server with mobile devices connected to Exchange ActiveSync, in which case you can enable the Exchange connector in Intune to discover and manage these devices.

3. **What security policies does your organization require?** Consider encryption compliance, rights management, and multi-factor authentication when accessing corporate data.

 EXAM TIP

Make sure you are familiar with the available device management solutions and understand which platforms are supported by each solution.

First, look at the management prerequisites from a device platform perspective. Refer to Table 2-2 for platform compatibility.

TABLE 2-2 Mobile device management compatibility chart

Device Platform	MDM for Office 365	Microsoft Intune	System Configuration Manager 2012 R2	System Center Configuration Manager 2012 R2 with Intune integration
Microsoft Windows	No	Windows 8.1 or later	Windows XP Professional SP3 or later	Windows XP Professional SP3 or later
Microsoft Windows Server	No	No	Windows Server 2003 SP2 or later	Windows Server 2003 SP2 or later
Windows Phone	Windows Phone 8.0 or later	Windows Phone 8.0 or later	No	Windows Phone 8.0 or later
Windows RT	No	Windows RT and 8.1 RT	No	Windows RT and 8.1 RT
iOS	iOS 7.1 or later	iOS 7.1 or later	No	iOS 7.1 or later
Android	Android 4.0 or later	Android 4.0 or later	No	Android 4.0 or later
Mac OS X	No	Mac OS X 10.9 or later	Mac OS X 10.9 or later	Mac OS X 10.9 or later
UNIX/Linux	No	No	Based on file dependencies	Based on file dependencies

With the variety of devices that Intune supports, you need to be familiar with the prerequisites for each platform. Some configurations are mandatory and others are optional. These settings are available in the Intune admin console by navigating to the Admin workspace and clicking the Mobile Device Management task. The following is a breakdown of the platform configuration options:

- **Windows** This section is used for configuring the management of devices running Microsoft Windows.
  - **Step 1: Enrollment Server Address** Assign a verified domain for user self-enrollment.
  - **Step 2: Add Sideloading Keys** Add sideloading keys for the installation of line-of-business apps.
  - **Step 3: Upload Code-Signing Certificate (Optional)** Upload code-signing certificates for line-of-business apps that are not already trusted.
- **Windows Phone** This section is used for configuring the management of devices running Windows Phone.
  - **Step 1: Enrollment Server Address** Assign a verified domain for user self-enrollment.
  - **Step 2: Obtain an Enterprise ID and Code-Signing Certificate (Optional)** Windows Phone 8 devices require a code-signing certificate from Symantec before they can be managed. Windows Phone 8.1 and Windows 10 Mobile can be managed without a code-signing certificate, but for sideloading apps one is required.
  - **Step 3: Download and Sign the Company Portal App** For Windows Phone 8 management you must download the Company Portal app and sign it with a Symantec code-signing certificate. Windows 8.1 and Windows 10 Mobile devices can download the Company Portal app from the Windows Store.
  - **Step 4: Upload and Deploy the Signed Company Portal App** For Windows Phone 8 management, upload, configure, and deploy the signed Company Portal app.
- **iOS** This section is used for configuring the management of devices running iOS.
  - **Step 1: Upload an APN Certificate** Assign an Apple Push Notifications (APN) certificate for managing iOS devices.
  - **Step 2: Device Enrollment Program** Assign a Device Enrollment Program (DEP) token for (over the air) enrollment of corporate-owned iOS devices. Devices enrolled through DEP cannot be un-enrolled by users.
- **Microsoft Exchange** This section is used for configuring the Microsoft Exchange Connector with your existing Microsoft Exchange infrastructure. Configuring this connector enables Microsoft Intune to discover and manage mobile devices that are currently connected to your environment using Exchange ActiveSync.

- **Multi-factor Authentication** This section is used for configuring multi-factor authentication within Microsoft Intune, requiring users to provide additional verification before accessing company resources.

- **Certificate Connector** This section is used for configuring the Certificate Connector, enabling you to deploy certificates to mobile devices managed by Microsoft Intune.

- **Enrollment Rules** This section is used to configure the device enrollment limit for your users. The default value is five.

Continuing with device-specific prerequisites, it's also important to note that Microsoft Intune does have web browser requirements. These requirements include the following:

- **Web Browser Prerequisites** Managed devices need to have a compatible web browser in order to access web services like the Company Portal. Intune administrators will also require a supported browser in order to access the Intune admin console. Refer to Table 2-3 for the list of compatible web browsers.

TABLE 2-3 Microsoft Intune web browser compatibility

Device Platform	Service	Supported Browser Version
Windows Phone, Windows RT, iOS, and Android	Company Portal	The default web browser is supported on each mobile device
Mac OS X	Company Portal	Safari browser on Mac OS X 10.9 or later
Microsoft Windows	Company Portal, Microsoft Intune account portal, Microsoft Intune admin console	Internet Explorer 9 or later Google Chrome 42 or earlier Mozilla Firefox Note: Microsoft Edge is not supported

Web Browser Plug-ins Client computers accessing the Company Portal using a web browser will require the Microsoft Silverlight plug-in. If it is not installed, the user will be prompted to download and install the plug-in before signing-in to the portal.

Now that you have reviewed device prerequisites, you can address infrastructure requirements. Microsoft Intune is a cloud-based management service, requiring you to address specific network infrastructure prerequisites. These prerequisites include the following:

- **Firewall Prerequisites** Devices require regular Internet communication with Microsoft Intune and related online services. Refer to the diagram in Figure 2-4 for a layout that describes the required services, ports, and domains that Microsoft Intune requires.

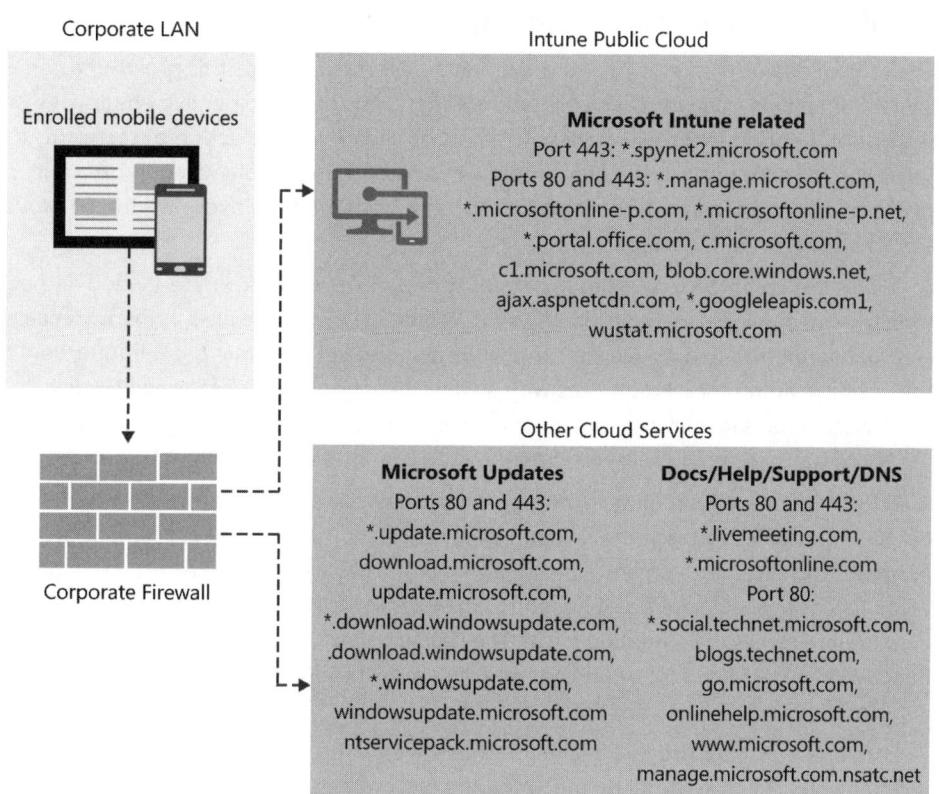

FIGURE 2-4 The firewall requirements for Microsoft Intune

- **Bandwidth Prerequisites** Your organization's Internet connection needs to have adequate bandwidth to accommodate the added client traffic. Depending on the platform, device enrollment requires a client download that is approximately 125 MB in size. Other traffic activities include daily client communication, malware definition updates, software updates, and software distribution.

NEED MORE REVIEW? **UNDERSTAND THE MICROSOFT INTUNE BANDWIDTH REQUIREMENTS**

For more information about the bandwidth considerations required to support Microsoft Intune, visit *https://technet.microsoft.com/library/dn646966.aspx#BKMK_NetworkBandwidth*.

Configure device enrollment profiles

The mobile device market is growing exponentially. The influence this growth has on consumers directly impacts the business world. From a business perspective employees want the option to bring their own devices to work, or have a wide selection of platforms that the company can support. If a new employee is more comfortable working on a particular platform, but the organization doesn't support it, both the employee and the employer will see an impact in performance.

Today, it is common to see organizations with a mixture of corporate-owned and personally owned devices working side by side. Personally owned devices in the workplace are commonly referred to as *BYOD*, or bring your own device. With Microsoft Intune, companies can start providing their employees with a wider selection of corporate-owned devices while still ensuring the same level of governance. Enrollment of these devices can be handled in a variety of ways, including self-enrollment by the user or direct enrollment by an administrator.

When it comes to enrolling corporate-owned devices, administrators can use device enrollment profiles to tailor the first-run experience and simplify the enrollment process. For example, you can create an enrollment profile that disables all of the setup assistant prompts on an iOS device, eliminating the need for the user to walk through the various options. Enrollment profiles are compatible with the iOS and Mac OS X platforms. You can create them directly from the Microsoft Intune admin console or through the Apple Configurator desktop application.

Enrollment profiles provide a solution that scales with larger deployments. For example, imagine you are the system administrator at a high school. Over the summer, the school orders 300 iPads for their students to start using during the fall semester. It is your job to enroll them with Microsoft Intune and set up the high school wireless network profile so the devices can get online. You need a streamlined method to enroll and configure these devices. You also want to customize the iOS Setup Assistant so students aren't prompted with a number of questions when they turn on the device. By using an enrollment profile, you as the administrator can achieve these requirements. Once you have created the profile you can use Apple Configurator to mass deploy it. The custom profile will provide a tailored first-run experience and ensure that the device is properly enrolled with Microsoft Intune.

The following steps describe how to create and deploy a device enrollment profile using Microsoft Intune.

STEP 1: ASSIGN AN INTUNE ACCOUNT AS A DEVICE ENROLLMENT MANAGER

1. Sign in to the Microsoft Intune admin console at *http://manage.microsoft.com/*.
2. Click the Admin workspace.
3. In the tasks list, click Device Enrollment Managers.

4. On the Device Enrollment Managers page, click Add.

5. On the Add Device Enrollment Manager dialog box, enter the user ID for the Intune user account. Click OK to add the user account and confirm it appears in the list of Device Enrollment Managers, as shown in Figure 2-5.

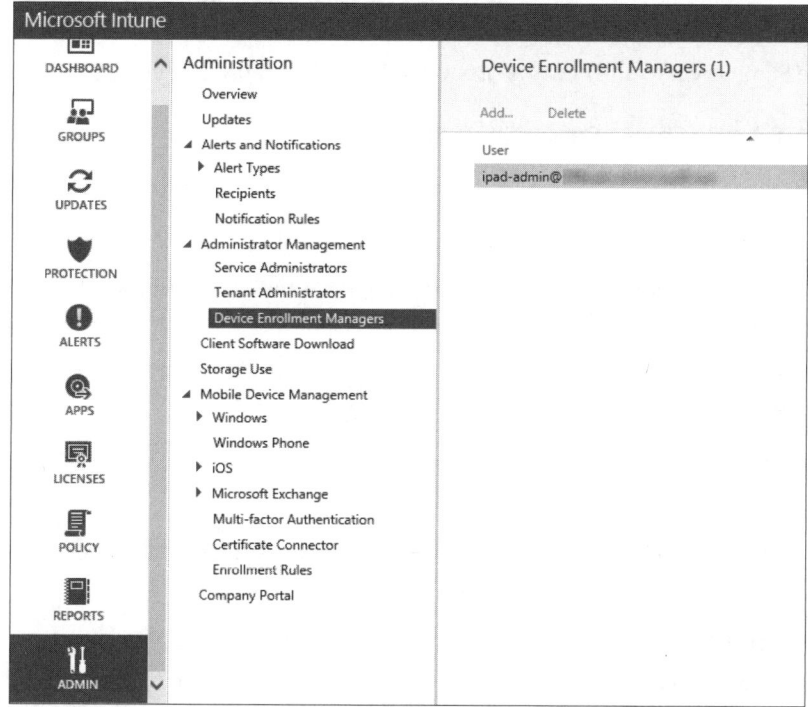

FIGURE 2-5 The Microsoft Intune admin console showing the Device Enrollment Managers page

STEP 2 – CREATE A NEW DEVICE GROUP

1. Click the Groups workspace.

2. In the tasks list, right-click All Devices and select Create Group. Groups are used for managing devices and scoping deployments.

3. On the General page, fill in the following criteria and click Next.

- **Group name** Enter a name for the group. In this example we will call our group Student iPads.

- **Description** Enter a description for the group.

- **Select A Parent Group** Select All Devices from the list. As your environment grows you can start choosing child groups to help scope future deployments.

4. On the Criteria Membership page, select the following settings and click Next.

- **Device Type** Select Mobile.

- **Specify The Mobile Devices To Be Included In The New Group** Click Empty Group. You are setting up a new group of devices so you want to avoid including any existing clients.

5. On the Direct Membership page, leave the fields blank and click Next. You have the option to selectively include and exclude devices.

6. On the Summary page, review the group configuration and click Finish.

STEP 3 – CREATE A NEW DEVICE ENROLLMENT PROFILE

In this step, you create a new device enrollment profile:

> **IMPORTANT   DEFAULT ENROLLMENT PROFILE**
>
> The first device enrollment profile you create is assigned as the default for all future pre-enrolled devices. The default profile cannot be deleted until a new profile is created and assigned as the default.

1. Click the Policy group.
2. In the tasks list, click Corporate Device Enrollment.
3. On the Corporate Device Enrollment page, click Add.
4. On the Create Profile page, fill in the following criteria, as shown in Figure 2-6:

- **Name** Enter a name for the device enrollment profile.

- **Description** Enter a description for the device enrollment profile.

- **Enrollment Details** User Affinity. Click the drop-down list and select Prompt For User Affinity. User Affinity associates the device with a user account, which is beneficial when running reports or looking at inventory. Self-enrollment handles this automatically when users enter their credentials for the first time. With enrollment profiles you have the option to not require User Affinity. In the high school scenario, you will require User Affinity to ensure students can access their applications and sign in with their Office 365 account.

- **Assign Devices To The Following Group** Select the Student iPads group from the dropdown list. Use Group assignments to assign different enrollment profiles to different groups of devices. For example, you can split students and faculty into two separate groups, each with their own enrollment profile and configuration.

- **Device Enrollment Program Settings**   Leave the slider in the Off position. DEP is a service Apple provides to their customers for over-the-air MDM enrollment. This is covered in more detail later in this section.

5. Click Save Profile.

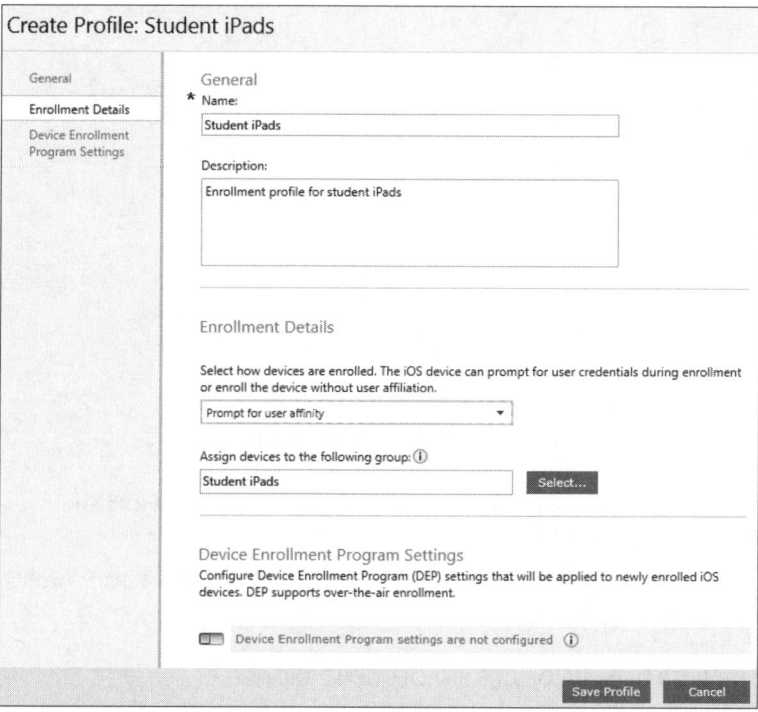

FIGURE 2-6   The Create Profile page

STEP 4 – IMPORT PRE-ENROLLED DEVICES AND ASSIGN AN ENROLLMENT PROFILE

1. Click the Groups workspace.

2. In the tasks list, click to expand Corporate Pre-enrolled devices. Click By iOS Serial Number.

3. On the By iOS Serial Number page, click Add Devices.

4. On the Select Method page, select Upload A CSV File Containing Serial Numbers And Details. You can add pre-enrolled mobile devices to Microsoft Intune using one of two methods:

   - The first is a comma-separated (.csv) file containing two columns. Column 1 is mandatory and must contain the list of mobile device serial numbers. Column 2 is optional and can contain any remarks, names, or other identifiable notes that you want to associate with the mobile device serial number.

- The second is to manually enter the serial number and remarks using the web interface.

5. Click Browse, locate the .csv file, click Open, and click Next to upload the .csv file. Figure 2-7 shows a sample .csv file.

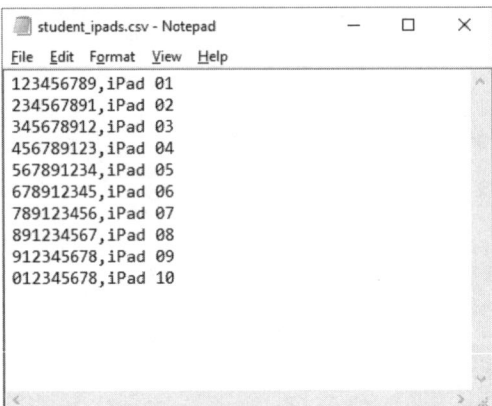

FIGURE 2-7 A comma-separated (.csv) file for bulk device import

6. On the Review Devices page, review any conflicts with existing devices and overwrite them as needed. Click Next to continue.

7. On the Assign Profile page, under Assign Profile, select the Student iPads profile from the dropdown list and click Finish.

STEP 5 – ENROLL DEVICES USING A DEVICE ENROLLMENT PROFILE

1. Click the Policy group.

2. In the tasks list, click Corporate Device Enrollment.

3. On the Corporate Device Enrollment page, select the Student iPads enrollment profile and click Export.

4. Copy the profile URL to a Mac running Apple Configurator. Add the URL to Apple Configurator with the iPad attached. Apple Configurator will connect to the URL, download the profile, and reset the iPad using the custom profile.

Enrollment profiles can also reduce administrative overhead in geographically dispersed environments. Microsoft Intune supports Apple's Device Enrollment Program. This service pre-registers new devices with Intune and enrolls them over the air, without any physical interaction from IT. For example, you are the system administrator for Tailspin Toys, and its headquarters are located in New York City, but you have sales staff all over the United States. A new iPhone has been requested for Jim Daly out of Chicago. During the procurement process you provide your DEP customer ID to the reseller for device registration. The iPhone is shipped directly from the service provider to Jim's home in Chicago. The first time Jim starts the iPhone it receives your enrollment profile over the air, enrolls in Microsoft Intune, and receives the associated device policies. The following steps will guide you through setting up DEP, creating a DEP enrollment profile, and deploying it to your pre-enrolled devices.

STEP 6 – ENABLE THE DEVICE ENROLLMENT PROGRAM

1. Click the Intune Admin workspace.

2. In the tasks list, click to expand iOS and Mac OS X. Click Device Enrollment Program. Note that this option will not appear until you have uploaded an APN certificate.

3. On the Upload A Device Enrollment Program Token page, click the button to Download Encryption Key. Save the encryption key (.pem) locally.

4. Click the Apple Device Enrollment Program Portal link and log in using your Apple ID. In order to log in to the Apple Device Enrollment Program, you must be a member of a business or educational institution with an active Apple ID that has been registered for DEP.

5. From the DEP portal, click to expand Device Enrollment Program. Click Manage Servers, and then click Add MDM Server.

6. In the Add MDM Server dialog box, enter a name to refer to the Intune server connection and click Next. The server name given is purely for identification.

7. In the Add <ServerName> dialog box, click Choose File. Locate the DEP encryption key and click Next to upload it.

8. In the Add <ServerName> dialog box, click the Your Server Token link to download the DEP server token (.p7m). Click Done to complete the process.

9. Back on the Microsoft Intune admin console, click Upload The DEP Token.

10. In the Upload The DEP Token dialog box, for the DEP Token, click Browse and locate the DEP server token. For the Apple ID, enter the Apple ID credentials associated with your DEP account. Click Upload to complete the process.

STEP 7 – UPDATE THE DEVICE ENROLLMENT PROFILE

1. Create a new device group to manage corporate-owned iPhones.

2. Click the Policy workspace.

3. In the tasks list, click Corporate Device Enrollment.

4. On the Corporate Device Enrollment page, click Add.

5. On the Create Profile page, complete the form. Under Device Enrollment Program Settings, toggle the slider to enable this feature. Fill in the following criteria, as shown in Figure 2-8 and Figure 2-9.

- **Department** Enter a department name associated with this enrollment profile. This field will be visible on the device during the Setup Assistance experience.

- **Support Phone Number** Enter a support phone number. This field will be visible on the device during the Setup Assistant experience.

- **Preparation Mode** Select Supervised from the dropdown list. Supervised devices give the administrator more granular control, including the ability to manage the device even after a remote wipe. Unsupervised devices are more open. Supervised mode is more tailored toward a classroom environment, whereas unsupervised mode would be more ideal for bring your own device.

- **Lock Enrollment Profile To Device** Select Enable from the dropdown list. Locking the enrollment profile to the device prevents removal.

- **Setup Assistant Panes** Set each of the values to Disable. These settings control the Setup Assistant experience through which the user navigates. In this example, you do not want students to set these options. They will be handled separately through the use of policies.

- **Enable Additional Apple Configurator Management** Select Disallow from the dropdown list. This setting enables an administrator to set up post-configuration as an option. You might enable this in a test scenario or in an environment where device requirements change regularly. You can also add additional security through the use of an Apple Configurator certificate.

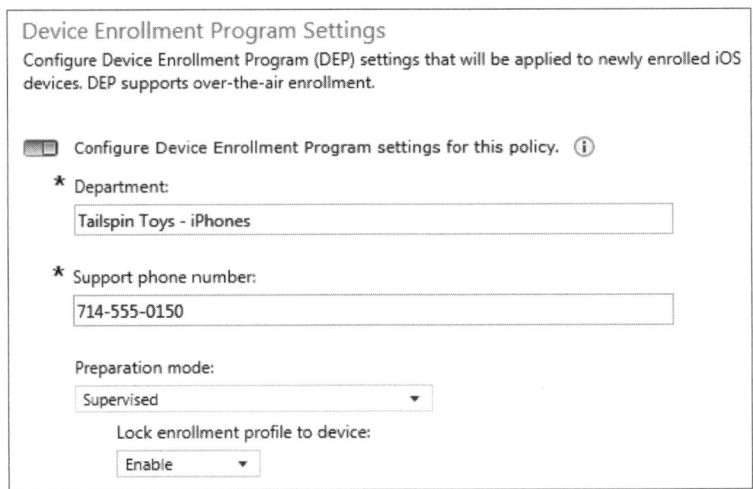

FIGURE 2-8 The DEP settings on the Create Profile page for Corporate Device Enrollment

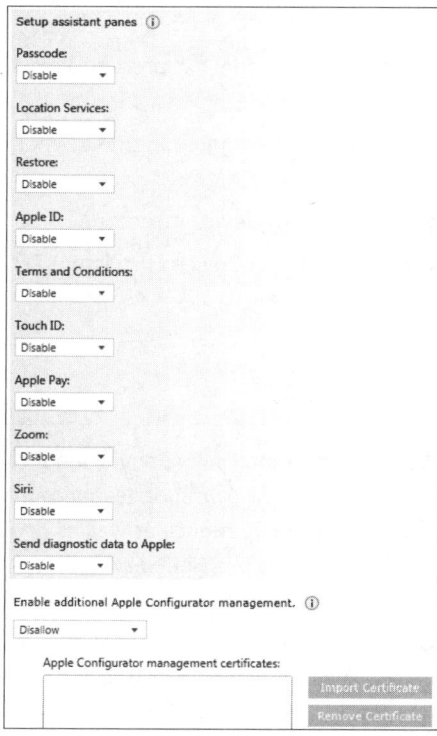

FIGURE 2-9 Additional DEP settings on the Create Profile page

6. Click Save Profile.

✔ **Quick check**

- You are a system administrator for Tailspin Toys, working from their San Francisco office. The marketing department manager for Tailspin Toys is planning to hand out iPads to his team at an upcoming conference in Chicago during the holidays. He needs to order 50 new iPads for the event, but he would prefer to have them shipped directly to Chicago. You need to insure these devices are enrolled and configured to access company resources. What steps would you take to achieve this?

Quick check answer

1. Create a configuration profile in Microsoft Intune that has DEP enabled and configured.

2. Sign up for the Apple device enrollment program.

3. Submit an order for the 50 iPads using your DEP account ID.

4. Retrieve a list of the iPad serial numbers and pre-enroll them into the Microsoft Intune admin console using a .csv file. Assign the new configuration profile to the pre-enrolled devices to automate their enrollment.

Summary

- Know which device platforms are supported by Intune, and what prerequisites a particular platform may have before it can be managed.
- Know what the system prerequisites are for Intune, like assigning Intune as your MDM authority.
- Know how to navigate the report workspace and review device inventory.
- Know the difference between an enrollment profile and a Device Enrollment Profile. Understand the use cases and deployment process for each.

Skill 2.2: Plan for the Company Portal

Many industry-leading MDM providers have adopted a common philosophy when it comes to device management: empowering your users to manage their own mobile device experience. In recent years a big emphasis has been placed on user experience as well as on finding ways to further reduce administrative overhead within IT. Terms like (self-service,) (app store,) and (one-click install) are all core features in today's MDM solutions.

With Microsoft Intune your users will have access to the Intune Company Portal, a custom-tailored access point that empowers them to manage their own mobile device experience. As an administrator, you still have all the tools you need to manage and secure these devices, but Intune lets users take advantage of the familiar app store experience using the self-service theme. For this skill, you will review how to customize the Company Portal, including the user terms and conditions, take a look at the available device policies, and close with some real-world use cases.

> **This section covers how to:**
> - Customize the Company Portal and company terms and conditions
> - Design configuration policies, compliance policies, conditional access policies, Exchange ActiveSync policies, and resolve policy conflicts

Customize the Company Portal and company terms and conditions

The Microsoft Intune Company Portal is accessible to all licensed and enabled Intune user accounts. Users are required to sign in using their account credentials. The Company Portal is available across all Intune supported devices. The portal is available in one of two formats:

- **Company Portal Website** Users of devices with a supported web browser can go to the following web address and sign in with their Intune credentials: https://*portal.manage.microsoft.com/*.

- **Company Portal App** The Company Portal app is available to all devices with a supported app store ecosystem. In Table 2-4, you can see the device types and their compatibility with the Company Portal.

TABLE 2-4 Company Portal app compatibility

Platform	Company Portal App Availability
Microsoft Windows	Windows 8.0 or later through the Microsoft Store
Microsoft Windows Server	No
Windows Phone	Windows Phone 8 or later through the Microsoft Store
Windows RT	Windows RT or later through the Microsoft Store
iOS	iOS 7.1 or later through the App Store
Android	Android 4.0 or later through the Google Play store
Mac OS X	No
UNIX/Linux	No

When users sign in to the Company Portal they will find multiple resources at their disposal. As the administrator, you have the ability to customize these settings to help enhance the user experience and provide more value to your users. The following is a look at the existing Company Portal features for Microsoft Intune.

- **Enroll Devices** This feature provides users with the capability to enroll their device with Microsoft Intune, whether the device is personally owned or corporate-owned. You can streamline the enrollment experience and any legal requirements by using the Terms and Conditions policy, discussed in more detail later in this section.

- **View Device Status** This feature provides an overview of your users' currently enrolled devices. Users can then select any of their devices by name and view information like model, manufacturer, operating system, and policy compliance.

- **Install Software** This feature provides users with a list of custom-tailored corporate applications. Users can navigate the available options and choose to install the software they need.

- **IT support** This feature provides users with the contact information for their organization's help desk or other support resource. Contact information can include the phone number, email address, and web address for environments that use a ticket system.

Each of the features mentioned previously is targeted at making the IT ecosystem more efficient and lowering overall costs. Let's review a real-world scenario that highlights some of these benefits: Imagine you are a system administrator for Tailspin Toys, a global organization with 4,500 employees. For the past several years the company has been providing their employees with email access on their mobile devices using Exchange ActiveSync. You have written documentation that accounts for the assortment of operating systems and Exchange ActiveSync settings.

Every time a new device is enrolled you are required to enter multiple pieces of information. Once connected the device is automatically quarantined as part of your company's mobile device security policy. Employees must then open a service desk ticket and attach a signed copy of the organization's mobile device terms and conditions policy. That ticket needs to be reviewed and approved by management before the device can be activated. It isn't uncommon for this process to take 2-3 days.

As the system administrator for Tailspin Toys, you recently implemented Microsoft Intune as the MDM solution. Using the Company Portal you have converted the existing terms and conditions form into an Intune policy, configured a series of device security policies, and enabled self-enrollment for employees. Users can now choose to self-enroll any of their mobile devices up to a maximum of five, pending their acceptance of the terms and conditions. Reports are run on an as-needed basis to audit device activity, security compliance, and accepted terms and conditions.

In Figure 2-10, you can see an example of the client computer experience when accessing the Company Portal from a web browser.

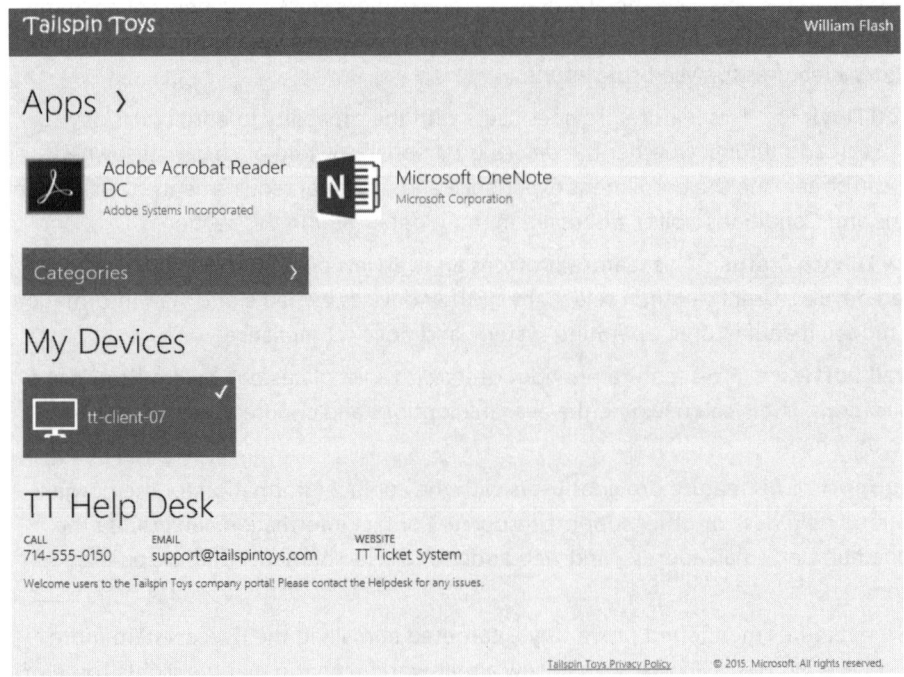

FIGURE 2-10 The Microsoft Intune Company Portal

You have control over several options that are available in the Microsoft Intune admin console, which are described in the following steps.

1. Sign in to the Microsoft Intune admin console via *https://manage.microsoft.com/.*
2. Click the Admin workspace.
3. In the tasks list, click Company Portal.
4. On the Company Portal page, fill in the options according to your organization. These options include:

   - **Company Name**   This name is displayed at the top of the Company Portal.
   - **IT Department Contact Name**   This name is displayed in the IT contact section.
   - **IT Department Email Address**   This address is displayed in the IT contact section.
   - **Additional Information**   This text is displayed under the IT contact section.
   - **Company Privacy Statement URL**   This URL is displayed at the bottom of the Company Portal.
   - **Support Website URL** (not displayed)   This URL is listed in the IT contact section.
   - **Website Name** (displayed to user)   This name masks the previous URL, keeping the presentation clean.
   - **Theme Color**   This setting adjusts the color theme for the Company Portal.
   - **Include Company Logo**   When set up, this setting enables you to upload a company logo in JPG or PNG format, no larger than 750 KB, with a maximum resolution of 400 x 100 pixels. You have the option to upload two versions of your logo: one for when the portal is using a white background and one for when the portal is using the selected theme color. The variation in background colors is isolated to the Windows 8 Company Portal app.
   - **Choose A Background For Windows 8 Company Portal App**   This setting enables you to choose between the default Company Portal app background and a solid white background.
5. Click Save to apply your changes to the Company Portal.

> **NOTE   REVIEW CHANGES TO THE COMPANY PORTAL**
> At the bottom of the Company Portal customization page, a link will open the Microsoft Intune Company Portal website. This can be helpful for reviewing recent changes.

Many organizations require their employees to accept a series of terms before allowing them to access company data from a mobile device. Terms can be more rigorous for a user who wants to access data from a personally owned device. It is not uncommon that this process also involves a physical printout and handwritten signature, resulting in lengthy paper trails. After customizing the Company Portal interface, you need to become familiar with the Terms and Conditions policy. This single feature can reduce IT overhead exponentially and administrators should use it whenever possible. The following items outline how this feature works and what scenarios it can support:

- **Reviewing the terms and conditions**   When a new terms and conditions policy is deployed users will be prompted to review it the next time they access the Company Portal, whether or not the connected device is enrolled with Intune.

- **Accepting the terms and conditions**   Accepting the terms will grant the user access to the Company Portal. Declining the terms will log the user out before accessing the Company Portal.

- **User deployment**   Terms and conditions are deployed to users, not devices. This means users will only need to accept the terms once from any of their devices, regardless of enrollment. Enrollment is not a requirement to access the Company Portal.

- **Multiple policies**   Terms and conditions are created as policies. You have the flexibility to create multiple policies and deploy them to different groups based on the organization's requirements. For example, you may have a global set of terms and conditions for accessing the Company Portal, in which case you would create the policy and deploy it to all users. You may also have a requirement that deals with user conduct when accessing the Google Play store on Android devices, in which case you would create a separate terms and conditions policy and deploy it to a group containing only Android users.

- **Versions**   Each terms and conditions policy includes a version number attribute. As the administrator, you can modify an existing policy and choose whether or not to update the version number. When you update the version number it will force all of your users in the deployed group to review the updated terms and choose to accept or decline them the next time they access the Company Portal. If you choose not to update the version number only new users will see the updated terms. The accepted version number is also included in the terms and conditions report.

The following steps demonstrate how to create a terms and conditions policy, deploy it to a group of users, update it, and generate a report that identifies who has accepted the terms.

STEP 1 – CREATE A TERMS AND CONDITIONS POLICY

1. Sign in to the Microsoft Intune admin console at *manage.microsoft.com/*.
2. Click the Policy workspace.
3. In the tasks list, click Terms And Conditions.
4. On the Terms And Conditions page, click Add.
5. On the Create Terms And Conditions page, fill in the fields according to your organization's needs. Refer to Figure 2-11 as an example. The fields that are available include the following:
   - **Name**   This field is used to identify the policy from within the Intune admin console. This field is only visible to administrators.

- **Description** This field is also used to identify the policy from within the Intune console and is also only visible to administrators.

- **Title** This field populates the title of the terms and conditions. Users will see this field at the top of the page when reviewing the terms from within the Company Portal.

- **Text For Terms** This field contains the terms you are requesting users to accept. All terms related to the new policy should be included in this field.

- **Text To Explain What It Means If The User Accepts** This field contains the explanation of acceptance to users. For example, "By clicking accept, you agree to abide by these terms and conditions."

6. Click Save to create the policy.

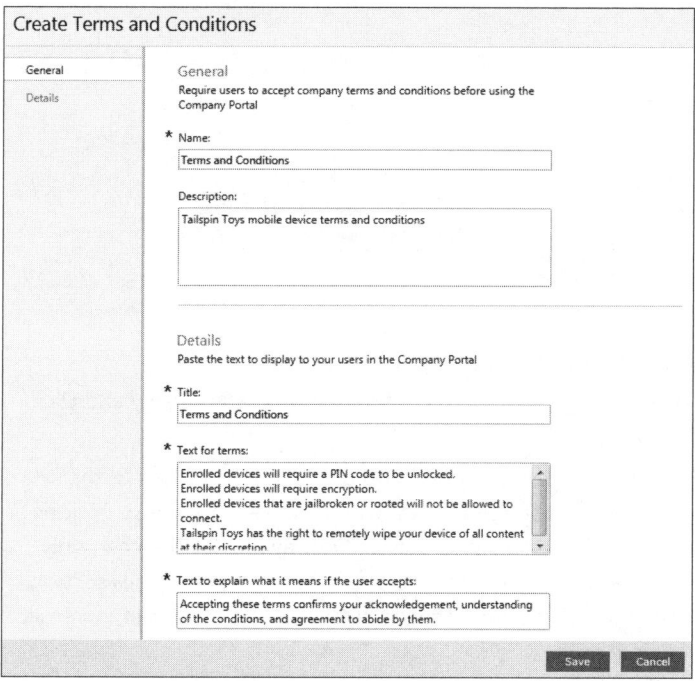

FIGURE 2-11 The Create Terms and Conditions page

STEP 2 – DEPLOY THE TERMS AND CONDITIONS

1. On the Terms And Conditions Policies page, select the newly created policy and click Manage Deployment.

2. In the Select The Groups To Which You Want To Deploy This Policy dialog box, select the target group from the left and click Add. If you want to apply this to all users, select the All Users group.

3. Click OK to create the deployment. The next user in the deployed group to access the Company Portal will be prompted to review the terms and conditions, as shown in Figure 2-12. Clicking Read Terms displays the contents of the agreement.

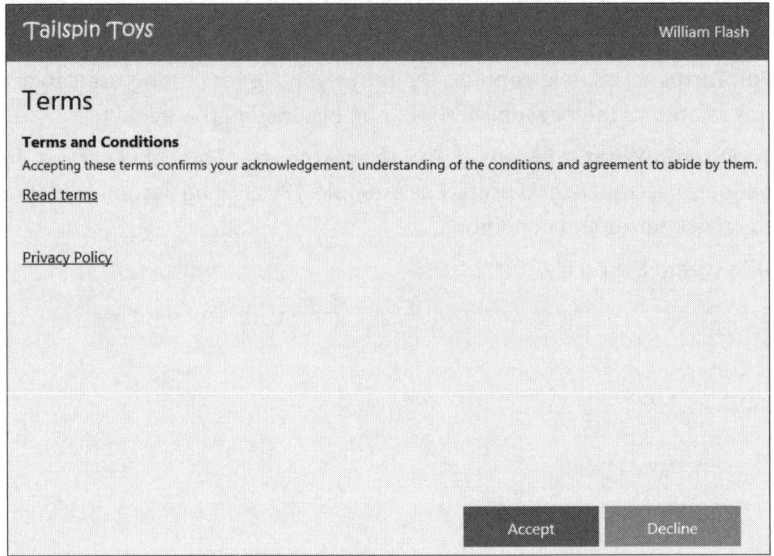

FIGURE 2-12   The new Terms and Conditions policy.

STEP 3 – UPDATE THE TERMS AND CONDITIONS POLICY

1. On the Terms And Conditions Policies page, select the newly created policy and click Edit.

2. On the Edit Terms And Conditions page, you will find a new set of options at the bottom of the page dealing with versioning, as shown in Figure 2-13. These options are dimmed or "grayed out" initially, meaning they are not available. They can only be adjusted after a change has been made to one of the fields in the Details section. Once a change is made to the text in one of these fields, you can choose to increase the version number or leave it as is.

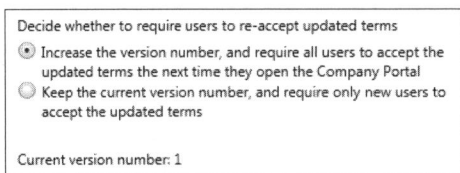

FIGURE 2-13   The Edit Terms and Conditions page

3. Click Save to update the policy.

STEP 4 – RUN THE TERMS AND CONDITIONS REPORT

1. Click the Reports workspace.

2. In the tasks list, click the Terms And Conditions Reports.

3. On the Terms And Conditions Reports page, you have two options available to generate this report.

   - **Option 1:** Click the Users option and click Edit under Select Users. Select the check box for each user you would like to include in the report and click OK. It is important to note that this option enables you to select a maximum of 10 users per report performance. If you need to see the details for more than 10 users, it is recommended that you run the report based on terms, as shown in option 2.

   - **Option 2:** Click the Terms option and click Edit under Select Terms And Conditions Policy. Select the check box for each policy you would like to include in the report and click OK.

4. Click View Report to see the results. In Figure 2-14, you can see an example of the final report using the Terms option. The output for this report shows all of the policies that were selected, the users that viewed the policy, which version they accepted, when they accepted it, and whether or not they accepted the latest version.

Terms and Conditions Name	User	Accepted Version	Date Accepted	Accepted Latest
Google Play Terms and Conditions	Dean Halmaert	1	11/30/2015 10:37:35	Yes
Google Play Terms and Conditions	Glen John	1	11/30/2015 10:32:34	Yes
Google Play Terms and Conditions	Erik Rucker	0	No date available	No
Google Play Terms and Conditions	David Pelton	1	11/30/2015 10:34:03	Yes
Google Play Terms and Conditions	Tom Perham	0	No date available	No
Google Play Terms and Conditions	Daniel Roth	0	No date available	No
Google Play Terms and Conditions	John Kane	1	11/30/2015 10:36:52	Yes
Google Play Terms and Conditions	Aaron Hulett	1	11/30/2015 10:31:27	Yes
Google Play Terms and Conditions	William Flash	0	No date available	No
Google Play Terms and Conditions	Julian Isla	1	11/30/2015 10:34:24	Yes
Google Play Terms and Conditions	Sarah Jones	1	11/30/2015 10:35:41	Yes
Terms and Conditions	Dean Halmaert	2	11/30/2015 10:37:35	Yes
Terms and Conditions	Glen John	2	11/30/2015 10:32:34	Yes
Terms and Conditions	Erik Rucker	0	No date available	No
Terms and Conditions	David Pelton	2	11/30/2015 10:34:02	Yes
Terms and Conditions	Tom Perham	0	No date available	No
Terms and Conditions	Daniel Roth	0	No date available	No
Terms and Conditions	John Kane	2	11/30/2015 10:36:52	Yes
Terms and Conditions	Aaron Hulett	2	11/30/2015 10:31:28	Yes
Terms and Conditions	William Flash	1	11/30/2015 10:09:18	No
Terms and Conditions	Julian Isla	2	11/30/2015 10:34:25	Yes
Terms and Conditions	Sarah Jones	2	11/30/2015 10:35:41	Yes

Terms and Conditions Report (22)   Search

FIGURE 2-14 The Terms and Conditions Report

Intune design policies and policy conflicts

Continuing through the device management lifecycle, you arrive at a critical step in the planning process: Microsoft Intune policies. Similar to Group Policy, Intune policies help administrators control certain aspects of the operating system and application behavior on enrolled mobile devices and client computers. Policies are administered from within the Intune admin console. A policy contains a group of administrative settings that you can set according to the needs of your environment. From within the admin console, policies are broken down into three main categories: configuration, compliance and conditional access. Refer to Table 2-5 for the available Intune policy types.

TABLE 2-5 Microsoft Intune Policy Types

Policy Type	Purpose	Example Use Cases
Configuration Policies	Apply device configurations to assist with administration and enhance the user experience.	Deploy a configuration policy that automates the setup of the user's email client, wireless network profile, and VPN profile.
Compliance Policies	Require a specified level of security compliance on the enrolled device.	Deploy a compliance policy that requires a password when users unlock their devices, requires device encryption, and validates the device is not jailbroken or rooted.
Conditional Access Policies	Require a certain level of compliance before accessing Exchange email and/or SharePoint Online.	Enable a conditional access policy for Exchange Online that requires the connected device to be domain joined before accessing email.

Configuration policies

Configuration policies contain the widest assortment of options for administrators. This is mainly due to the wide range of devices that Intune supports, and each one having its own method for controlling system settings. For example, you will find a separate wireless profile policy for Android, iOS, Mac OS X, Windows Phone, and Windows desktop. To help navigate these options, Microsoft Intune provides policy templates. These templates mask much of the administrative overhead of creating individual configuration files that contain all of the necessary settings. Instead, when you select a policy template you are presented with a web form that contains empty fields, checkboxes, and dropdown menus.

When you create a new configuration policy you will be prompted to select from an arrangement of pre-generated templates based on the content you need to manage, as shown in Figure 2-15.

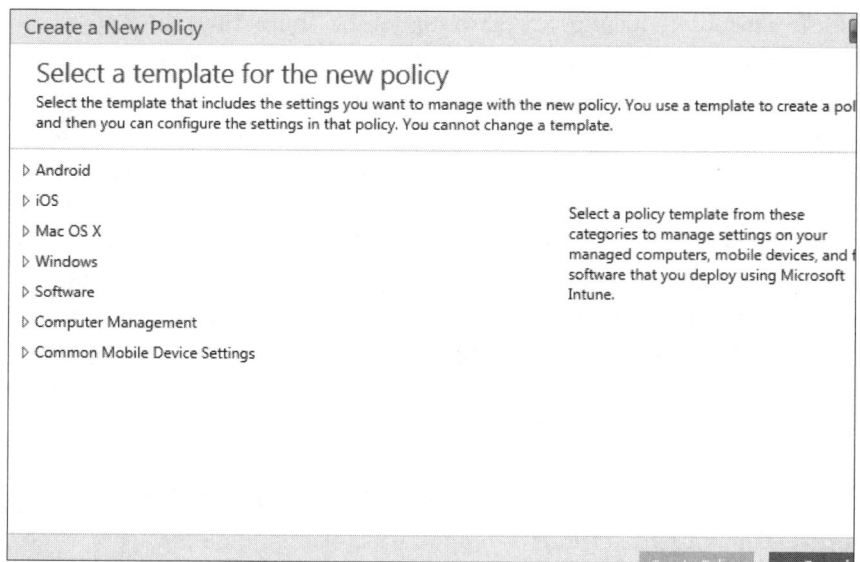

FIGURE 2-15 The Microsoft Intune admin console showing the available options for creating a new configuration policy.

The policy web interface is split into different categories based on device or task. The following list provides additional details for each of these categories:

- **Android** This section contains policy templates for the Android mobile device platform. Administrators have the option to configure wireless profiles, VPN profiles, certificate profiles, email profiles, and general device configuration profiles using a series of pre-defined options or by uploading a custom configuration based on the Open Mobile Alliance Uniform Resource Identifier (OMA-URI) standard.

- **iOS** This section contains policy templates for the iOS mobile device platform. Administrators can configure wireless profiles, VPN profiles, certificate profiles, email profiles, mobile app configurations, and general device configuration profiles using a series of pre-defined options or by uploading a custom configuration file created using the Apple Configurator.

- **Mac OS X** This section contains policy templates for the Mac OS X device platform. Administrators can configure wireless profiles, VPN profiles, certificate profiles, and general device configuration profiles using a series of pre-defined options.

- **Windows** This section contains policy templates for the Windows Phone mobile device platform and the Windows desktop platform. Administrators can configure wireless profiles, VPN profiles, certificate profiles, email profiles, and general device configuration profiles using a series of pre-defined options or by uploading a custom configuration file based on the OMA-URI standard.

- **Software** This section contains policy templates for the Android and iOS mobile device platforms. Administrators are presented with two template options:

  - **Microsoft Intune Managed Browser** This is a dedicated web browser app that can be deployed to enrolled devices. The corresponding policy template enables you to configure restrictions associated with this browser. For example, you can restrict users from browsing certain URLs when using the Microsoft Intune Managed Browser.

  - **Mobile Application Management** This policy template enables administrators to control the behavior of mobile device apps by applying the policy when an app is deployed using the Company Portal. For example, you can create a mobile application management policy that prevents the data of an associated app from being backed up using iTunes or iCloud.

- **Computer Management** This section contains policy templates for the Windows desktop platform. Administrators are presented with three template options:

  - **Microsoft Intune Agent Settings** This policy template controls the agent settings for Microsoft Intune Endpoint Protection and Microsoft Intune update services. These settings enables administrators to control Endpoint Protection scans and exclusions, as well as the software update schedule and behavior.

  - **Microsoft Intune Center Settings** This policy template enables administrators to populate their support contact information in the Microsoft Intune Center. The Microsoft Intune Center is a desktop application that is installed as part of the Intune agent.

  - **Windows Firewall Settings** This policy template enables administrators to control the Windows Firewall, including its enablement and active exclusions.

- **Common Mobile Device Settings** This section contains policy templates for common mobile device settings. Administrators are presented with two template options:

  - **Exchange ActiveSync** This policy template contains a number of security settings similar to what you would see in the ActiveSync policy found on an Exchange server. Options include password requirements, encryption requirements, and email behavior.

  - **Mobile Device Security** This policy template contains common mobile device security settings that work across multiple platforms. For example, when you use this policy it enables you to enforce password and encryption requirements across iOS, Android, and Windows Phone devices. This can be useful in situations where you need to deploy the same settings across multiple devices and want to avoid creating an individual policy for each one.

Now that you have reviewed the different configuration policy templates, take a look at a real-world scenario where you need to create a configuration policy from scratch and deploy it to some enrolled devices. In this example, you are preparing to enroll some devices into Microsoft Intune for Tailspin Toys. The security team has reviewed the Endpoint Protection client and would like you to adjust some of the security settings. You start by creating a Microsoft Intune Agent Settings policy. Typically, this is one of the first policies you will create before enrolling devices, because this policy controls the deployment and configuration for the Endpoint Protection agent, as well as the configuration for Windows update services. You will create the policy and then deploy it to a group of devices. Note that you can only deploy configuration policies to devices.

1. Sign in to the Microsoft Intune admin console at *https://manage.microsoft.com/*.

2. Click the Policy workspace.

3. In the tasks list, click Configuration Policies.

4. Click the Add button.

5. In the Create A New Policy dialog box, click to expand Computer Management.

6. Click Microsoft Intune Agent Settings. In the right pane, select Create Deploy a Custom Policy. This gives you the ability to make changes to the policy before it is created. Note that if you select instead the Create and Deploy a Policy with the Recommended Settings, the policy will immediately be created and you will be prompted to deploy it. If you use this option, you do have the option to cancel out of the deployment window and still edit the policy before deploying it.

7. Click Create Policy.

After completing these steps, look at the web form for the Intune agent policy. As you scroll through the list of options, notice that many of the settings are already predefined, but can be changed to meet the needs of your environment. Also notice that several of the settings have an information icon to the right of their item description, as shown in Figure 2-16. Clicking this icon will provide additional administrative details that can assist you in understanding how a particular setting is applied. From here, you start making modifications to the policy to meet the requirements of Tailspin Toys.

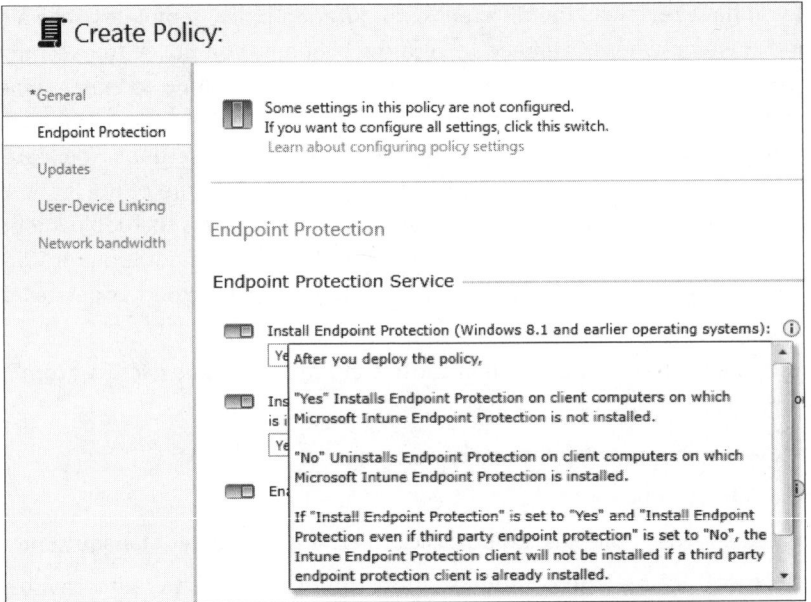

FIGURE 2-16 The Microsoft Intune admin console showing the Create Policy web form

8. On the Create Policy page, apply the following configuration changes:

   ■ **Name**   Enter a name for the policy. This field is used to identify the policy from within the Intune admin console. This is visible only to administrators.

   ■ **Description**   Enter a description for the policy. This field is used to further identify the policy from within the Intune admin console. This is visible only to administrators.

   ■ **Endpoint Protection Service**   In this section, locate the setting Submit File Samples Automatically When Further Analysis Is Required. Select Never Send from the drop-down list. This setting will prevent clients from submitting any file samples to Microsoft, without prompting the user.

   ■ **Microsoft Active Protection Service**   In this section, locate the setting Join Microsoft Active Protection Service. Select No from the drop-down list. This setting will prevent clients from sending data about possible malicious software to the Microsoft Active Protection service.

9. Click Save Policy to save the changes.

10. After saving the policy you will be prompted to deploy it. Click Yes.

11. In the Manage Deployment dialog box, select a device group on the left and click Add to add it to the deployment.

12. Click OK to deploy the policy.

After completing these steps, you should have a better understanding of how to manage configuration policies. From an exam perspective, be sure to familiarize yourself with the supported device platforms and understand that configuration policies are used for managing client settings and software that you deploy.

> **NEED MORE REVIEW?** **MICROSOFT INTUNE CONFIGURATION POLICIES**
>
> For more information about configuration policies in Microsoft Intune, visit *https://technet. microsoft.com/library/dn743712.aspx.*

Compliance policies

After working with configuration policies, the first thing that stands out when you create a compliance policy is the simplistic set of options. To be more specific, the Microsoft Intune admin console contains only one policy template for compliance. A compliance policy defines the security settings that your organization requires in order for a connected device to be identified as compliant. Compliance is checked at a regularly scheduled interval (30 days by default). It can be monitored using a report, and in some cases, it can be automatically remediated. From a deployment perspective it is important to note that unlike configuration policies, compliance policies are deployed to groups of users. Every device used by a targeted user will receive the corresponding policy. Compliance policies serve two important purposes:

- **Compliance** Compliance policies govern whether or not an enrolled device is compliant based on the settings defined in your policy. For example, your policy might be set to require device encryption. A device that is not encrypted would be identified as noncompliant. A device that is identified as noncompliant will be placed in one of the following two categories:

  - **Quarantine** Devices that cannot be remediated will be quarantined. This prevents the user from accessing other services that rely on conditional access—a topic that is described in more detail later in this chapter. The user will also be notified about compliance issues when accessing the Company Portal.

  - **Remediate** Devices that can be remediated will be addressed automatically. For example, if your policy requires all Windows 10 client computers to have a password length of 15 characters, and a device has a password of 10 characters, the user will be prompted to update their password.

- **Conditional access** When configured, conditional access policies can prevent your users from accessing services like email if a device is identified as noncompliant. The conditional access policy will refer to your active compliance policies for this ruling. You will take a deeper look at conditional access policies in the next section.

At this point you should have a better understanding of how compliance policies can be used, and what the process is when a device is identified as noncompliant. Now take a look at the rules that a compliance policy can govern. The web form used to edit a compliance policy includes several options, broken down into four sections:

- **Passwords**   This section deals with password requirements. The options include, but are not limited to, enforcement, length, complexity, expiration, and history.

- **Encryption**   This section deals with device encryption. The only available option is to require, or not require, device encryption.

- **Jailbreak**   This section deals with device tamper detection and is directly associated with the iOS (jailbreak) and Android (rooted) platforms. The only available option is to require, or not require, jailbreak detection.

- **Email profiles**   This section deals with local email profile detection. For example, when Intune is enabled it will define the device as noncompliant if it is unable to deploy an email profile because the user has manually created one. The only available option is to enable or disable email profile detection.

As you progress through the various policy types, you will see some settings that overlap. For example, requiring a password to unlock a device is an option you will see across multiple policy templates. From a compliance policy perspective, you need to be familiar with its capabilities (the four rules mentioned previously) and how they work in conjunction with configuration policies.

> **NEED MORE REVIEW?   MICROSOFT INTUNE COMPLIANCE POLICIES**
>
> For more information about compliance policies in Microsoft Intune, visit *https://technet. microsoft.com/library/dn705843.aspx.*

Conditional access policies

Conditional access is a framework that Microsoft Intune uses to help customers protect corporate resources. The guidance for conditional access is built based on the rules you define in your compliance policies. These two policy types work in conjunction to ensure data security. For example, imagine you have a compliance policy deployed to all of your mobile devices. You then create a conditional access policy for Exchange Online. In this scenario, when a user accesses Exchange Online the conditional access policy will be enforced. If the user's device is compliant, access will be approved. If the user's device is not compliant, access will be restricted.

In the Microsoft Intune admin console you have three services that can use conditional access. These services include:

- **Exchange Online policy**   This access policy protects connections to the Exchange Online service. Connected devices must be managed and compliant in order to access email. The controls available to administrators include:

  - **Platform targeting**   The access policy can be targeted at all of the supported device platforms, or only Windows 10 Mobile, for example.

- **Mobile device compliance**   The access policy can restrict connections based on the device platforms that Intune supports. Approving non-supported Intune devices would approve connections from devices like Blackberry.
- **User targeting**   The access policy can be targeted at all users or specific security groups. You also have the option to assign groups that are exempt from the policy altogether.

- **Exchange on-premises policy**   This access policy protects connections to Exchange on-premises. Connected devices must be managed and compliant in order to access email. For this policy to work there is a prerequisite. The Microsoft Intune Exchange Connector must be set up and working with your organization's infrastructure. The controls available to administrators include:
  - **User targeting**   You can target the access policy at all users or select specific security groups. You also have the option to assign groups that are exempt from the policy altogether.
  - **Exchange settings**   The access policy supports advanced Exchange ActiveSync settings for devices that are connected to Exchange. This includes devices that are not directly managed by Intune. Many of these options are pulled directly from Exchange's native ActiveSync settings. You have the option of adding custom device rules for models that are outside the Intune support list. You also have the option to block or quarantine devices that fail the compliance check.

- **SharePoint Online policy**   This access policy protects connections to the SharePoint Online service. Connected devices must be managed and compliant in order to access documents. The controls available to administrators include:
  - **Platform targeting**   The access policy can be targeted at all of the supported device platforms, or only Windows 10 Mobile, for example.
  - **User targeting**   You can target the access policy at all users or select specific security groups. You also have the option to assign groups that are exempt from the policy altogether.

EXAM TIP

Be sure to familiarize yourself with how conditional access policies work in conjunction with compliance policies, and be sure to note which services support these policies.

This concludes the process for managing conditional access policies.

NEED MORE REVIEW?   **MICROSOFT INTUNE CONDITIONAL ACCESS POLICIES**

For more information about conditional access policies in Microsoft Intune, visit *https://technet.microsoft.com/library/dn818907.aspx.*

Exchange ActiveSync Policies

Many organizations run Microsoft Exchange for their email delivery system, and among those companies, many have Exchange ActiveSync (EAS) enabled to accept connections from a multitude of mobile devices. This is something that Microsoft Intune has accounted for and supports. Depending on the environment, administrators have one or two options to accommodate the existing collection of connected devices. For Exchange on-premises, you can install the Microsoft Intune Exchange Connector. For Exchange online, you can configure the service-to-service connector. Using either of these options will enable you to manage your existing EAS devices from within the Intune admin console, without requiring them to enroll with Intune. This solution gives administrators the ability to start leveraging Intune immediately, versus the possibility of a lengthy enrollment process, as is the case with other MDM solutions.

> ***NEED MORE REVIEW?*** **MICROSOFT INTUNE EXCHANGE ACTIVESYNC POLICIES**
>
> **For more information about connecting Exchange with Microsoft Intune, visit**
> *https://technet.microsoft.com/library/dn646988.aspx.*

When you have connected your Exchange environment with Microsoft Intune, you will have the ability to manage your connected EAS devices using the EAS policy. The EAS policy was also referred to in the configuration policies section. It resides under the common mobile device settings category, as shown in Figure 2-17.

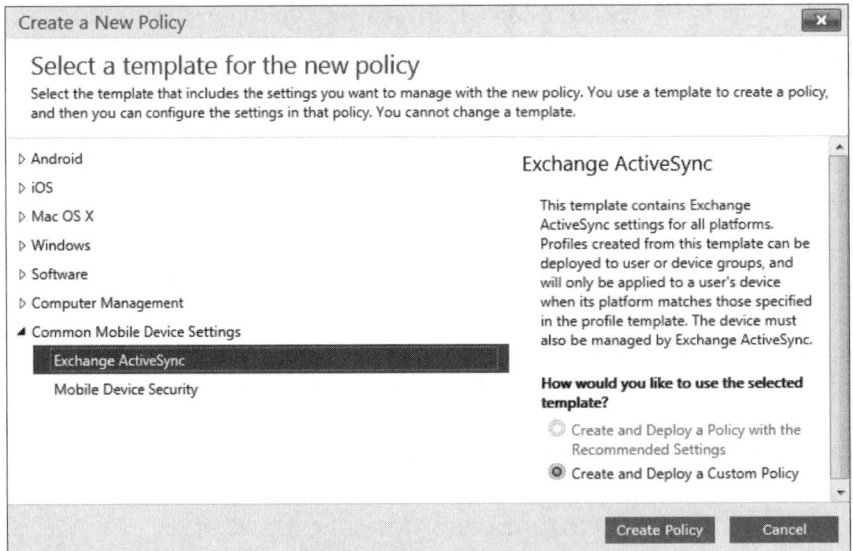

FIGURE 2-17 The Microsoft Intune admin console showing the Create A New Policy dialog box

Policy conflicts

The more advanced your policy structure becomes, the more likely you are to run into a policy conflict when managing devices. In today's infrastructure, devices can receive instruction from numerous sources. Understanding the priority system for these instructions is a critical step to insuring that your security and configurations policies are applied correctly. For the exam you might see questions dealing with policy conflicts and resolution. As an administrator, you don't want to hear that users cannot receive email on their devices, and then discover it is due to a policy conflict.

With Microsoft Intune in your environment, you have four policy sources that can manage device settings. You need to be aware of these sources and the settings that they can apply.

- **Group Policy**   From a client computer perspective, many of the settings that Intune policies can control are also addressed in Group Policy. For example, you can apply a GPO that controls the Windows Firewall settings for a Windows 10 client computer, and you can also manage and deploy those same settings using the Intune Windows Firewall settings policy. If a conflict is detected, domain-level Group Policy will always win. The only exception to this rule applies to computers that are unable to log on to the domain.

- **Intune compliance policies**   Following Group Policy you have Intune compliance policies. In terms of hierarchy, these policies have the next priority. Policy conflict for compliance policies include the following:

  - **Compliance policy vs. Group Policy**   Group policy settings will always win, unless the device cannot log on to the domain.

  - **Compliance policy vs. compliance policy**   Conflicts are judged on a per-setting basis, where the most restrictive setting wins.

- **Compliance policy vs. configuration policy**   The compliance policy setting will always win.

- **Intune configuration policies**   Next in the chain is the Intune configuration policy. Policy conflicts for configuration policies include the following:

  - **Configuration policy vs. Group Policy**   Group policy settings will always win, unless the device cannot log on to the domain.

  - **Configuration policy vs. compliance policy**   The compliance policy setting will always win.

  - **Configuration policy vs. configuration policy**   Conflicts are judged on a per-setting basis, where the most restrictive setting wins.

- **Intune application management policies**   The Intune application management policy separates itself from the rest of the pack. These policies are specific to Intune management, and only apply to apps that operate in the app store ecosystem. The only possibility for a conflicting application management policy would be against another application management policy. Policy conflicts for application management policies include the following:

  - **Application management policy vs. application management policy**   For apps that are deployed with two or more policies that contain conflicting settings, conflicts are judged on a per-setting basis, where the most restrictive setting wins. For apps that are deployed with one policy, and then a conflicting policy is deployed at a later date, the original setting will take precedence, but the policy conflict will be reported to the Intune admin console.

Now that you've reviewed the various sources that can configure settings on your devices, and how those settings overlap, take a look at how Intune reports a policy conflict to the admin console. As an administrator, you have plenty of things to keep your eye on, and Intune does a great job at alerting you when things are not set up properly.

In the following example, you have configured a Windows Firewall policy that is reporting conflicts. For testing, you have deployed this policy to a test group that contains a single Windows 10 client computer. Figure 2-18 shows the Intune dashboard, your first view when you log in to the admin console.

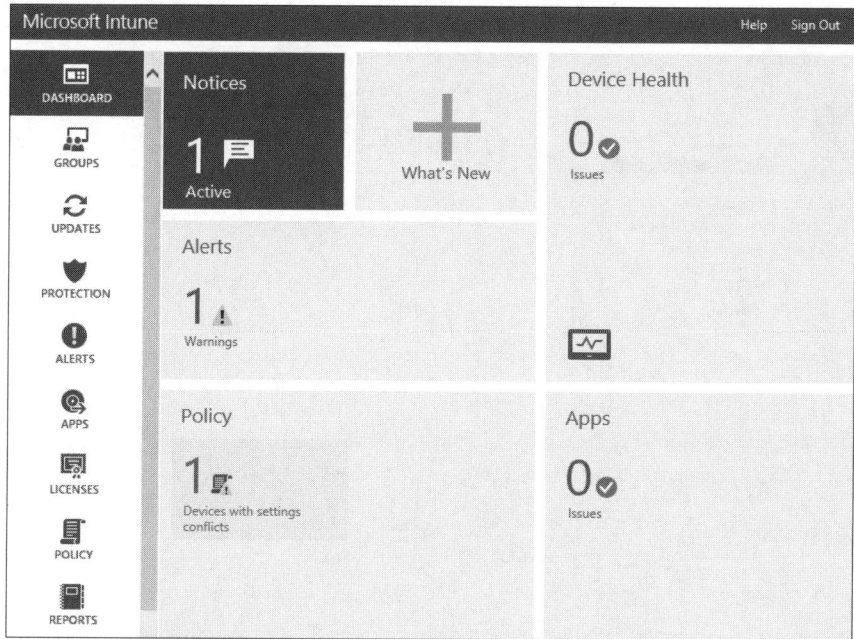

FIGURE 2-18 The Microsoft Intune admin console with dashboard tiles

Notice that the policy tile is reporting one device with settings conflicts. In the following steps, you navigate the Intune admin console and identify the conflicting policy:

1. From the Dashboard page, click the Devices With Settings Conflicts link inside the Policy tile. This action takes you directly to the Groups workspace with the All Devices group highlighted. The list of devices will automatically be filtered, as shown in Figure 2-19. In this example, the filter is set to With Group Policy Setting Conflicts. The page that is displayed identifies the devices that are reporting conflicts, along with a link to the conflicting policies.

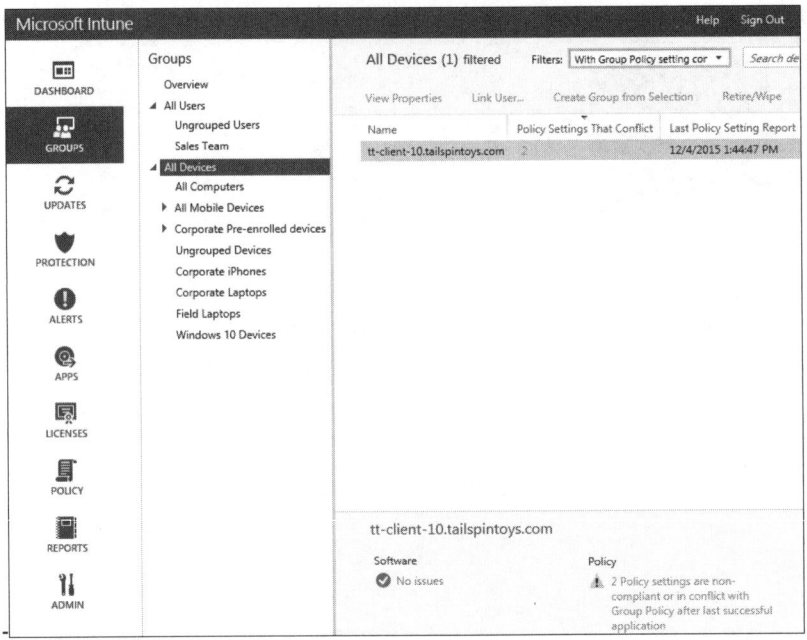

FIGURE 2-19 The Microsoft Intune admin console showing a device with a policy conflict

2. On the All Devices page, with the troubled device selected, click the number 2 link
    under the Policy Settings That Conflict column. This loads the device's property page
    with the Policy tab selected, as shown in Figure 2-20. On this page you can expand the
    reported conflicts and identify which policies are not in agreement. You also have links
    to each of the conflicting policies so you can easily open them. Once identified, you
    need to decide which source is accurate and rectify the other. After the conflict has
    been remediated, the device will need a policy refresh, at which time the notification
    on the Intune admin console dashboard will be cleared.

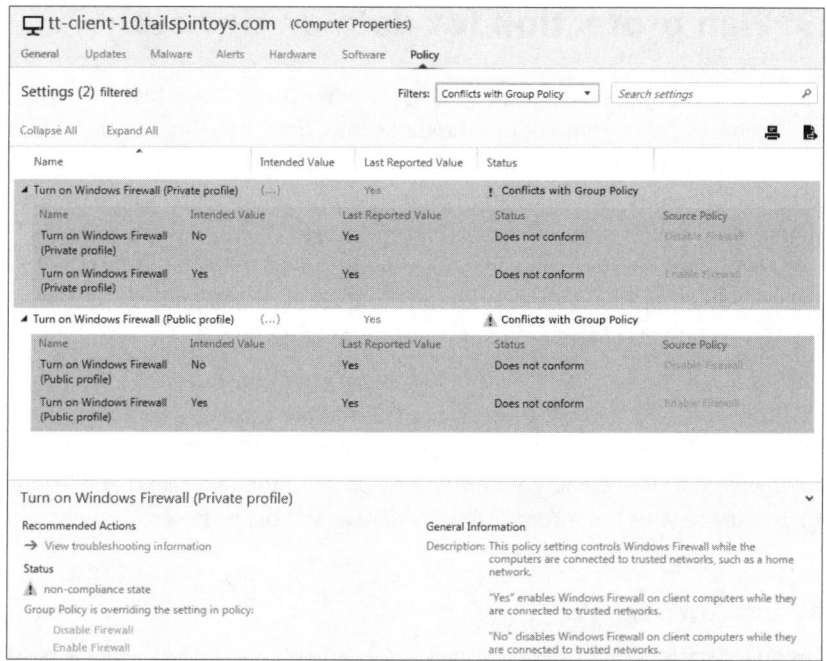

FIGURE 2-20 The Microsoft Intune console showing multiple policy conflicts for a device

EXAM TIP

Spend a few extra minutes studying the policy conflict priority system. This is an ideal topic on which to stage a troubleshooting question. Understanding which policy has priority in a given situation is a strong representation of a real-world troubleshooting scenario. You should also continue to familiarize yourself with the Intune admin console.

Summary

- Know how to customize the company portal, including the different elements that can be altered.

- Know how to setup terms and conditions, and how to view the results using the corresponding report.

- Know how to create and deploy configuration, compliance, and conditional access policies. Understand the different use cases for each.

- Know how Exchange Active Sync policies are utilized with Microsoft Intune.

- Know how to detect and resolve policy conflicts.

Skill 2.3: Plan protection for data on devices

Mobile devices in the enterprise introduce an entirely new attack vector for IT security professionals. Taking corporate data out of the datacenter and enabling it to be wirelessly transmitted to pocket size devises that are frequently lost or stolen is a major concern for many organizations. Data protection on mobile devices is a fundamental requirement for any industry-leading MDM solution, and Microsoft Intune operates with this expectation. Moving beyond Intune for a moment, the parent product, Enterprise Mobility Suite (EMS), offers additional layers of protection. These include identity and access management using Azure AD Premium and Azure Rights Management Services (Azure RMS).

For this skill, you will review the data protection needs of mobile devices in the enterprise. You will look at the data and services that require heightened security and how to accommodate those needs. This includes designing a layer of protection around email and SharePoint services, implementing a device encryption requirement for mobile devices, and leveraging full and selective device wipes to insure company data is not lost or stolen.

> **This section covers how to:**
> - Design for protection of data in email and Microsoft SharePoint when accessing them from mobile devices
> - Design for protection of data of applications by using encryption
> - Design for full and selective wipes

Design for protection of data in email and Microsoft SharePoint

Imagine that you just started a new job. Today is your first day in the office. What would you generally expect to have access to on day one? In many cases the answer to this question is email and possibly the company SharePoint site. Depending on the circumstances, these access rights may even be provided before you have physical access to the facility through services like Office 365. Given the importance of these services, it makes sense that our users expect to have access to them on the go, whether that be on a smartphone or tablet. The problem is, many of these services store company data on the local device. From a device management perspective, you need to insure that this local data is protected on those mobile devices. In this section, you will look at the methods that Microsoft Intune provides for protecting data on devices.

With Microsoft Intune you have the ability to control how data is stored and shared on an enrolled mobile device through the use of configuration policies and published applications. This framework gives you the tools to protect company data that is downloaded to mobile devices. For example, imagine you are the system administrator at a financial institution. The majority of employees carry a smartphone with access to company email. Recently, a security vulnerability was identified when users started copying sensitive email attachments to cloud-based file sharing services on their mobile devices. To resolve this vulnerability, you create a new mobile device security policy that disables the option for users to download email attachments on their mobile devices. Setup and deployment of the policy takes only minutes, and now you can report to management that the problem has been resolved.

Understanding your options is the first step in designing a solution for data protection. Intune has two approaches for achieving data protection. These include the following:

- **Application management** Apps that you publish to the Company Portal can optionally be assigned an application management policy. These policies provide a series of data relocation options that enable you to control the way data is preserved in the associated app. For example, you could publish Microsoft Word for iOS, but prevent users from backing up the app data to non-company sources. In this scenario you can prevent iTunes and iCloud backups for the specified app. Refer to Figure 2-21 for additional options associated with data relocation control.

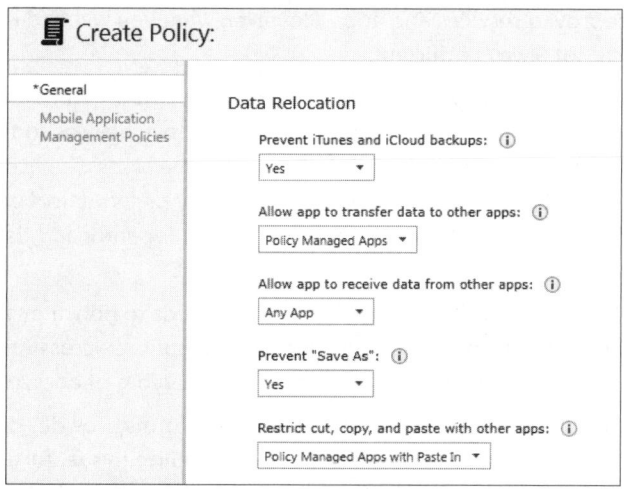

FIGURE 2-21 The Create Policy dialog box for an iOS device

Data relocation is a great tool for managing how data is stored on mobile devices, but you should know the prerequisites for the exam:

- **Microsoft Intune App SDK** For an app to acknowledge the configuration policy, it must support the Microsoft Intune App SDK. For a list of apps that support the Intune App SDK, visit *https://technet.microsoft.com/library/dn708489.aspx*.
- **Wrapped app** Microsoft provides the Intune App Wrapping Tool, which enables you to repackage an existing app to include the Intune App SDK. This is only supported on in-house apps or through separate arrangements with other app developers. Apps downloaded from the App Store or Google Play store cannot be repackaged.

- **Mobile device security**   The mobile device security policy provides an extensive list of security options. From a data protection standpoint, you have the ability to block cloud services, restrict copy and paste, restrict screen captures, prevent the downloading of email attachments, and much more.

NEED MORE REVIEW?   **UNDERSTANDING APPLICATION MANAGEMENT POLICIES**

For more information about the application management policies supported by Microsoft Intune, visit *https://technet.microsoft.com/library/dn878026.aspx*.

Design for protection of an application's data by using encryption

The previous section discusses the protection of data on mobile devices. More specifically, it identifies ways to control the data that is downloaded by users, and how to prevent that data from being shared with non-approved services and apps. However, what you will find is that controlling how data is stored is not always sufficient.

In a world filled with mobile devices, encryption at rest is becoming a new requirement among organizations. You need to encrypt the data stored on your mobile devices so that data isn't at risk in the event that the device is lost or stolen, and in some cases it's necessary to encrypt data at an application layer in situations where a device may be left unlocked. In this section, you look at the methods that Microsoft Intune provides for enforcing data encryption on devices.

All mobile device manufacturers have different standards with regards to how they implement security features like encryption. From a policy perspective, the options across Android, iOS, and Windows Phone will differ slightly. Take a look at the different types of encryption.

- **Device encryption**   Each of the supported mobile device platforms provides device-level encryption. However, some prerequisites are required to utilize this feature. First, the owner must configure a device-specific password. Second, some devices require the owner to manually enable encryption. The encryption keys for the device are protected by the password. As long as the device remains locked and encryption is enabled, all data is encrypted. If a device does not have a password configured or encryption was not enabled, it is effectively unencrypted. If a user's device password is compromised, all of the data on the mobile device is accessible once a perpetrator bypasses the lock screen. Based on this information, the following steps can be taken with Microsoft Intune to enforce device-level encryption:

    - Create and deploy a configuration policy that requires enrolled devices to use a password. For added security, you can require a complex password.

    - Depending on the device, enable the encryption enforcement settings in the previously mentioned configuration policy. Not all devices will enforce encryption automatically when a password is enabled, so this setting is important to remember.

- Create a compliance policy that checks devices for encryption. This will help govern enrolled devices and remediate when possible.

- **App encryption**   Through the use of device-level encryption, Microsoft Intune gives you the capability to encrypt data at an application layer. This means if a device is left unlocked, the data contained in a particular app can remain encrypted. The encryption of data in apps is controlled though the mobile application management policies. You have reviewed these policies in previous sections, and their ability to control data relocation for apps. In the case of encryption, you have the option to encrypt application data and require a PIN to access the encrypted app. Refer to Figure 2-22 for a look at the available access and encryption settings.

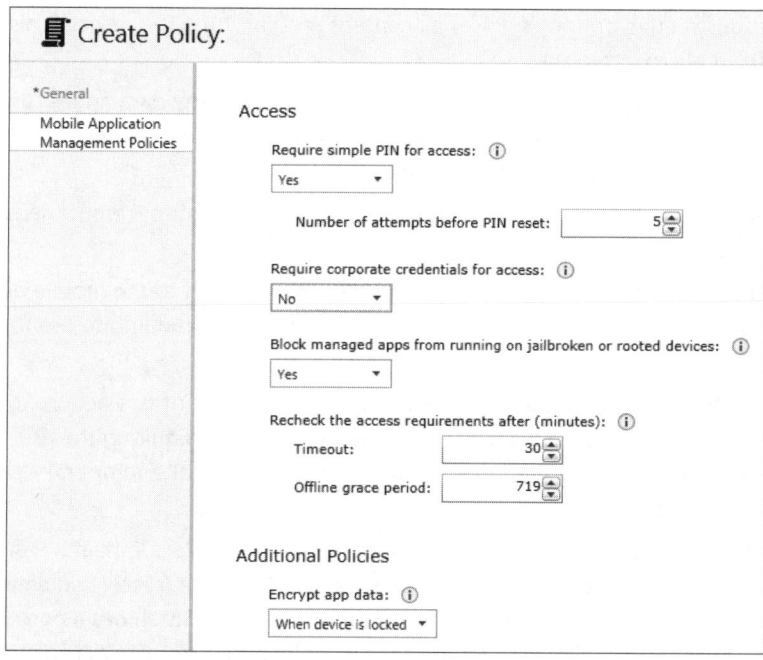

FIGURE 2-22   The Create Policy dialog box for an iOS device

EXAM TIP

For the exam, you should be familiar with the different types of encryption, both device-level and application-level encryption. It is also important to remember that apps must support the Intune App SDK in order for application management policies to apply. This is generally not an issue for in-house app development, but apps downloaded from an app store will need to be reviewed.

Design for full and selective wipes

In many ways, the ability to securely erase a remotely managed device has gotten infinitely and rightfully better over the years. IT departments around the world are supporting more mobile devices than ever before. Therefore, they need the capability to remotely and securely wipe these devices. In a very short time span, we have progressed from manually wiping devices through the on-screen display, to fully wiping a remote device over the Internet, and now having the ability to perform a selective wipe that intelligently erases company data.

Microsoft Intune supports the ability to send a full wipe or a selective wipe through the Intune admin console. Review what each of these commands does from a device standpoint:

- **Full wipe** Performing a full wipe restores the target device to factory settings. This is the most destructive option; it erases all content and settings. The device is also removed from Microsoft Intune.

- **Selective wipe** Performing a selective wipe erases all company data and settings, preserving the user's personal data and settings. After completing a selective wipe, the device is also removed from Microsoft Intune.

Two additional options that are provided, and are equally important for remote security, include the following:

- **Remote Lock** Performing a remote lock will send a command to the mobile device that requires the user to enter their device passcode before continuing to use the device.

- **Passcode Reset** Performing a passcode reset will initiate one of two actions, depending on the platform. The passcode will either be cleared, requiring the user to enter a new passcode depending on your compliance policies, or a temporary passcode will be applied.

Now that you have a better understanding of these commands, take a look at a real-world scenario. Imagine you are the system administrator at Tailspin Toys. You receive an email from your HR department stating that Sarah Jones in marketing is retiring. Sarah has a personally owned Surface Pro that she enrolled with Intune so she could access company email. You need to send a selective wipe to insure her personal data is unharmed, and all company data is securely erased. The following steps detail the process of accomplishing a selective wipe:

1. Sign in to the Microsoft Intune admin console at *https://manage.microsoft.com/*.
2. Click the Groups workspace.
3. In the Groups list, click All Devices.
4. On the All Devices page, locate and click Sarah Jone's Surface Pro.
5. In the menu bar, click the Retire/Wipe link. This opens the Retire Device dialog box, as shown in Figure 2-23. In this example, the targeted device is running Windows 8.1 Enterprise, which does not support the option for a full wipe.

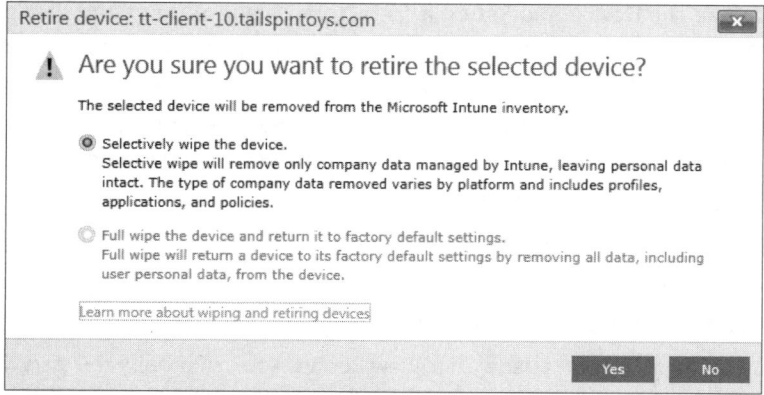

FIGURE 2-23 The Retire Device dialog box with a selective wipe

6. Select Selectively Wipe The Device and then click Yes. This command will be sent to the remote device and the device will be removed from the Intune admin console.

   Next you receive a call from the finance manager, John Kane. John just got home from a business trip and realized that he lost his corporate-owned iPad. The device has not shown up, so you need to send a full wipe to ensure all company data is securely erased. The following steps walk you through the process of accomplishing a full wipe:

1. Sign in to the Microsoft Intune admin console at *https://manage.microsoft.com/*.

2. Click the Groups workspace.

3. In the Groups list, click All Devices.

4. On the All Devices page, locate and click John Kane's iPad.

5. In the menu bar, click the Retire/Wipe link. This opens the Retire Device dialog box. The iPad supports both the full and selective wipe commands. This dialog box provides you with both options, as shown in Figure 2-24.

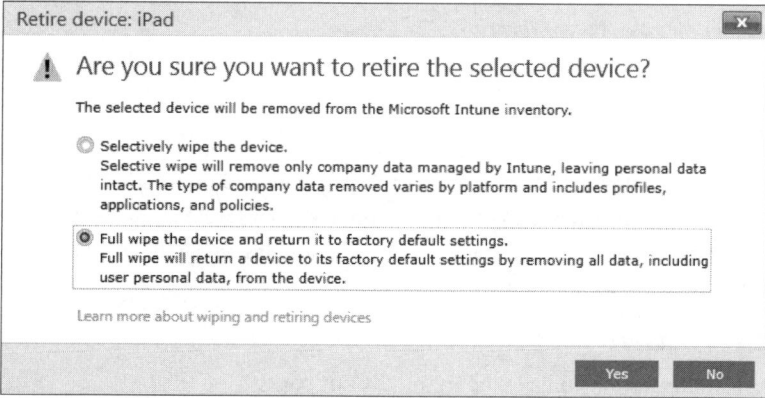

FIGURE 2-24 The Retire Device dialog box with a full wipe

6. Select Full Wipe The Device And Return It To Factory Default Settings and then click Yes. The command will be sent to the remote device and the device will be removed from the Intune admin console.

From within the Intune admin console, after you have issued a wipe or retire command, you as the administrator can run a report to audit these activities. The following steps walk you through running a device history report:

1. Sign in to the Microsoft Intune admin console at *https://manage.microsoft.com/.*

2. Click the Reports workspace.

3. In the tasks list, click Device History Reports.

4. On the Device History Reports page, click View Report. You can modify the start and end date range to meet your needs. The results are displayed in a new window, as shown in Figure 2-25. This report contains a historical log for all retire, wipe, and delete actions. The results shown are current. In this example, the device state is at Delete Pending. After these devices are wiped, the device state will be updated the next time the report is run.

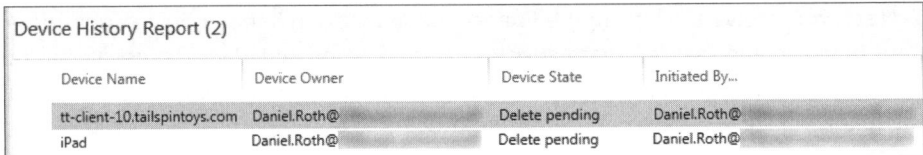

Device History Report (2)			
Device Name	Device Owner	Device State	Initiated By...
tt-client-10.tailspintoys.com	Daniel.Roth@	Delete pending	Daniel.Roth@
iPad	Daniel.Roth@	Delete pending	Daniel.Roth@

FIGURE 2-25 The Microsoft Intune admin console showing the results of a Device History Report

For the exam, familiarize yourself with the process of wiping a device, and the differences between a full wipe and a selective wipe. Review the device history reports and the data they provide administrators.

NEED MORE REVIEW? **MICROSOFT INTUNE DATA PROTECTION**

For more information about remote wipe, selective wipe, remote lock, and passcode reset, visit *https://technet.microsoft.com/library/jj676679.aspx.*

 Quick check

- You are a system administrator for Tailspin Toys. Today you received a call from your manager telling you that the toy train division has been sold. You need to securely wipe the 35 personally owned devices that this team has enrolled with Microsoft Intune. What steps would you take to achieve this?

Quick check answer

1. From the Intune admin console, open the Groups page.

2. Locate each of the devices, preferably using a dedicated group for the team, and issue a selective wipe using the Retire/Wipe option.

Summary

- Know how to protect applications using mobile application management policies.
- Know how to identify and resolve policy conflicts using the Intune admin console.
- Know how to use policies to achieve data protection, both on device-level and an application-level encryption.
- Know how to navigate devices in the Intune admin console, send full and selective wipes, and review the results of those actions using the device history report.

Thought experiment

In this thought experiment, demonstrate your skills and knowledge of the topics covered in this chapter by creating a new MDM solution. You can find the answer to this thought experiment in the next section.

You are a system administrator for Tailspin Toys, a global organization with 1,800 employees. The management team at Tailspin Toys has a strict policy preventing email on personally owned devices. Employees with corporate-owned devices are approved to receive email using Exchange ActiveSync, but they must first print and sign a copy of the company's mobile device terms and conditions. Currently, the IT department does not have an MDM solution in place to manage the growing request for mobile device access. Your manager has asked you to implement Microsoft Intune to support these mobile devices, but first he has asked the following questions:

1. What mobile device platforms does Microsoft Intune support?
2. You need a way to manage terms and conditions. How can you achieve this with Microsoft Intune?
3. All of your mobile devices today are connected using Exchange ActiveSync. How will these devices receive email with Microsoft Intune?
4. You need a way to view device inventory, like model and operating system. How can you achieve this with Microsoft Intune?
5. The company has a new requirement to support personally owned devices. How can you securely protect company data and insure it is erased if needed?

Thought experiment answer

This section contains the solution to the thought experiment.

1. Microsoft Intune supports Windows 8.1 or later, Windows Phone 8.0 or later, Windows RT and 8.1 RT, iOS 7.1 or later, Android 4.0 or later, and Mac OS X 10.9, or later.

2. Microsoft Intune supports a terms and conditions policy, along with a corresponding report that you can use to audit user acceptance of those terms.

3. Microsoft Intune provides the Exchange Connector, a solution that syncs existing Exchange ActiveSync devices with the Intune admin console. Users continue to receive their email the same way as they did before.

4. Microsoft Intune includes a computer inventory report, a mobile device inventory report, and a Mac OS X hardware and software report. You can leverage the data in these reports as needed.

5. Microsoft Intune provides configuration and compliance policies that enable you to enforce secure passwords and device encryption on enrolled devices. When you remove a device from Intune you can choose to issue a full or selective wipe based on the situation.

CHAPTER 3

Design for data access and protection

Today, some data is stored on-premises and some data is stored in the cloud. Storing data in the cloud, especially in bulk, is a new strategy that is expected to continue to grow. Some organizations, especially smaller organizations, store all of their data in the cloud. Other organizations are still storing all of their data on-premises. No matter where your data is stored, you need to have sensible data protection solutions in place to ensure that your data is safe from unauthorized access and protected in the event of a theft. Additionally, you need to know who has accessed files or attempted to access files that they do not have permission to access.

Skills covered in this chapter:

- Plan shared resources
- Plan advanced audit policies
- Plan for file and folder access

Skill 3.1: Plan shared resources

As a system administrator managing technologies to monitor and protect devices and data, you need to routinely design and implement file and disk encryption technologies and secure devices on your network. BitLocker for protecting data is covered in this section, along with the Network Unlock feature for securing your devices, and Encrypting File System (EFS) for encrypting data.

> **This section covers how to:**
> - Design for file and disk encryption and BitLocker encryption
> - Design for the Network Unlock feature
> - Configure BitLocker policies
> - Design for the Encrypting File System (EFS) recovery agent
> - Manage EFS and BitLocker certificates including backup and restore

Design for file and disk encryption and BitLocker encryption

When you decide to implement a file and disk encryption solution in your organization, you need to look at the available technologies and match up the technology's features with your requirements. In addition to documenting your company requirements, you also need to understand the technology you are implementing. In this section, you review two technologies: EFS and BitLocker. This section reviews their features, compares them for various scenarios and then walks through some design considerations. To get started, take a look at EFS.

EFS

EFS is a file and folder encryption technology built into Windows client and server operating systems. Because EFS is built in, it is easy to use and easy to deploy. Because the components are built in to Windows, you only need to focus on policies and recoverability. EFS provides the following features:

- **Encrypts files and folders stored on NTFS partitions**   Only authorized users can access encrypted files. Authorized users are those users who have been granted explicit access to the files.

- **Generates a user certificate to use for encryption and decryption automatically**   Because EFS automatically generates a certificate, users can begin using it immediately. In most enterprise environments, managing automatically generated user certificates requires a great deal of administrative overhead.

- **Integrates with an on-premises public key infrastructure (PKI) such as Active Directory Certificate Services (AD CS)**   While you can use automatically generated certificates with EFS, it is good practice to use an on-premises PKI instead. This is because an on-premises PKI centralizes the certificate distribution and management, which reduces administrative overhead and increases security.

- **Integrates with Group Policy for administrative-authorized decryption and recovery**   A data recovery agent is a special account that is granted access to decrypt all files encrypted with EFS, no matter which employee encrypted the files.

When you encrypt a file with EFS, EFS generates a symmetric encryption key. That key is encrypted with your public key and then stored as part of the encrypted file. Figure 3-1 shows the encryption process in a flowchart.

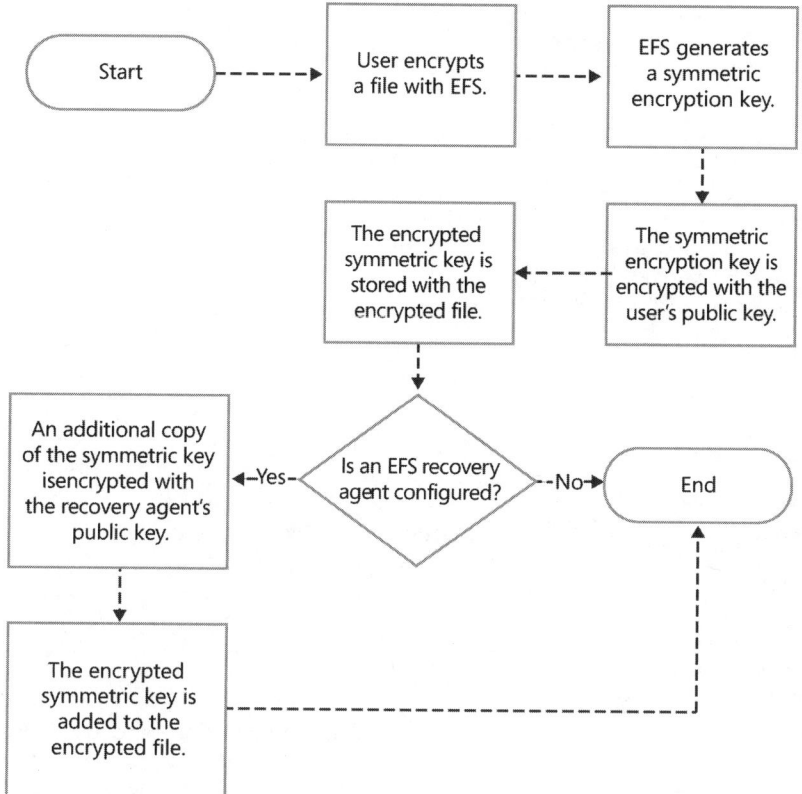

FIGURE 3-1 A flowchart shows the EFS file-encryption process

When you attempt to decrypt a file encrypted with EFS (such as by opening a Microsoft Word document encrypted with EFS), your private key is inspected to see if it corresponds to the public key used to encrypt the file. If it does, you gain access to the file transparently. If your private key does not correspond to the public key used to encrypt the file, the only way you can decrypt the file is by having access to a recovery agent certificate. If you don't, you are denied access to the file. Figure 3-2 shows the decryption process in a flowchart.

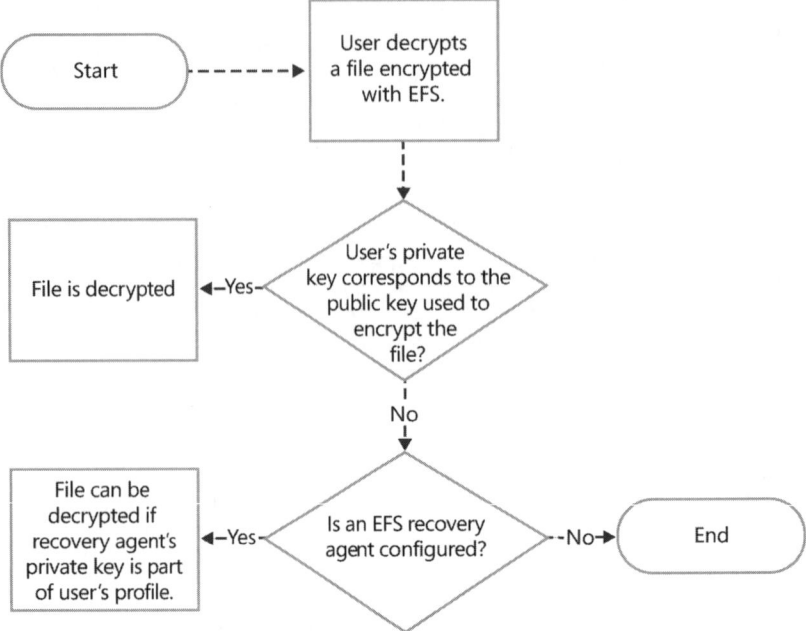

FIGURE 3-2 A flowchart shows the EFS file-decryption process

When thinking about the design considerations with EFS, you should perform the following actions:

- **Ascertain whether you need file and folder encryption, volume encryption, or both.** To do this, you need to gather the requirements and understand what your organization is trying to protect against. For example, if you have computers that are used by multiple users, EFS is a good choice to ensure that each user's data isn't accessible by the other users. On the other hand, if you are trying to protect against a stolen computer, EFS is probably not the best choice. To maximize protection for a stolen computer, you should encrypt all hard drive volumes. When working in a high-security environment, where security is the most important design consideration, consider using a combination of EFS and a volume-encryption solution. BitLocker is such a solution, and it works well in combination to maximize security.

- **Create a plan for recovering encrypted files.** Imagine a scenario where a key developer suddenly leaves your company. The desktop team attempts to export his data from his computer but can't because they don't have access to the encrypted files. If you don't have a recovery agent configured and you don't have access to the user's credentials, you lose all of the files. As the administrator, you need to look at this situation and similar situations and be sure that you are properly designed and configured to minimize data loss.

- **Create a plan for managing EFS certificates, which are stored in a user's profile.**   If the EFS certificate is lost or damaged (such as in a disk crash scenario), a user cannot gain access to encrypted files. While a recovery agent could gain access to the files, many organizations have not configured a recovery agent. To minimize the problem with lost or damaged certificates, you should plan to back up certificates. Backing up EFS certificates is often referred to as *key archival*. With key archival, if a certificate is lost or damaged, the key can be recovered from the archive and then used to decrypt files, as though the certificate was never lost or stolen.

✔ **Quick check**

- Which key do you use to encrypt the symmetric encryption key when you encrypt a file with EFS?

Quick check answer

- You use your public key to encrypt the symmetric encryption key when you encrypt a file with EFS.

BitLocker

BitLocker is a drive-encryption technology that is built into the Windows client and Windows Server operating systems. On the Windows Server operating system, BitLocker is a feature named BitLocker Drive Encryption that you can add to a server like any other role or feature. On Windows client operating systems, such as Windows 10, you can enable BitLocker by using the BitLocker Drive Encryption feature in Control Panel.

BitLocker provides the following features:

- **Encrypts an entire volume**   Whereas EFS works at the file and folder level, BitLocker works at the volume level. You can opt to have BitLocker encrypt an entire volume or just the used part of the volume.

- **Combinable with EFS**   If you have a shared computer used by multiple users (such as on different work shifts), BitLocker doesn't protect the user's data from another user because once one user is authenticated, that user has full access to the disk volumes. You can use NTFS in such a scenario, but if the users have administrative access to the computer, then they can bypass NTFS by taking ownership of a file or folder and resetting the permissions. EFS can protect data for different users sharing the same computer. Thus, if you have high-security requirements, consider deploying EFS and BitLocker.

- **Protects the integrity of the Windows startup process**   BitLocker protects the critical Windows startup files from tampering to ensure they remain uncorrupted. Additionally, Windows does not start if BitLocker detects that the critical Windows startup files have been tampered with, such as by a rootkit or other malware.

- **Controllable by using Group Policy**   You can use Group Policy to manage BitLocker across your domain-joined computers. Group Policy has 42 settings available, with some settings being valid for specific hard drive types. You will look at the available policies in more detail in the section titled "Configure BitLocker policies" later in this chapter.

Compared to EFS, BitLocker has more prerequisites when you start using it, depending upon the features you plan to use. For example, a Trusted Platform Module (TPM) is required on a computer if you plan to use system integrity verification or multifactor authentication. A TPM is a hardware chip that is used on a computer for encryption and decryption. As part of your BitLocker design, you need to consider the following questions:

- Do all of the target computers have a TPM? In many cases, the answer is no. Thus, you need to plan how to handle computers without a TPM. Will you use BitLocker with those? And if so, which settings will you use?

- Do you require multi-factor authentication? If so, don't forget that a TPM is required.

- Which drive-encryption method will you use? In many cases, it is a good idea to use the strongest encryption method available for your computers, even if they have minor performance degradation in some situations.

- Will you enable users to choose any of the BitLocker settings for their devices? In a high-security environment, it is most common to have IT enforce BitLocker settings with Group Policy. In environments that aren't dealing with sensitive data, enabling users to choose some BitLocker settings can improve the user experience.

Whereas EFS is an encryption solution for files and folders, BitLocker is an encryption solution for entire volumes. BitLocker also offers additional enhancements such as multi-factor authentication, where you can require supplemental authentication when a computer starts up. Another enhancement BitLocker offers is the Network Unlock feature, which improves the user experience for computers protected with BitLocker. In the next section, you look at the Network Unlock feature in detail.

Design for the Network Unlock feature

When you add security to the environment, it can sometimes degrade the user experience. For example, if users have to enter a pin or a USB key each time they turn on their devices, but they also have to type a username and password to sign in to Windows, the experience isn't as seamless as it could be. This is the case with BitLocker, when you require additional authentication at startup. To maintain the extra security of additional authentication at startup, while remediating the degraded user experience, you can implement the Network Unlock feature. The Network Unlock feature automatically unlocks a BitLocker-encrypted drive at startup, if your computer is on a trusted subnet.

To use the Network Unlock feature, the following prerequisites must be met:

- Client computers must run Windows 8, Windows 8.1, or Windows 10 and servers must run Windows Server 2012, Windows Server 2012 R2, or Windows Server 2016.
- Computers must have a TPM.
- You must have a Windows Deployment Services (WDS) server.
- The BitLocker Network Unlock feature must be added to Windows.
- You must have a DHCP server (separate from the WDS server).
- You must obtain an SSL certificate for the WDS server.
- Group Policy must be configured to enable Network Unlock and to specify the SSL certificate in use.

An optional feature of Network Unlock is to require clients to be on a specific subnet. This adds another layer of security to your environment. The Network Unlock process is shown in Figure 3-3 and described in more detail following the figure.

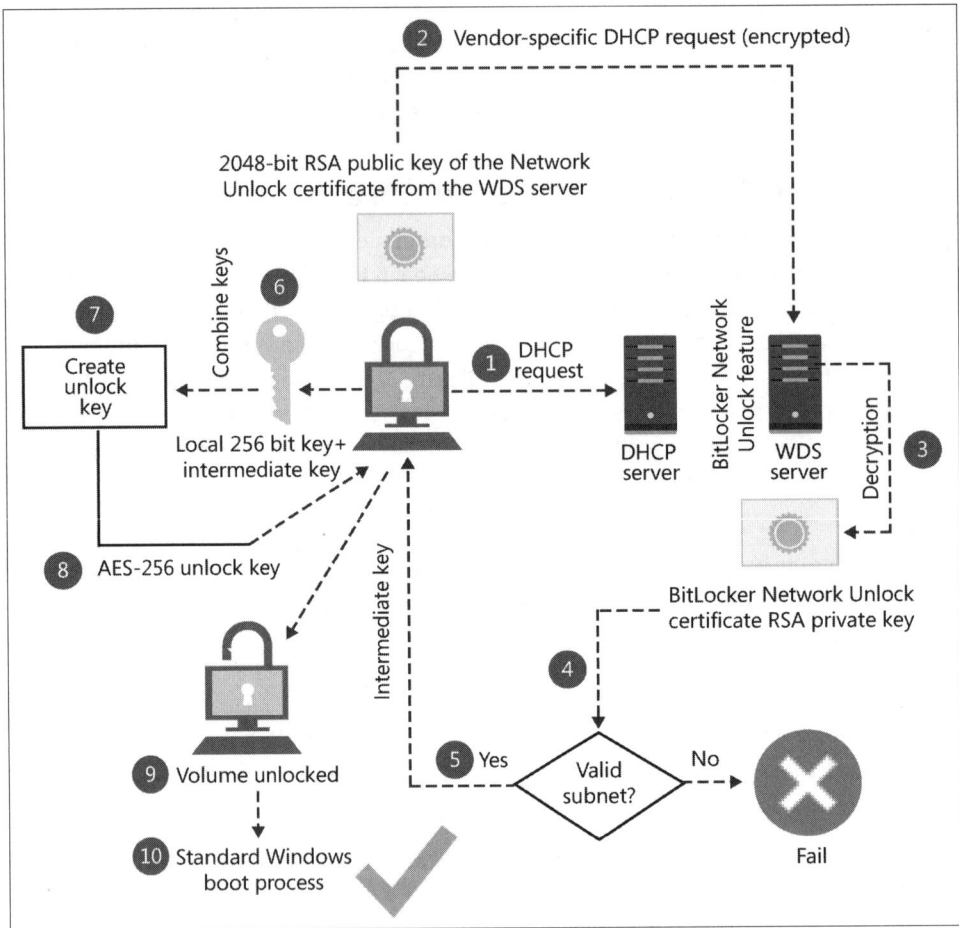

FIGURE 3-3 Flowchart shows the BitLocker Network Unlock process

The following steps describe the process shown in Figure 3-3:

1. In the first step, the client computer broadcasts a standard DHCP request. A DHCP server issues an IP address and the client computer can then communicate on the network.

2. After obtaining an IP address from the DHCP server, the client computer sends out a vendor-specific DHCP request. Within the request is a 256-bit intermediate key and an AES-256 session key, which are used for the reply. The 2048-bit RSA public key of the Network Unlock certificate encrypts the request on the WDS server.

3. The WDS server processes the vendor-specific DHCP request. First, it decrypts the request by using the BitLocker Network Unlock private key.

4. The WDS server checks to see if the client computer is on a trusted subnet. If it isn't, the process fails immediately.

5. If the client computer is on a trusted subnet, then the WDS server sends the network key to the client computer. The network key is encrypted with the session key and becomes an intermediate key. Note that verifying subnets is a valid and optional configuration on the WDS server.

6. The client computer takes the intermediate key from the WDS server and combines it with another local 256-bit intermediate key.

7. An unlock key is created from the combined keys.

8. The AES-256 unlock key is sent to the client computer.

9. The client computer unlocks the operating system volume with the AES-256 unlock key.

10. The client computer continues with a standard Windows boot process.

EXAM TIP

Be sure you memorize the prerequisites for BitLocker Network Unlock. Prerequisites fall into the design category and the Network Unlock feature has some not-so-obvious prerequisites that could trip you up on the exam.

This section covers the BitLocker Network Unlock feature. In the next section, BitLocker policies in Group Policy are covered, where you can gain some insight about Network Unlock from a policy perspective.

 Quick check

- You have an internal PKI. You issue a certificate for BitLocker Network Unlock. Which server should have the certificate and the private key installed?

Quick check answer

- The WDS server must have the certificate installed, along with the private key. The client computers only require the public key.

Configure BitLocker policies

You configure BitLocker policies in Group Policy. With 42 BitLocker settings, you can configure policies based on your requirements. For the exam, you don't need to memorize all of the settings. You should browse through all of them, however, and be comfortable with the most commonly used settings. The following settings represent some of the most commonly used BitLocker settings:

- **Choose Drive Encryption Method And Cipher Strength**   This setting allows you to choose an encryption method based on the drive type. For example, you can select AES 256-bit encryption for operating system volumes and AES 128-bit encryption for removable drives. To maximize security, you should opt to use this setting and set the encryption level at the highest supported level for your operating system versions. For Windows 10 and Windows Server 2016, you can opt to use a new encryption method named XTS AES 256-bit.

- **Allow Network Unlock At Startup**   This setting must be enabled to have a computer use the Network Unlock feature. If disabled, Network Unlock cannot be used.

- **Allow Enhanced PINs For Startup**   When you configure BitLocker to require a PIN at startup (such as for computers that don't have a TPM), you can use this setting to enable the use of complex PINs, such as case-sensitive letters, symbols, numbers, and spaces.

- **Require Additional Authentication At Startup**   This GPO setting has several key settings, including the one for the TPM policy, which dictates whether a TPM is required. Optionally, you can enable the setting to allow BitLocker without a TPM. In that case, a startup key or USB flash drive is required. You can also opt to use a startup key and a PIN with TPM, which is useful in a high-security environment where you must maximize security and minimize risk.

- **Enforce Drive Encryption Type On Operating System Drives**   You can configure this setting to enforce BitLocker to always fully encrypt a volume when protecting it. Optionally, you can configure this setting to encrypt only used space. Encrypting only used space is less secure because unused space can contain data that can be recovered. You also can enable users to choose the option instead.

- **Allow Secure Boot For Integrity Validation**   This feature is used on client computers running Windows 8 or later, and on servers running Windows Server 2012 or later. Secure Boot ensures, for example, that authorized software vendors digitally sign the firmware in the startup environment.

- **Deny Write Access To Fixed Data Drives Or Removable Drives Not Protected By BitLocker**   You can use this setting to prevent users from using fixed data drives or removable drives unless they are protected with BitLocker. You can enable or disable this setting for each drive type (fixed or removable).

Many of the BitLocker settings that you configure in Group Policy are stored in the registry in the HKEY_LOCAL_MACHINE\SOFTWARE\Policies\Microsoft\FVE key. For the exam, note the following information about configuring BitLocker with Group Policy:

- **Policies are located in the Computer Configuration container.** This means that Policies are computer settings. The BitLocker GPOs need to target computer objects, not user objects.

- **There are operating system, version-specific.** The newest version of Windows supports the latest and most secure encryption options. But, in an environment with older operating systems, you might need to use multiple policies to target different settings based on the operating system version. For example, Windows 10 and Windows Server 2016 support the new XTS-AES 256-bit encryption. Older operating systems do not support it. Thus, if you want to mandate XTS-AES 256-bit encryption, you need a GPO that targets only client operating systems that are Windows 10 or later and servers that run Windows Server 2016 or later.

> *NEED MORE REVIEW?* **USING WMI FILTERS TO TARGET OPERATING SYSTEM VERSIONS**
>
> You can use a WMI filter with a GPO to target specific operating system versions. To refresh your knowledge, see the article at *https://technet.microsoft.com/library/ cc758471(v=ws.10).aspx* for more information.

Design for the Encrypting File System (EFS) recovery agent

Every user of EFS has their own private key that they use to decrypt data encrypted with EFS. The private key is stored in the user's profile. If the key becomes corrupt or is lost, the data encrypted with EFS cannot be decrypted. For most organizations, this situation is unacceptable. You have a couple of options that you can use to avoid a situation where you cannot decrypt data after a private key is lost:

- **Back up the user's certificate, which includes the private key.** You can use any backup method. The most important part of the backup is to send it to another computer, to another datacenter, or to the public cloud. If a user's private key becomes corrupt or is otherwise lost, you can restore the user's certificate from backup. This solution works but it requires a large amount of administrative overhead.

- **Use an EFS recovery agent.** A data recovery agent (DRA) is a user account with authorization to decrypt all EFS data, no matter who encrypted it. By default, the Administrator account in the AD DS domain is the EFS recovery agent (although you need to create the data recovery agent to add the Administrator User as a recovery agent). It is a good practice to have a dedicated EFS recovery agent account instead because the Administrator User account has more permissions than necessary for performing EFS data decryption.

Before you deploy EFS in your environment, you need to design the solution. EFS might not have many design considerations, but digital certificates are a big consideration. You need to decide how you want to use EFS certificates in your environment. While you can purchase a third-party certificate from a provider, it isn't cost effective or easy to manage if you have more than a few users. Most organizations opt to use an internal public key infrastructure (PKI) because it keeps costs low and certificate management generally can be automated.

The following list outlines the high-level steps that you must perform to configure a PKI for EFS. You do not need a more detailed step-by-step representation of these steps because they are outside the scope of the exam. However, you'll be expected to understand the information contained in the following steps; they provide a thorough understanding of how EFS works and which components are part of the environment.

1. Install a Certification Authority.

2. Duplicate the EFS Recovery Agent certificate template.

   This step is optional. By default, as long as your DRA account is a member of the Domain Admins or Enterprise Admins group, you can use the default EFS Recovery Agent template (because both of those groups have the rights to enroll for the template). However, the job of the DRA account is data recovery. Thus, the account does not need rights to the AD DS environment. It is a good practice to assign only the least permissions necessary to perform a task or job. Instead, modify the template so that the DRA user account has rights to read and enroll.

3. Sign into a computer as the DRA account and then enroll for the custom EFS Recovery Agent certificate.

Now that you have a DRA account with a recovery certificate, you can add an EFS recovery agent to a GPO so that all computers can access the recovery agent. The GPO must be linked to a location that contains all of the computers that use EFS. Perform the following steps:

1. Open the Group Policy Management Console. Create a new GPO named EFS and then edit the GPO.

2. In the Group Policy Management Editor, navigate to Computer Configuration\Windows Settings\Security Settings\Public Key Policies\Encrypting File System.

3. Right-click Encrypting File System and then click Add Data Recovery Agent.

4. In the Add Recovery Agent Wizard, on the Welcome To The Add Recovery Agent Wizard page, click Next.

5. On the Select Recovery Agents page, click Browse Directory.

6. In the Find Users, Contacts, and Groups dialog box, type the name of the DRA user account in the Name text box and then click Find Now.

7. On the Search results page, double-click the DRA user account.

8. On the Windows Security page, verify that the certificate displayed is the DRA user account's EFS recovery certificate. Then, click OK.

9. On the Select Recovery Agents page, your DRA user account should be listed. Click Next to continue.

10. On the Completing The Add Recovery Agent Wizard page, click Finish.

Now that you have a dedicated DRA account and it has been added to a GPO, you might be wondering how you can figure out if it is functional. Once computers have refreshed Group Policy, you can sign in as a standard user account and encrypt a file or folder with EFS. Once complete, you can look at the properties of the file or folder. From the properties window, click Advanced, then click Details. You should see your standard user account's certificate in the top of the results dialog box and the DRA user account's recovery certificate in the bottom part of the dialog box, as shown in Figure 3-4.

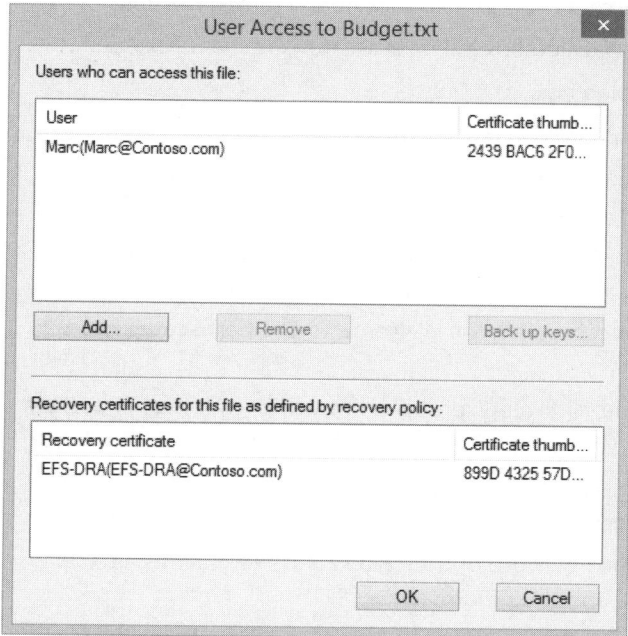

FIGURE 3-4 An encrypted file's properties show whether a DRA has been correctly configured

There is one important detail you should know about EFS recovery agents. A recovery agent can only recover EFS-encrypted files if the encrypted files were encrypted after the recovery agent was created and added to a GPO (and technically after computers refreshed their Group Policy). Files encrypted before a recovery agent was configured cannot be decrypted (and thus recovered). From a design perspective, this bit of information is critical because you can plan to have a DRA configured and ready before you enable users for EFS.

Manage EFS and BitLocker certificates including backup and restore

You need to back up EFS and BitLocker certificates. With EFS, users always have certificates. With BitLocker, users often do not have certificates, although there is a use case for certificates (which is using virtual smart cards with BitLocker). In either case, you can use the same backup and restore methods because these methods exist outside of EFS and BitLocker.

The following high-level steps represent the certificate backup process:

1. Right-click the Start button, click Run, type **mmc**, and then click OK.

2. If User Account Control displays a window, click Yes.

3. In the Console1 MMC snap-in, click File and then click Add/Remove Snap-ins.

4. In the Add Or Remove Snap-ins dialog box, click Certificates and then click Add.

5. In the Certificates Snap-ins dialog box, click Finish.

6. In the Add Or Remove Snap-ins dialog box, click OK.

7. In the Console1 snap-in, expand Certificates – Current User, expand the Personal container, and then click the Certificates container.

8. Right-click the certificate that you want to back up, click All Tasks, and then click Export.

9. In the Certificate Export Wizard page, on the Welcome To The Certificate Export Wizard page, click Next.

10. On the Export Private Key page, click Yes, Export The Private Key. This is an important step. Without the private key, you cannot use the certificate for decryption. Click Next to continue.

11. On the Export File Format page, click Next.

12. On the Security page, click the Password option, type a password, and type the password again to confirm it, and then click Next.

13. On the Files To Export page, click Browse to browse to the location where you want to save the file. Type a file name and then click Save. You should browse to a network location so that you can have a copy of the certificate off of the computer. That way, if the computer has a hardware problem, you won't lose the original certificate and exported certificate.

14. On the Files To Export page, click Next.

15. On the Completing The Certificate Export Wizard page, click Finish.

16. On the Certificate Export Wizard dialog box, click OK.

After a user has a certificate and it has been backed up, you don't need to continue to back it up regularly because unlike regular data, the certificate doesn't change. So, while the initial backup of all of your users' certificates is a lot of work, the work isn't ongoing.

The following steps represent the certificate recovery process. During a recovery, you might be restoring a certificate to the original computer (such as if the original certificate was corrupt or accidentally deleted) or you might be restoring the certificate to a new computer (if the old computer had to be replaced). In either situation, the steps are the same.

1. Right-click the Start button, click Run, type **mmc**, and then click OK.

2. If User Account Control displays a window, click Yes.

3. In the Console1 MMC snap-in, click File and then click Add/Remove Snap-ins.

4. In the Add Or Remove Snap-ins dialog box, click Certificates and then click Add.

5. In the Certificates Snap-ins dialog box, click Finish.

6. In the Add Or Remove Snap-ins window, click OK.

7. In the Console1 snap-in, expand Certificates–Current User, and then click the Personal container.

8. Right-click the Personal container, click All Tasks, and then click Import.

9. On the Welcome To The Certificate Import Wizard, click Next.

10. On the File To Import page, click Browse to browse to the location where the certificate is backed up.

11. In the Open dialog box, click the drop-down menu and then click Personal Information Exchange (*.pfx,*.p12). Next, click the name of the file that contains the backed-up certificate and then click Open.

12. On the File To Import page, click Next.

13. On the Private Key Protection page, type the password to the backed up certificate and then click Next.

14. On the Certificate Store page, click Next.

15. On the Completing The Certificate Import Wizard page, click Finish.

16. On the Certificate Import Wizard dialog box, click OK.

Backing up and restoring certificates is important. While you don't need to memorize all of the steps, you should plan to try the steps in a lab environment as part of your exam preparation.

Summary

- EFS is a file and folder encryption technology built into Windows operating systems.
- EFS relies on certificates for encryption. As part of your disaster recovery planning, you should plan to back up EFS certificates.
- BitLocker is a volume encryption technology built into Windows operating systems.
- BitLocker encrypts the entire volume.
- BitLocker can be combined with EFS to maximize data security.
- The BitLocker Network Unlock feature enables you to configure multiple factors of authentication for BitLocker startup. But instead of having a user enter multiple factors of authentication, a computer's subnet is checked and if the subnet is a corporate LAN, the authentication is automatic.
- You can configure BitLocker for all of your domain's computers by using Group Policy and the 42 available settings.

Skill 3.2: Plan advanced audit policies

Skill 3.1 covers ways to secure your devices and data. While those are key foundational technologies, you also need a way to figure out if anyone attempts to gain access to unauthorized data or succeeds in actual unauthorized access. This section focuses on auditing, especially the latest advanced auditing policies that provide granular audit controls for your data and devices. It focuses primarily on how to leverage Group Policy for advanced auditing.

> **This section covers how to:**
> - Design for auditing by using Group Policy and AuditPol.exe
> - Create expression-based audit policies
> - Design for removable device audit policies

Design for auditing by using Group Policy and AuditPol.exe

Many administrators are familiar with the standard auditing policies:

- Audit account logon events
- Audit account management
- Audit directory service access
- Audit logon events

- Audit object access
- Audit policy change
- Audit privilege use
- Audit process tracking
- Audit system events

The exam, though, focuses on advanced audit policies. You need to understand the difference between standard audit policies and advanced audit policies. Table 3-1 breaks down the key differences.

TABLE 3-1 Standard vs. advanced auditing

Characteristic	Standard audit policy	Advanced audit policy
Granular control of auditing	No	Yes
Best for reducing size of logs	No	Yes
Simple setup	Yes	No
Requires a minimum server version of Windows Server 2008	No	Yes

To implement advanced audit policies, you can use Group Policy or the AuditPol.exe command-line tool. When applying advanced audit policies to multiple computers (such as all of the computers in your organization), you should use Group Policy because it is more efficient than AuditPol.exe. AuditPol.exe configures auditing on a single computer at a time. Thus, if you try to use it to configure several computers or more, it becomes time consuming and error prone. However, AuditPol.exe has some important uses:

- **Implement advanced auditing on individual computers.**   For domain-joined computers, you should use Group Policy. For computers that are not joined to a domain, such as computers in a perimeter network, AuditPol.exe enables you to configure them with the same advanced audit policies.

- **Query computers for their current advanced audit policy settings.**   You can query computers whether they have been configured with Group Policy, or AuditPol. exe, or any other method. For example, if you want to find out the account management audit settings on a local computer, you can run the **AuditPol.exe /get / category:"Account Management"** command.

- **Migrate auditing settings from one computer to another.**   You can back up auditing settings and then restore those settings to a different computer. For example, if you have several computers in a perimeter network and need to use AuditPol.exe to configure auditing, you can configure a single computer and then migrate the settings to the other computers. This reduces the amount of work it takes to configure auditing on all of the computers.

- **Set per-user auditing policies.** As mentioned earlier in this section, advanced audit policies are more granular than standard audit policies. You can also target auditing policies at individual users. You can reduce the amount of log data by setting up user-targeted policies.

With advanced audit policy, you have 10 categories of auditing. Inside each category, you have granular audit settings. For example, if you want to audit account management events, but you do not want to audit distribution group management events, you can use advanced audit policy settings to accomplish that. Figure 3-5 shows that configuration using the Group Policy Management Editor.

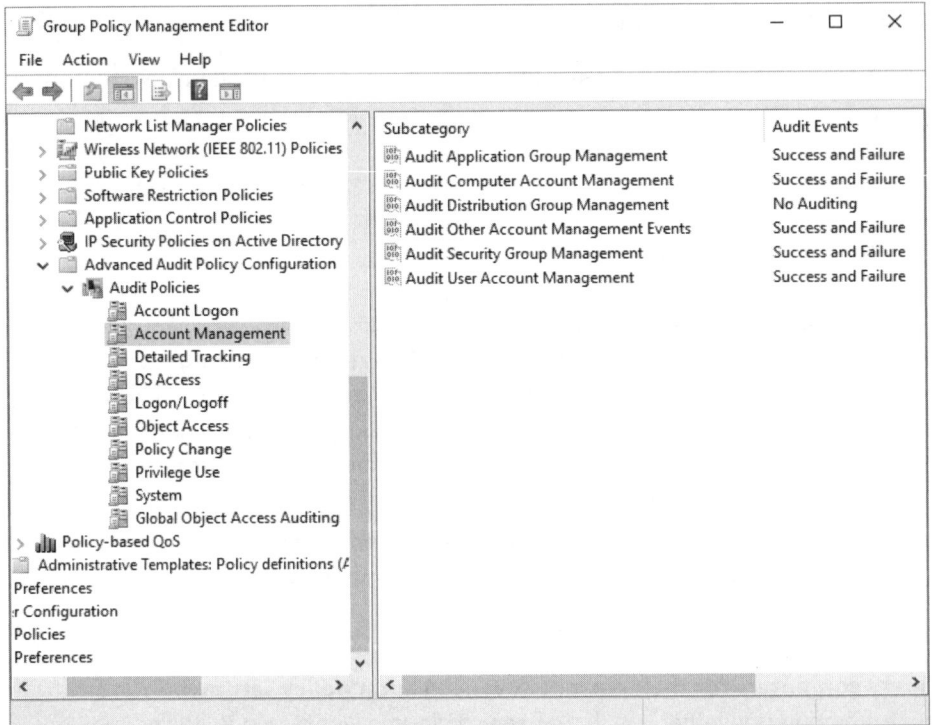

FIGURE 3-5 Advanced audit policy settings in a GPO

When you use Group Policy to implement advanced audit policies, remember that the settings are located in the Computer Configuration node. Thus, your GPOs should be linked to the locations where your computer objects are located. There are 60 advanced audit policy settings. If you configure both standard and advanced auditing settings and the settings conflict, you can have inconsistent results. Instead, you should ensure that you have enabled the following GPO setting: Audit: Force Audit Policy Subcategory Settings (Windows Vista Or Later) To Override Audit Policy Category. This setting ensures that the advanced audit policy settings take precedence.

Create expression-based audit policies

You configure standard auditing and advanced auditing so that activities are logged if a user performs audited actions such as reads a file, attempts to navigate a folder structure, or deletes a file. Expression-based audit policies provide additional granularity. Instead of having to audit specific users or groups, you can audit users based on attributes that you define. For example, you can audit a folder named Payroll and audit any activity from users who are not in the Payroll department. Or, you can audit HR users if they access sensitive HR information but only if they do so from a smartphone. You can mix combinations of attributes too. Expression-based audit policies work in conjunction with Dynamic Access Control (DAC, which is covered later in this chapter. But for now, you need to know that your expression-based auditing is extremely limited without DAC. In Figure 3-6, a condition has been added for auditing so that only users who are not members of the specified group are audited.

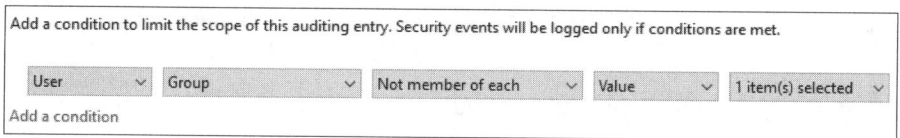

FIGURE 3-6 A simple expression-based audit condition

In Figure 3-7, auditing occurs only if a user's country is set to Germany and the computing device used is a member of the High Security Devices group.

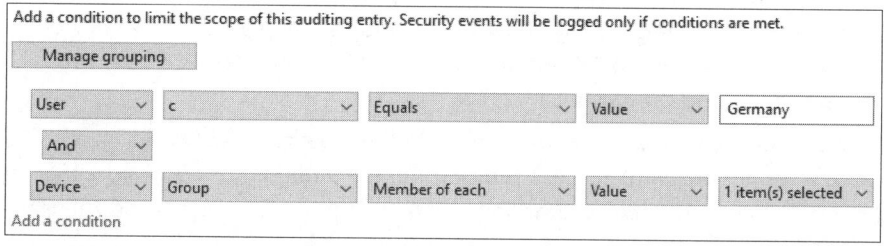

FIGURE 3-7 Two expression-based audit conditions

From a design perspective, you should know the following facts about expression-based auditing:

- **Expression-based auditing reduces the amount of data you log.**   Because you are filtering the amount of auditing you are capturing, less data is captured.

- **Expression-based auditing reduces the administrative overhead of auditing.** Managing auditing on a per user or per group basis is time consuming. Managing auditing with expressions automates much of the auditing and reduces the amount of administrative overhead.

- **You use DAC elements such as resource properties to create expression-based auditing.**   Thus, you need DAC for expression-based auditing.

 Quick check

- In your environment, all servers run Windows Server 2008 R2 and all client computers run Windows 7. You want to deploy expression-based audit policies. What should you do first?

Quick check answer

- Deploy a domain controller running Windows Server 2012, Windows Server 2012 R2, or Windows Server 2016. On the exam, you might be tested indirectly on your knowledge. For this question, while your goal is to create expression-based audit policies, your environment does not meet the prerequisites. DAC requires at least 1 domain controller that runs Windows Server 2012 or newer. Thus, that is your first step. Thereafter, you would deploy DAC and then deploy expression-based audit policies.

Design for removable device audit policies

In addition to auditing data access in your environment and on your servers, you might also encounter situations that require you to audit access to data on removable storage devices. For example, if somebody brings in their personal USB external hard drive and connects it to your environment, you want to know if data was copied to it.

Microsoft added auditing functionality for removable storage devices in Windows Server 2012. The Audit Removable Storage policy for auditing removable storage devices is functional for Windows Server 2012, Windows Server 2012 R2, Windows Server 2016, Windows 8, Windows 8.1, and Windows 10.

The configuration of the policy is straightforward. You enable the audit policy in a GPO and choose whether you want Success auditing, Failure auditing, or both. Note that the auditing policy is a computer configuration setting. In Figure 3-8, Success and Failure auditing have been enabled.

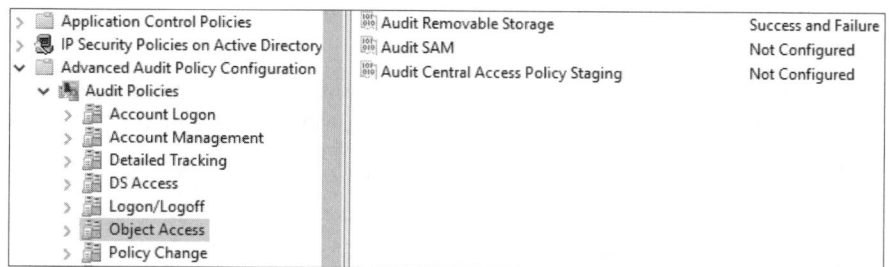

FIGURE 3-8 The Audit Removable Storage GPO setting

After you configure the GPO, you need to link it to the organizational units (OUs) that contain the computers you want to audit. After the GPOs are in place, the Security event log maintains log entries related to removable storage devices. In Figure 3-9, a snippet from the Security event log shows several removable storage entries.

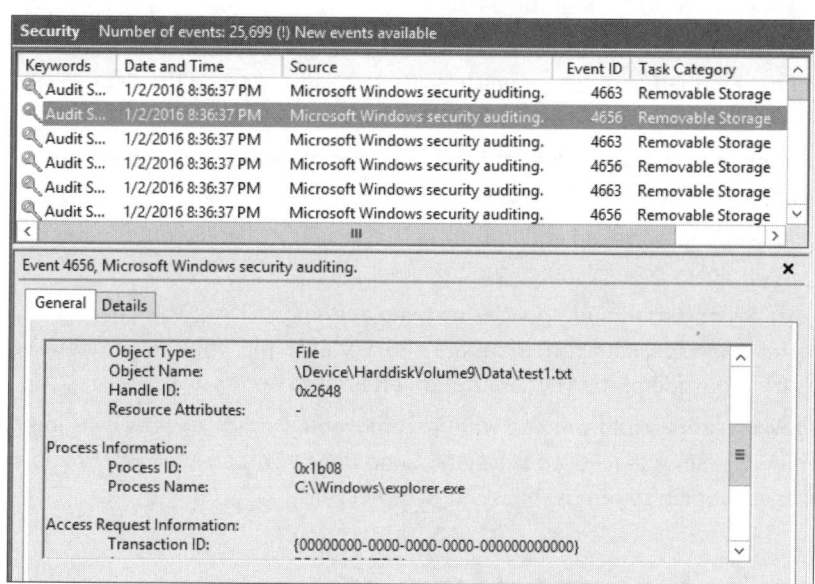

FIGURE 3-9 The Security event log showing removable storage entries

Be aware that auditing removable storage can generate a large amount of log entries. Thus, you should pilot the auditing with a small group of users before you deploy the Audit Removable Storage policy company-wide. Piloting the policy first enables you to find any problems before the impact is widespread.

✔ **Quick check**

■ In your environment, all client computers run Windows 7. The management team wants to enable auditing of removable storage devices. What should you do?

Quick check answer

■ Upgrade the computers to a minimum of Windows 8, because Windows 7 does not support the auditing of removable storage devices.

Summary

In this section, you looked at auditing with a focus on advanced audit policies. You should remember the following key points:

■ Standard audit policies provide auditing capabilities but often result in increased data and administrative overhead to filter the data.

■ Advanced audit policies provide auditing capabilities at a granular level that reduces the amount of data logged and the administrative overhead of finding the data you need.

■ You can implement and manage advanced audit policies on standalone servers by using the AuditPol.exe command-line tool.

■ To reduce administrative overhead and maximize the accuracy of your configuration, you can use Group Policy to implement audit policies.

■ You can use expression-based audit policies by tying auditing to DAC. This enables you to only audit when specified conditions are met. For example, you can audit an HR folder if users from outside of the HR department attempt to access it.

■ You can audit when users read from and write to removable storage devices by using the Audit Removable Storage GPO advanced auditing policy. This enables you to know when data is being copied to removable storage devices.

Skill 3.3: Plan for file and folder access

Now that you know how to protect your data and audit your environment, you need to investigate data-access technologies that you can use to both restrict access to data and provide access to data remotely. This section talks about Dynamic Access Control (DAC), the Web Application Proxy role service in Windows Server, and Azure Rights Management Services (Azure RMS). Your review includes exploring methods that enable you to automate access, expand access outside of your organization's network, and implement document security features to maintain control of data access even when documents are sent outside of the company.

Design for Windows Server Dynamic Access Control

DAC was first introduced with Windows Server 2012. DAC offers a new way to grant access to resources. Instead of relying on a user's group memberships, DAC enables you to grant access based on a wide variety of attributes, such as the department, country, and current location. At a high level, DAC policies can ascertain whether a user is given access to a resource based on who the user is, which device the user is using to access the data, and which data or resource is being accessed. DAC is a complex technology. For the exam, you don't need to focus on every component of DAC but you should be familiar with the design considerations, prerequisites, and integration scenarios.

First, review the following key technologies in DAC:

- **Central Access Policy (CAP)**   A CAP is a policy stored in AD DS that dictates the level of access specified users receive, based on the associated central access rules. You deploy CAPs by using Group Policy. A CAP named Contoso CAP is shown in Figure 3-10.

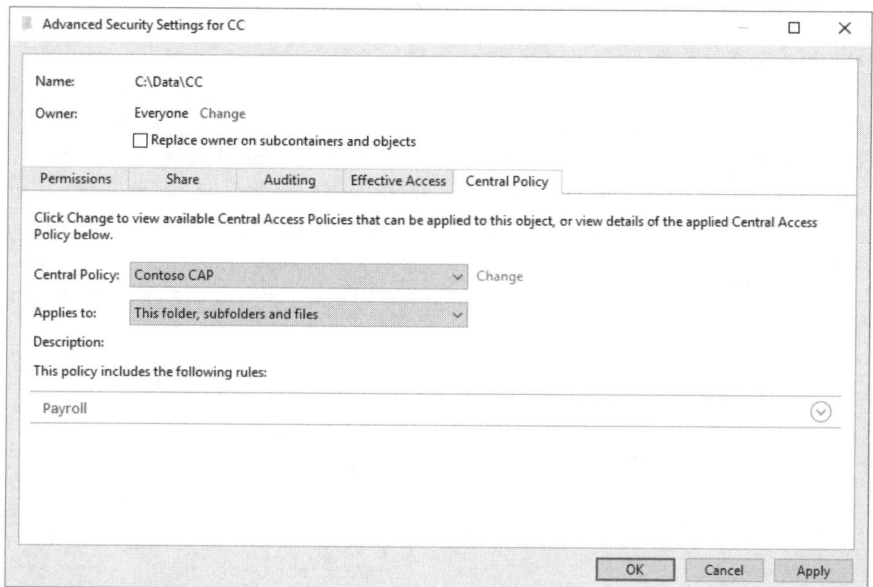

FIGURE 3-10 A Central Access Policy named Contoso CAP

- **Central Access Rules** A central access rule, shown in Figure 3-11, defines the permissions based on the target resources. In the figure, a central access rule named Payroll targets Payroll resources (resources with a department of Payroll) and assigns permissions to members of the Payroll group if their country is Canada. Central access rules are added to CAPs.

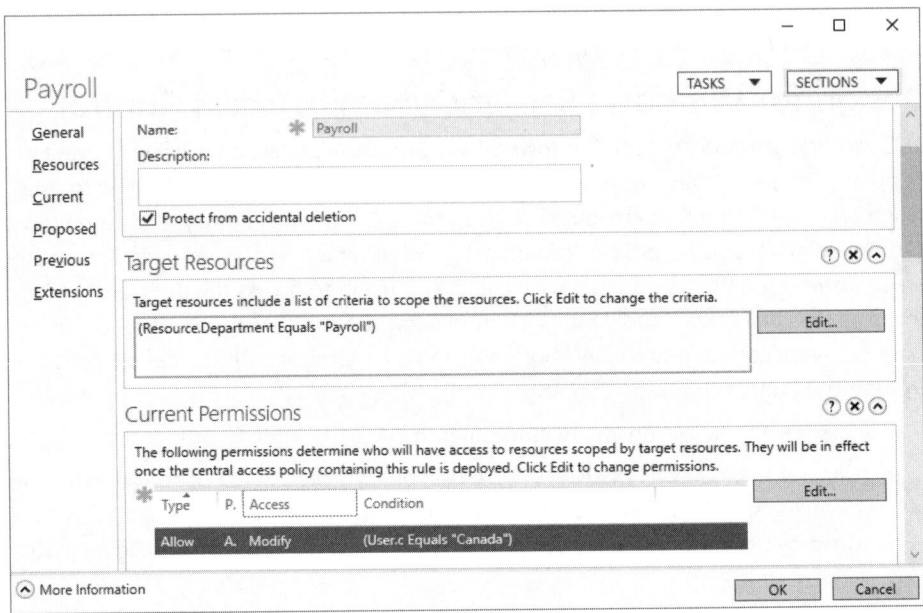

FIGURE 3-11 An example of a central access rule

- **Claim Types** A claim type is effectively a specified AD DS attribute. In identity management lingo, claim types are often called assertions. A claim type is tied to an attribute. For example, you can create a new claim type for the country of a user. You add claim types to central access rules. A claim type for the AD DS attribute C (which is used to indicate a user's country) is shown in Figure 3-12. Note that some claim types can be associated with both users and computers, which you can also see in Figure 3-12.

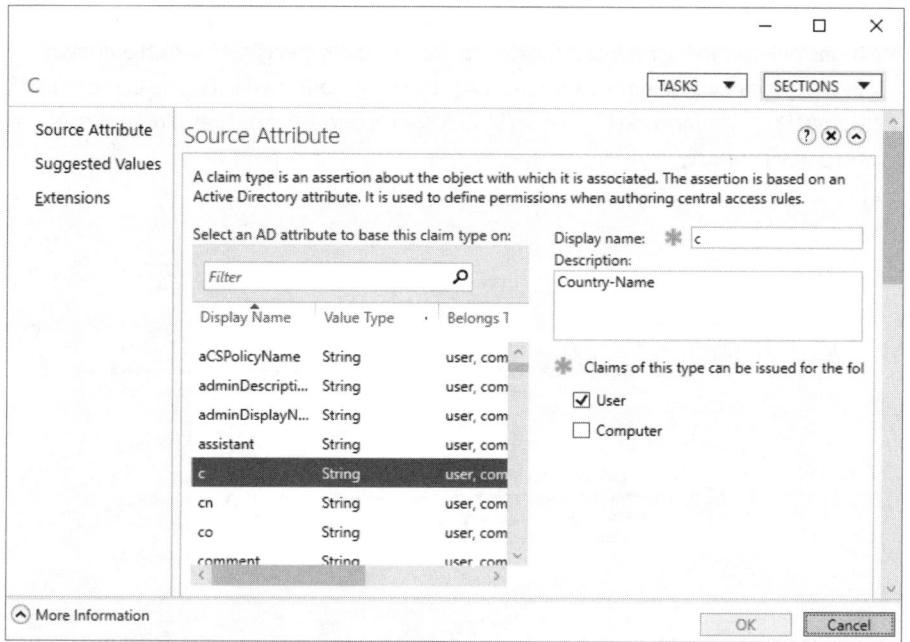

FIGURE 3-12 A claim type used to indicate a user's country

- **Resource Properties** A resource property is a description of a resource. For example, the Department AD DS attribute can be added as a resource property on a departmental printer. If you classify a printer in the HR area as an HR printer, then you can use DAC to automatically give HR department employees access to print to the printer. Figure 3-13 shows the Department resource property.

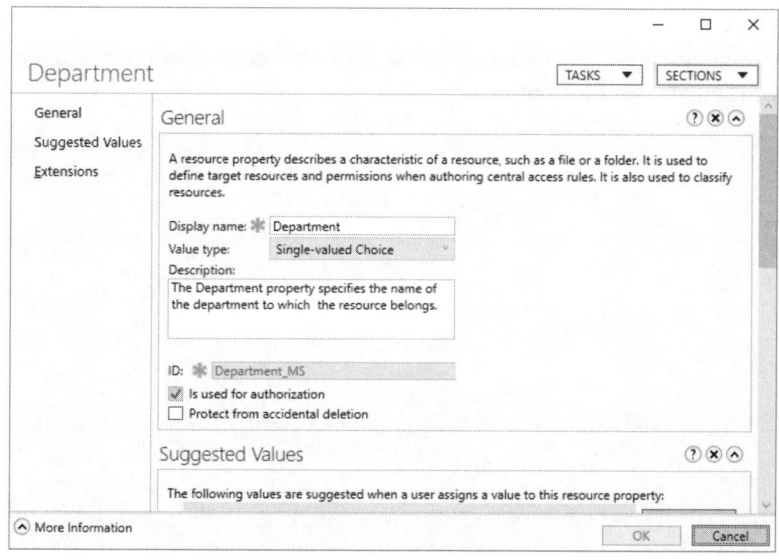

FIGURE 3-13 The Department resource property used to assign resources such as printer access

- **Resource Property Lists** A resource property list groups together resource properties. It has one default list named the Global Resource Property List, which is shown in Figure 3-14. You can also create additional lists. By using your own lists, you can ensure that department administrators see only relevant resource properties (and not every available resource property).

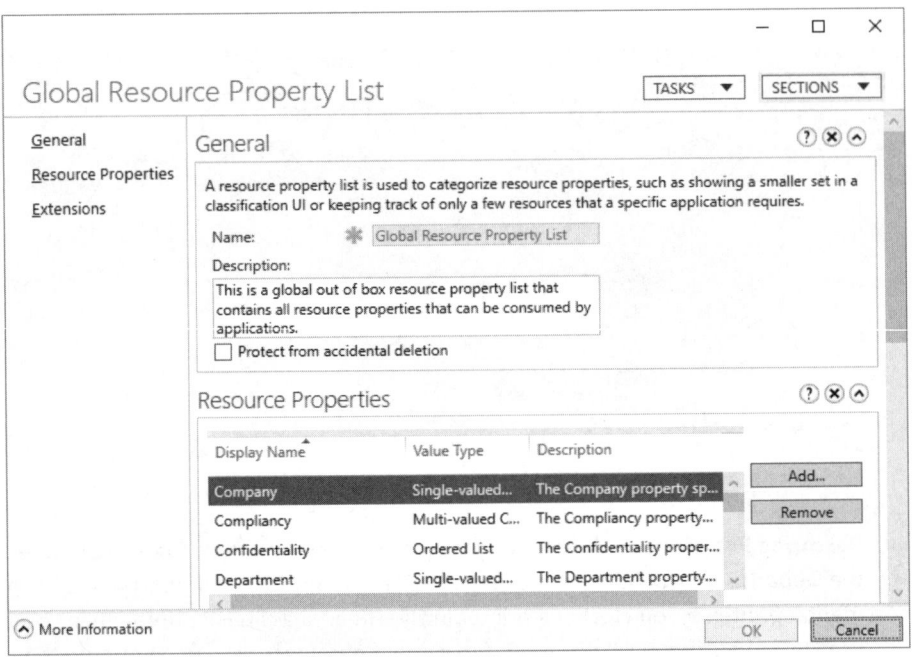

FIGURE 3-14 The default resource property list named Global Resource Property List

Now that you have reviewed DAC components, take a look at the Figure 3-15, which shows many of the components together. The figure can help you visualize how the components work together for Dynamic Access Control.

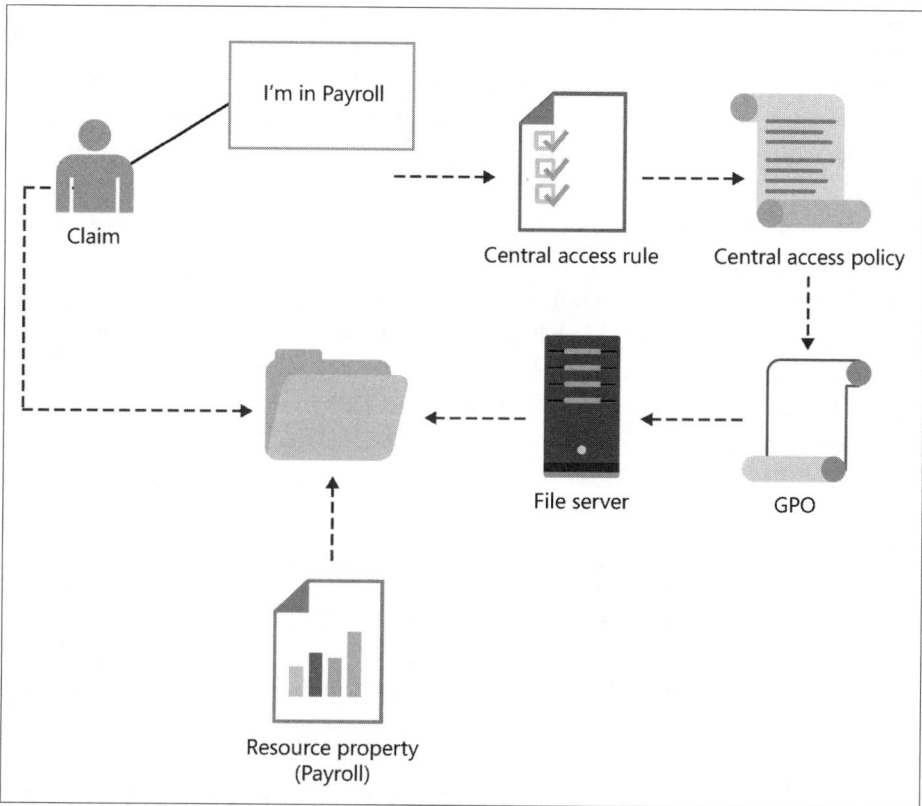

FIGURE 3-15 Dynamic Access Control components working together

Design considerations

From a design perspective, DAC provides the following enhancements:

- **Reduces administrative overhead**   Managing group memberships and group nesting requires a large amount of administrative overhead. However, DAC requires most of the administrative overhead during planning and implementation. Thereafter, you don't need to manage group memberships for access control because DAC can base access decisions on other attributes.

- **Reduces the use of AD DS security groups**   With DAC, you can grant access to resources without using any groups. Or, you can combine groups and attributes. In both cases, you can reduce the total use of groups. This minimizes your chances of experiencing "token bloat" in your environment. Token bloat occurs when so many groups are used that a user's access token can't hold all of the group SIDs. In such a case, some SIDs are not added to the token, which is similar to removing the user from those groups. Often, it leads to access issues, which are difficult to troubleshoot.

- **Automates some resource access**   Imagine that you are a system administrator for a healthcare company. Your job is to make sure that Canadian health data is accessible only to doctors in Canada, and U.S. health data is accessible only to doctors in the U.S. In other words, health data must remain in the source country. With DAC, you can automate access so that while doctors are in Canada, they can access Canadian health data but whenever they leave Canada, they cannot. Without DAC, you have to update group memberships each time a doctor travels to another country.

Before you opt to recommend Dynamic Access Control, you need to figure out if the organization can take on the implementation and support of a new technology. In addition to understanding the new technology, moving to DAC requires a bit of work, especially work tied to removing group and user access outside of DAC. You also need to consider the other technologies that can be required, such as Windows Server file services technologies, AD RMS, and Azure RMS.

Prerequisites

Because DAC was released for Windows Server 2012 and newer, you need to have at least 1 domain controller running Windows Server 2012 or newer. Client computers (or server-based clients) must also be running Windows 8 or Windows Server 2012 or newer. To maximize performance with DAC, you should have at least 1 domain controller running Windows Server 2012 or newer in close proximity to the DAC users. Because Windows Server 2012 or newer can handle all DAC-related authentication tasks, you need to have enough domain controllers running Windows Server 2012 or newer to maintain adequate performance levels. When entire organizations use DAC, it is a good practice to only use Windows Server 2012 or newer for all domain controllers.

Integration scenarios

You can integrate DAC into other technologies to enhance or expand the functionality. For the exam, you should be comfortable working through scenarios that involve multiple technologies and DAC. The following technologies are commonly used with DAC:

- **File Classification Infrastructure (FCI)**   This term refers to file server technologies built into Windows Server. The primary tool is the File Server Resource Manager (FSRM). FSRM performs the following tasks:
  - Classifies files based on the classification data you specify. You can individually classify files using File Explorer, enable FSRM to classify files, or use a combination of both methods. The file characteristics that you specify are referred to as classification

rules. You can specify simple characteristics such as specifying that data in the Payroll folder is Payroll data. Or, you can use complex regular expressions to narrow down classifications. For example, if a file contains numerical data in the U.S. Social Security format, classify the file as personally identifiable information (PII).

- Performs automatic scans of specified file locations and classifies files that are not already classified. You can scan the entire D:\ volume on a daily basis to ensure that files are classified. These automatic tasks are referred to as file management tasks.

- Takes action on classified files, based on classification. You can automatically encrypt files classified as "Confidential" with AD RMS or Azure RMS.

- **Active Directory Rights Management Services (AD RMS) and Azure Rights Management Services (Azure RMS)** AD RMS integration with DAC and FCI has been around since the introduction of these two technologies. However, integration with Azure RMS is new. It requires the installation of an RMS connector. Thereafter, you can configure specific services (Exchange, SharePoint, or FCI) and specific servers (such as your file servers) to use Azure RMS. In a DAC central access rule, you can configure the rule to apply only to files encrypted with AD RMS or Azure RMS by using a condition.

Design for Web Application Proxy

The Web Application Proxy role service, part of the Remote Access role, is the new name for the Active Directory Federation Services (AD FS) Proxy role service. The Web Application Proxy was released with Windows Server 2012 R2. At a high level, the Web Application Proxy server enables you to publish HTTP and HTTPS applications to the Internet so that remote clients can gain access to the applications. The Web Application Proxy server works in conjunction with AD FS servers in your internal network.

Web Application Proxy servers are most commonly deployed to a perimeter network. This maximizes security because it doesn't expose internal servers (those on the LAN) to the Internet. AD FS servers are most commonly deployed to the LAN. This allows them to efficiently service internal users directly and external users through a Web Application Proxy. In Figure 3-16, an example network is shown with Web Application Proxy servers, AD FS servers, and a single web server.

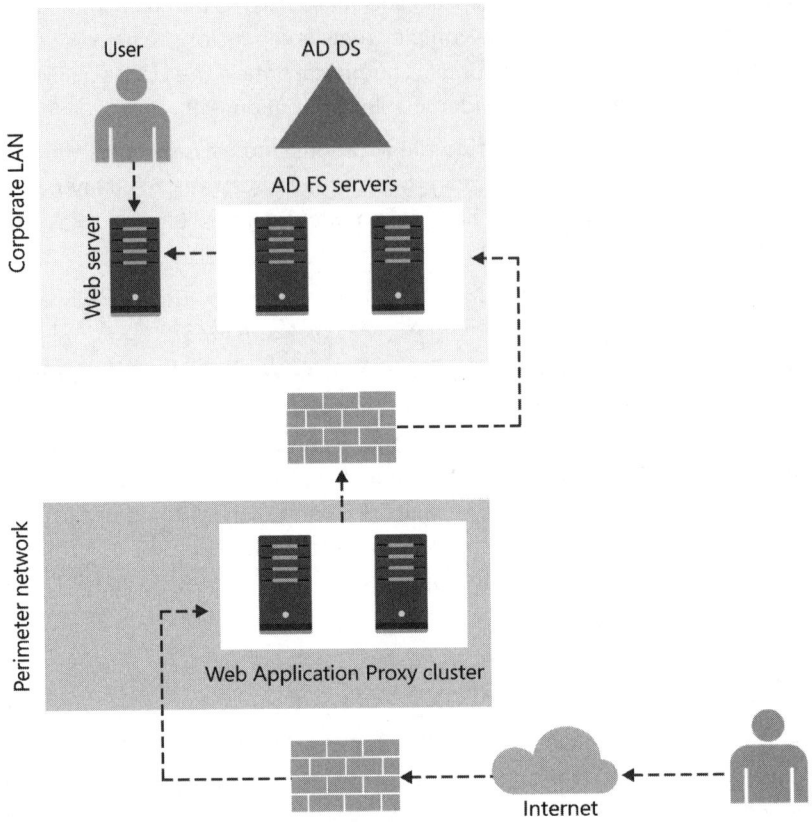

FIGURE 3-16 A typical implementation of Web Application Proxy servers

The Web Application Proxy server has two primary roles. One is providing a proxy service for an internal AD FS implementation. The other is making internal web applications available to Internet users. When you use a Web Application Proxy server to make an internal web application available to the Internet, you are publishing the application. In Figure 3-17, an application request from the Internet is shown going through the Web Application Proxy cluster. As the illustration shows, the initial connection is terminated and the cluster establishes a new connection to the web application. All communication to the web application goes through the proxy.

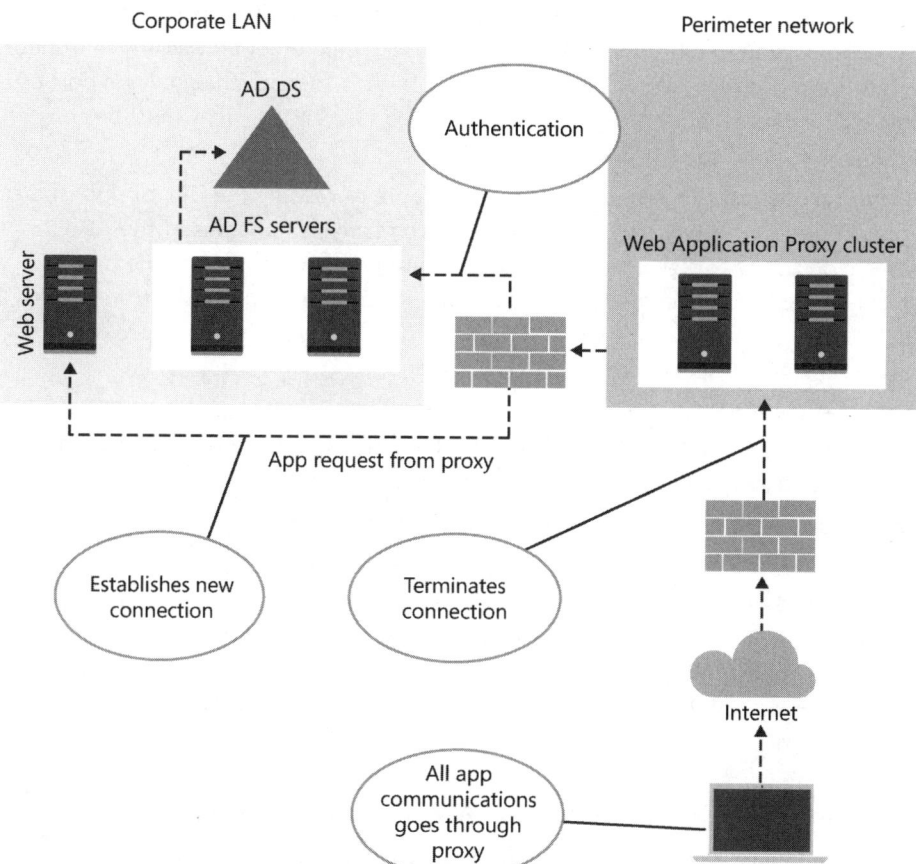

FIGURE 3-17 Web application request from the Internet

After you install the Web Application Proxy role service, the first thing you must do is open the Web Application Proxy Wizard. Then, you configure it with an existing SSL certificate installed on the server. Thereafter, you publish applications by using the wizard. For the exam, you need to be prepared to answer questions from a design aspect though. The following list contains the key areas you need to know:

- **Prerequisites** To use the Web Application Proxy, you must have AD DS and AD FS. At least 1 AD FS server must be running on Windows Server 2012 R2 or newer. Thus, from an order perspective, you install and configure the Web Application Proxy last. The Web Application Proxy role service must run on Windows Server 2012 R2 or Windows Server 2016 (although, at the time of this writing, Windows Server 2016 is in Technical Preview 4 and is not likely part of the exam).

- **Networking** You have many network configuration options for the placement of the proxy. For example, you can place the proxy behind an Internet firewall and in front of a LAN firewall. In all cases, Internet traffic must be able to reach the proxy on port 80 and 443 (or whichever port or ports you've chosen to use). The proxy must be able to communicate with the AD FS servers on the LAN.

- **Name resolution** Internet clients must be able to resolve the URL of published applications. The proxy server must be able to resolve the names of the AD FS servers on the LAN. Because servers in a perimeter network (where the proxy is usually deployed) often are not joined to a domain, you might need to rely on DNS conditional forwarding or HOSTS file entries to ensure that the proxy can resolve the AD FS server names.

- **Proxy configuration** The configuration of the proxy is stored on the AD FS servers on the LAN. This simplifies the deployment of additional proxy servers because the configuration can be automatically applied to the second proxy server.

- **Certificates** You need one SSL certificate for the proxy server. It must be configured for Server Authentication in the enhanced key usage.

As with all of the technologies on which the exam tests, it is helpful to have hands-on experience before you take the exam. Many of you who are reading this book work with only a subset of the technologies on a regular basis. Thus, you should plan to work on the rest in a lab environment. Deploying and configuring the technologies can help you remember key points.

> **NEED MORE REVIEW?**   **WEB APPLICATION PROXY WALKTHROUGH GUIDE**
>
> Microsoft published a guide to working with the Web Application Proxy role technology. See *https://technet.microsoft.com/library/dn280944.aspx* for more information.

Design for Azure Rights Management Service (Azure RMS)

Microsoft released Active Directory Rights Management Services (AD RMS) many years ago when Windows Server 2003 was the newest server-based operating system. Today, that technology still exists in Windows Server 2016. But a new cloud-based version has also been released. The cloud-based version is named Azure RMS. It is a subscription-based service that you can subscribe to through Office 365 and Microsoft Azure.

Initial setup

To begin using AD RMS, you need to acquire a license or a subscription that includes Azure RMS. Then, you need to enable Azure RMS. You can do that from the Office 365 admin portal or the Azure portal. In the Office 365 admin portal, you can see quickly that Azure RMS is activated, as shown in Figure 3-18.

FIGURE 3-18 The Office 365 admin portal showing that Azure RMS is activated

From the Office 365 admin portal, you can use the link to manage advanced features. However, that only takes you to the Azure portal. Thus, you must have an Azure subscription to configure Azure RMS advanced features.

EXAM TIP

With a design exam, you need to have a clear understanding of prerequisites. Having an Azure subscription to manage advanced features of Azure RMS is one such prerequisite. Such knowledge might be tested in a troubleshooting scenario that doesn't feel like design or planning. For example, an item writer can write a question where a company has enabled rights management in Office 365 and you need to add a new rights policy template. The question might ask which action you should take first. The answer would be to obtain an Azure subscription. But, be aware that seeing an answer choice that says to go to the portal and add a new rights policy template would be quite compelling.

You can also enable Azure RMS from the Azure portal. Once activated, it displays the activated status, as shown in Figure 3-19.

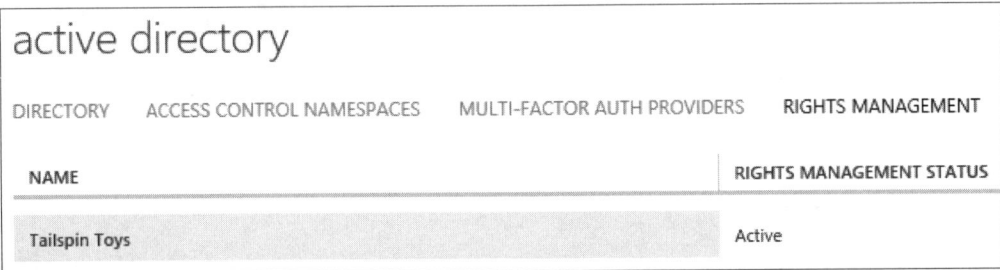

FIGURE 3-19 The Azure portal showing that Azure RMS is activated

Preparing Azure RMS for use

Once you have activated rights management, you can begin using it. However, without some further setup, you can end up degrading the user experience. In this section, you should examine some of the initial preparation tasks you need to take to have Azure RMS ready for use.

First, you need to create Rights Policy templates. A Rights Policy template provides users with a template to refer to when they add rights management protection to documents. For example, Figure 3-20 shows the available Rights Policy templates for Tailspin Toys in Microsoft Word.

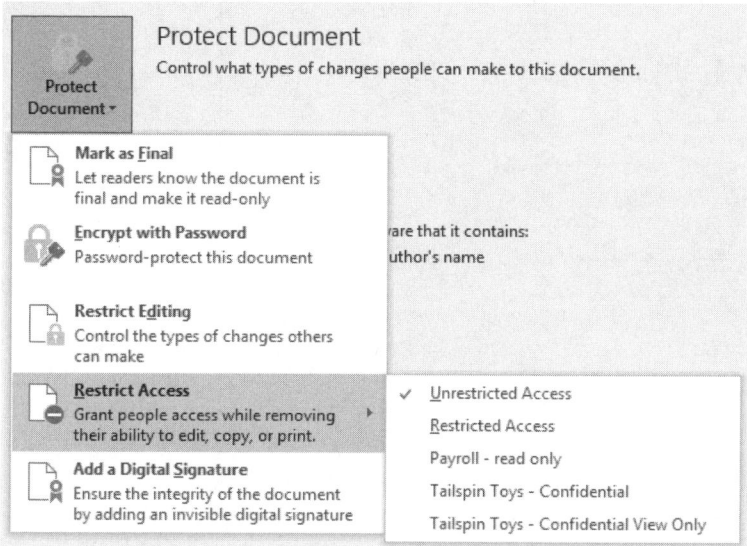

FIGURE 3-20 Rights Policy templates in Microsoft Word

In a typical organization, you should create templates for the most common use cases. Often, you need to create a few templates per department or business unit. Without templates, users have to individually add users to each document. Templates greatly improve the user experience. Without templates, users can opt out of protecting documents due to the degraded user experience of manually adding users.

The following instructions show the process to create a Rights Policy template and explain some of the design considerations. Note that these instructions are specific to the Azure management portal at the time of writing. Further, at the time of writing, Rights Management was available only on the standard portal – manage.windowsazure.com. Sometime in the near future, the new portal (*http://portal.azure.com*) can offer Rights Management functionality so these steps might change slightly.

1. From the Azure standard portal, navigate to Active Directory.

2. On the Active Directory page, click Rights Management at the top of the page.

3. Click the name of your Rights Management instance.

4. On the page that displays the current templates, click Add at the bottom of the page.

5. In the dialog box, shown in Figure 3-21, select English – United States in the Language list. Type a name and description for the template and then click the check mark to complete the creation process.

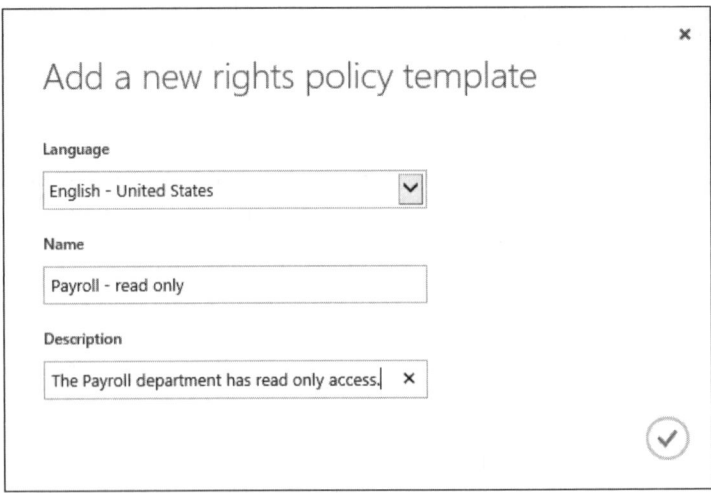

FIGURE 3-21 Creating a new Rights policy template

Note that the name and description are important because they are the only information the users see when they protect content. Users must understand what the template is used for and in which circumstances they should use it. In Figure 3-22, the ribbon in Microsoft Word is shown for a protected document. The name of the template is shown along with the description of the template. While you can click to view the permissions, a description is important because users won't always be able to make conclusions based on the permissions.

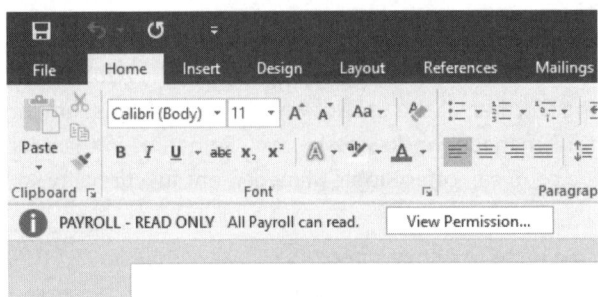

FIGURE 3-22 A Microsoft Office Word document protected by Azure RMS

Importantly, after you create a template, you have a few more steps left before it is ready for use. The next set of steps walks you through the access rights, scope, and configuration of the template.

1. Click the name of your Rights Policy template.

2. On the Rights page, click Add at the bottom of the page.

3. In the dialog box, shown in Figure 3-23, click the group for which you want to set document usage rights and then click the arrow on the right.

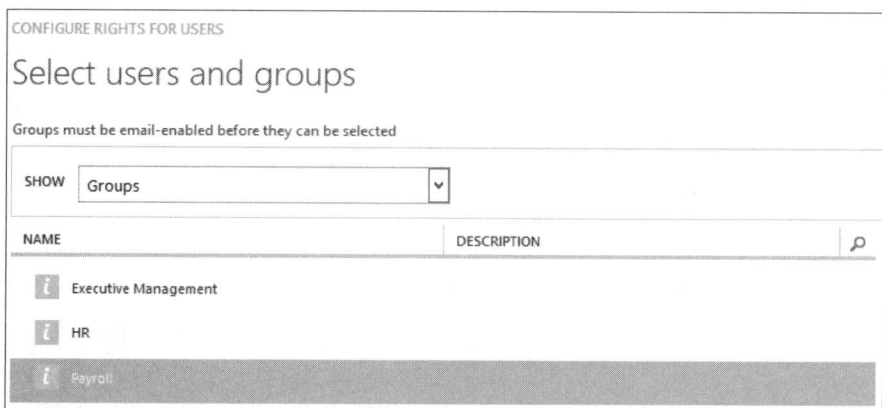

FIGURE 3-23 Choosing which users or groups have permissions to use a template

4. On the User And Group Rights page, shown in Figure 3-24, click the Reviewer option to give your group rights to view and edit documents. Then click the check mark.

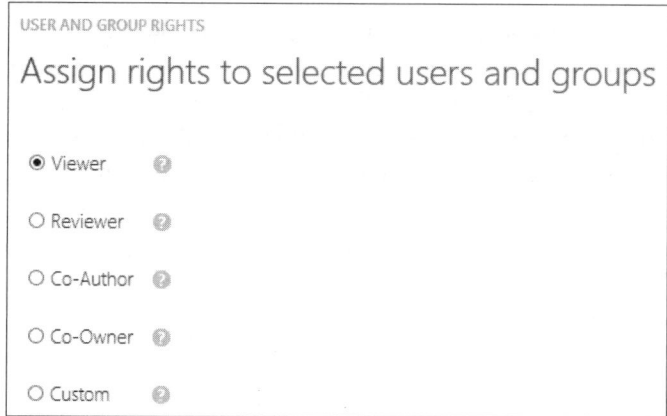

FIGURE 3-24 Assigning rights to users and groups so that they have specific permissions on documents

5. Click the Scope button at the top of the page.

6. Click the Add button at the bottom of the page.

7. On the Template Visibility page, shown in Figure 3-25, click a group to limit the template usage to that group. By default, all authenticated users can use any template. In

many cases, you might not want all users to see all templates. For example, if the HR department is going to have 10 templates for HR use, it is a good idea to configure the template so that only the HR employees can use the HR templates. Click the check mark to complete the process.

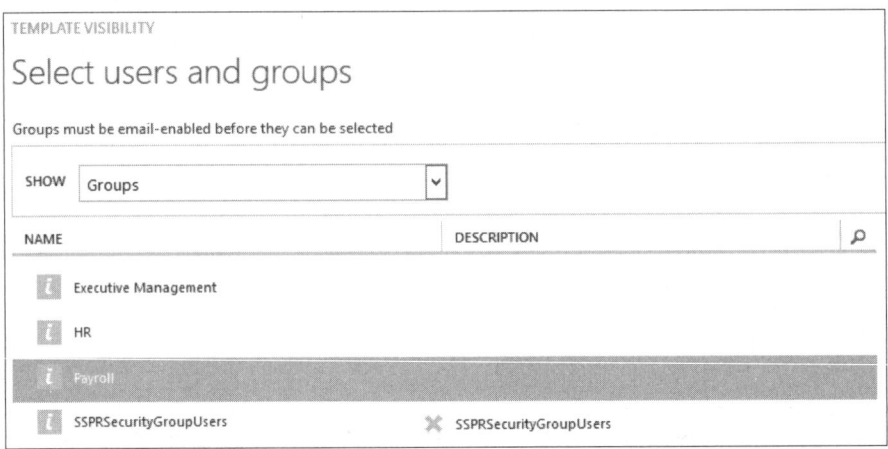

FIGURE 3-25 The Template Visibility page where you can configure users and groups to use a template

8. Click the Configure button at the top of the page.

9. Under the Content Expiration section, shown in Figure 3-26, click a desired expiration setting. By default, protected content never expires. For high-security scenarios, setting content expiration is a good idea. For example, if you needed to send credential information in a protected document, you should use a template that expires content to ensure that the information is not accessible after it has been consumed (read) once.

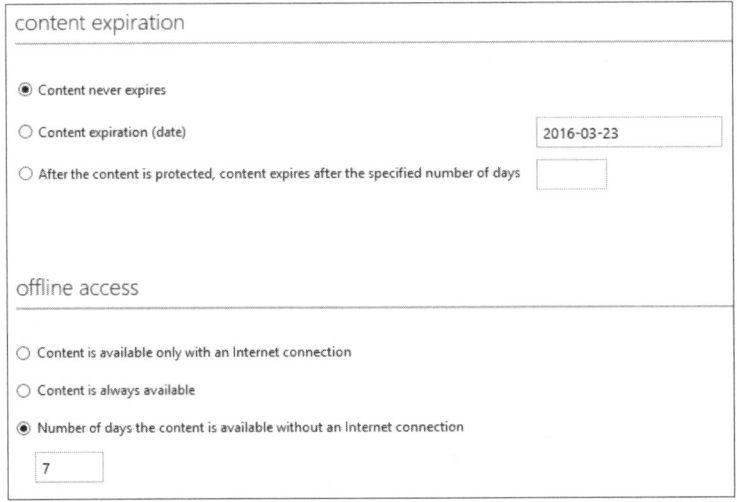

FIGURE 3-26 Adjusting content expiration and offline access for a template

10. Under the Offline Access page, shown previously in Figure 3-26, click a desired Offline Access setting. For high-security environments, you should disable this setting. In such a case, users must authenticate successfully and establish connectivity to Azure RMS to verify access before they can view or edit protected content.

11. The last step in the process is to publish the template. Until you publish a template, it isn't viewable or usable. At the top of the page, shown in Figure 3-27, click Publish. Then, click Save to save all of your changes on the Configuration page.

FIGURE 3-27 Publishing a template

After setting up templates, groups, and permissions, you still have a couple of configuration items left to do. The next section discusses integrating Azure RMS with the rest of your environment.

> **NEED MORE REVIEW? CONFIGURING CUSTOM TEMPLATES FOR AZURE RIGHTS MANAGEMENT.**
>
> Review more information about custom templates. See *https://technet.microsoft.com/library/dn642472.aspx* for more information.

Integrating Azure RMS with your environment

By default, Azure RMS integrates with Microsoft Office. But it integrates a little bit differently than an on-premises AD RMS. With an on-premises AD RMS, you are usually working within an AD DS domain, where all client computers and servers are joined to the domain. In such an environment, AD RMS works seamlessly. With Azure RMS, you can have similar functionality, but only if your client computers are joined to an Azure AD domain or if users are signing into Office. But Microsoft Office is just one part of integrating Azure RMS with your environment. Azure RMS also has integration to the following applications:

- **Microsoft Exchange Server**   For the exam, you only need to be aware of the high-level functionality associated with integrating with Exchange. Azure RMS integration with Exchange allows users to protect email messages with Azure RMS templates and use some built-in Outlook functionality, such as an option to disable forwarding and replying to all. Besides manual protection by users, you can use transport rules to automatically protect messages based on specific metadata. For example, you could create a transport rule so if the subject of an email contained the word "Encrypt," the email would be protected with Azure RMS using a predefined template. Azure RMS can be integrated with an on-premises Exchange organization, Office 365, or a hybrid environment where some users have a mailbox on-premises and some users have a mailbox in the cloud.

- **Microsoft SharePoint Server**   Similar to Exchange, you just need to understand the high-level integration for the exam. When you integrate SharePoint with Azure RMS, you enable document protection. You can configure document libraries to automatically protect documents with Azure RMS so that each time a document is downloaded, it is protected automatically based on an Azure RMS template.

- **File Classification Infrastructure**   For your file services environment, you can add a file management task to automatically encrypt documents based on specified criteria. For example, you could create a file management task to look for any files classified as sensitive in the E:\Data folder. That task could run daily and automatically apply an Azure RMS template to such files. To take advance of Azure RMS integration with your FCI, you need to also use file classification to classify files. Then, you can encrypt files automatically based on their classification.

Before you can integrate Azure RMS with your environment, you need to perform the following high-level tasks:

1. Install the Microsoft Rights Management connector. This is the service that communicates with Azure RMS for your on-premises resources.

2. Install the Microsoft Rights Management connector administration tool. This tool enables you to grant on-premises resources access to Azure RMS.

3. Run the server configuration tool script (GenConnectorConfig.ps1) on your Exchange and/or SharePoint servers. This tool prepares applications for integration with Azure RMS.

First, the following steps walk you through the installation of the connector that installs the administration tool, in addition to the connector.

1. Download the Microsoft Rights Management connector from *http://www.microsoft.com/download/details.aspx?id=40839*. The download includes 3 parts: the connector the administration tool, and the script.

2. Run the RMSConnectorSetup.exe installation file on a server. Ensure that the server isn't a domain controller, SharePoint server, or Exchange server.

3. On the Welcome To Microsoft Rights Management Connector Setup page, shown in Figure 3-28, ensure that the option to install the connector on the computer is selected. Click Next.

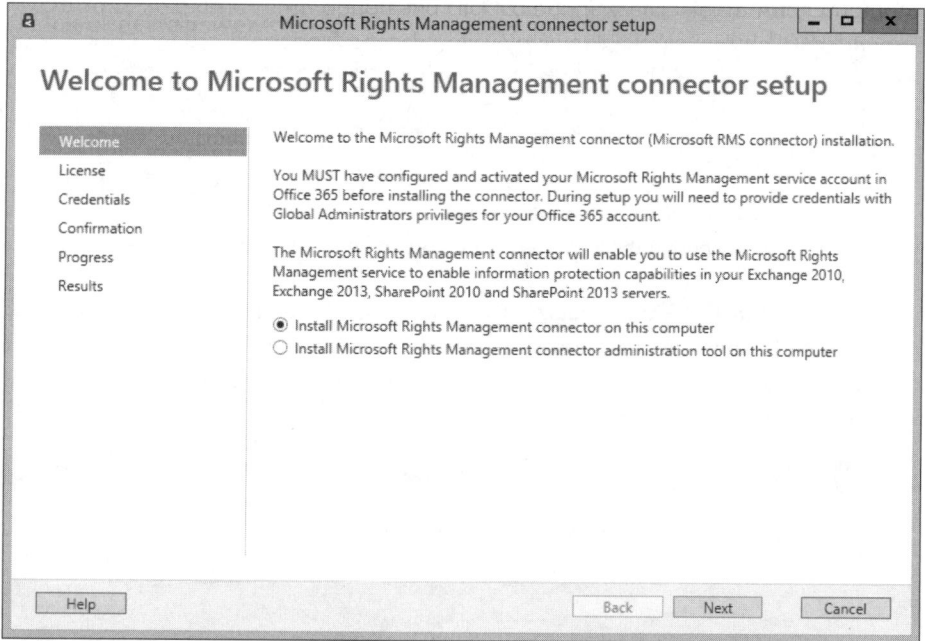

FIGURE 3-28 The Welcome page displayed when you run the connector installation

4. On the End-User License Agreement page, shown in Figure 3-29, review the license terms and if you agree, select the I Accept The Terms In The License Agreement check box, and click Next.

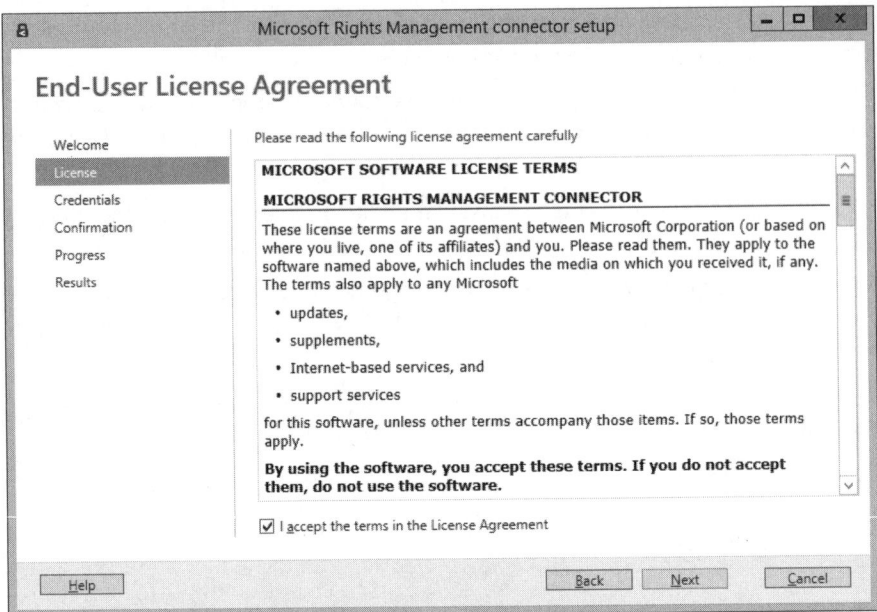

FIGURE 3-29 Accepting the End-User License Agreement as part of the connector installation

5. On the Microsoft RMS administrator credentials, shown in Figure 3-30, type your Azure RMS administration user name and password and then click Next. Note that the credential you specify here is used to communicate with Azure RMS so you must have administrative rights for Azure RMS to proceed.

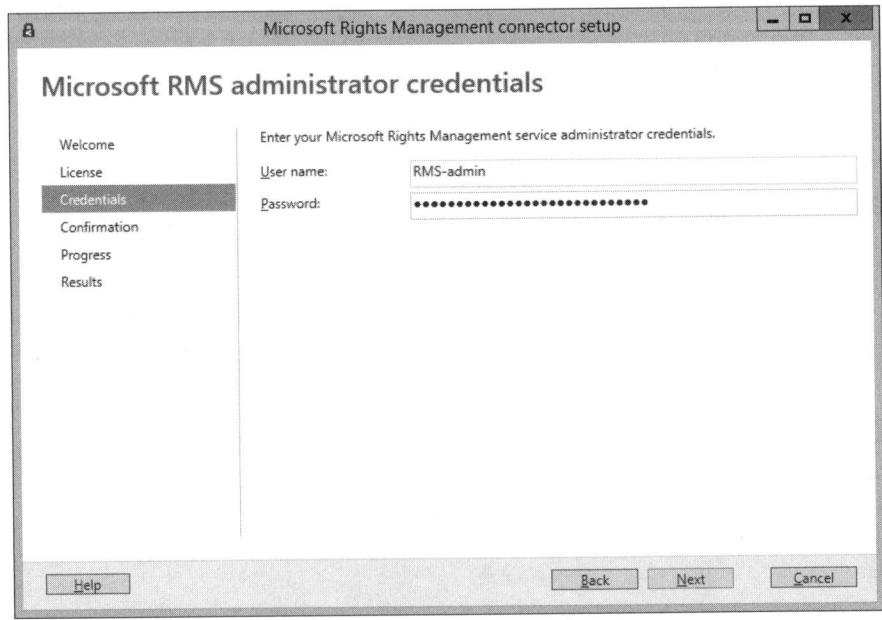

FIGURE 3-30 Specifying your Azure RMS credentials as part of the connector installation

6. On the Ready To install Microsoft Rights Management Connector Setup page, shown in Figure 3-31, click Install.

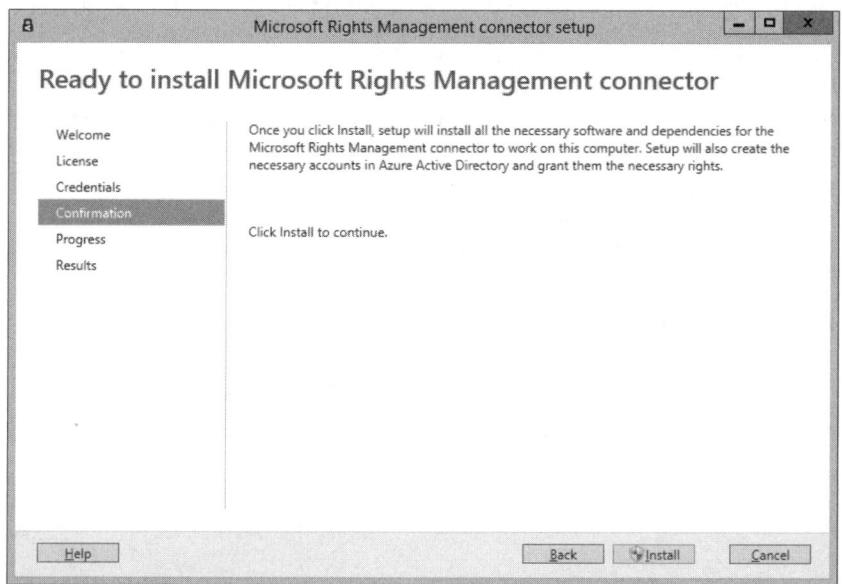

FIGURE 3-31 Proceeding with the actual installation as part of the connector installation

7. On the Installation Of Microsoft Rights Management Connector Completed page, shown in Figure 3-32, clear the check box to launch the administration tool because the next section walks you through that setup. Click Finish to complete the installation.

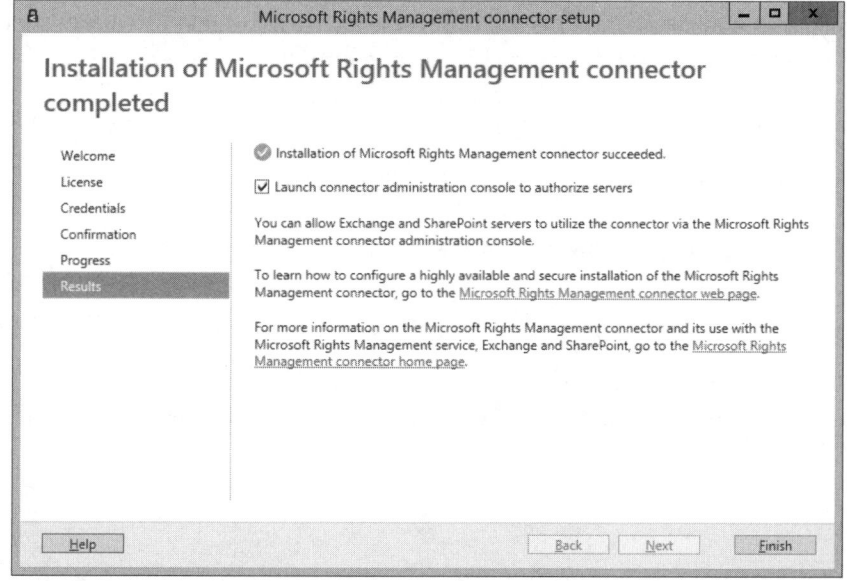

FIGURE 3-32 Completing the installation of the connector

Next, review the configuration steps with the administration tool. The tool is automatically installed as part of the connector installation. However, it is a good practice to install the administrator tool on at least one other computer, preferably a computer that isn't running the connector.

1. Run the tool by clicking Microsoft RMS Connector Tool on your Start screen. The first thing you need to do after running the tool is to authenticate with your Azure RMS administrative credentials. Figure 3-33 shows the authentication dialog box. Type your credentials and then click Sign In.

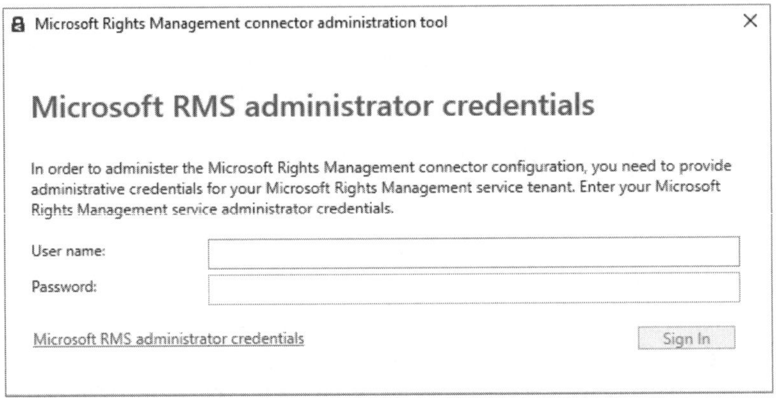

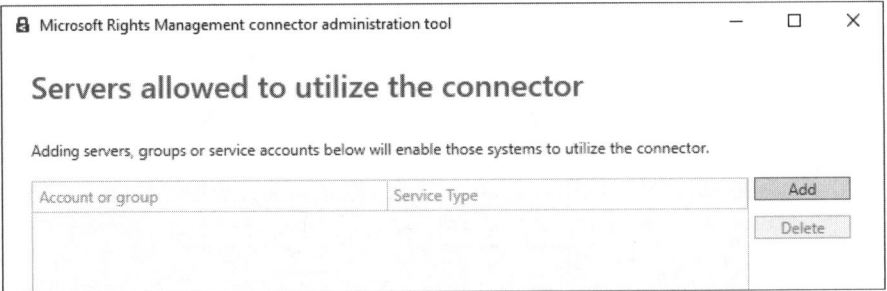

FIGURE 3-33 Signing in to the Microsoft Rights Management Connector Administrator tool

2. On the Servers Allowed To Utilize The Connector page, shown in Figure 3-34, you can see a list of configured resources (servers, groups, or user accounts). If this is the first time the connector has been installed and configured, the list is empty. If you configure resources and run the tool sometime later, a list of the currently configured resources displays. To add servers, click the Add button.

FIGURE 3-34 Configuring the Azure Rights Management Connector for integration with on-premises technologies

3. On the Allow A Server To Utilize The Connector page, shown in Figure 3-35, you click the drop-down menu to select Exchange Server, SharePoint Server, or FCI Server. Then, you specify a user, server, or group. When finished, click OK. You need to be careful on this step because it isn't intuitive. The following list contains the uses of a user, server, or group:

   - **User**   This can be any user account. However, it is usually a service account responsible for running Exchange services or SharePoint services. But you only specify user accounts if your services are running under a specified account, and not using Local System. To minimize administrative overhead, you should create a group and add the necessary service accounts to the group. Then, add the group for connector usage.

   - **Server**   When you add a server, it should be running Exchange, SharePoint, or file services. By not using a service account and by using Local System for the services, you should specify servers. Note that you should create a group for servers, add the servers to the group, and then configure the group for connector usage. This minimizes administrative overhead.

   - **Group**   You should, when possible, always use groups when assigning rights and permissions. For connector usage, you should plan to create a group for service accounts and a group for servers.

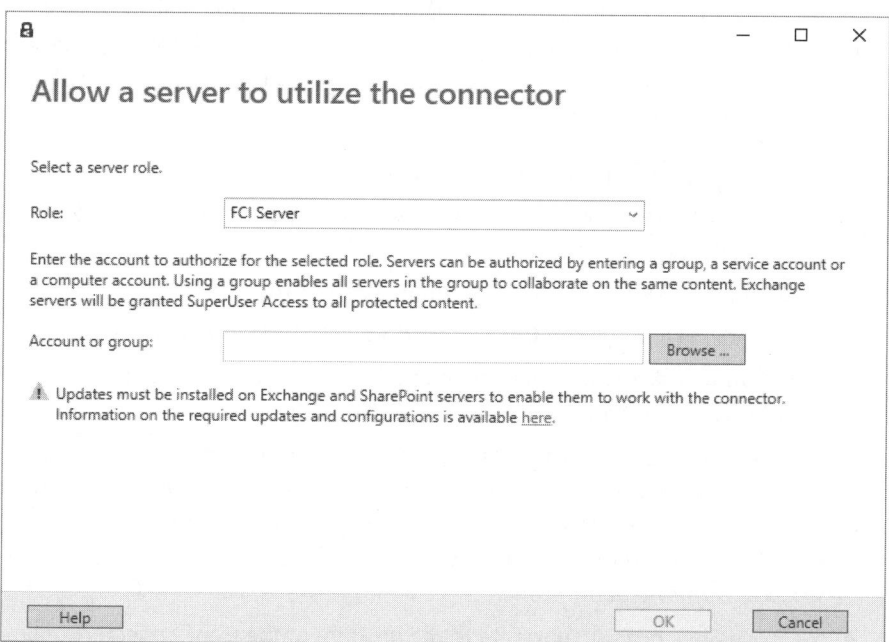

FIGURE 3-35 Specify a server role and add resources to enable them to use the connector

At this point, you have the connector installed and you have configured at least one resource for access. Next, look at the server configuration tool script. Before any of your servers can use Azure RMS, they must be configured to point to Azure RMS. With an on-premises installation of AD RMS, this step happens automatically because of the AD DS integration. For Azure RMS, you have the following options:

- **Manually set registry keys**   For small environments, such as having a single SharePoint server that uses Azure RMS, you can manually set registry keys on the SharePoint server.

- **Create registry files to set registry keys**   If you have multiple servers that use Azure RMS, you should consider creating registry files to reduce the administrative overhead of manually setting registry keys on multiple servers. A registry file is a file ending in .reg that can be run to automatically update the registry based on the content of the .reg file.

- **Use a GPO to set the registry keys on applicable servers**   In a large enterprise environment, you should use a GPO to configure servers that are going to use Azure RMS. This method requires the least administrative overhead and produces the most consistent results because administrators do not individually configure servers or manually run registry files.

The server configuration tool script eases the burden of performing all of the configuration options. Thus, to minimize administrative overhead, you should always use the script. The script is a PowerShell script file named GenConnectorConfig.ps1. The following examples show how to use the script.

UPDATE THE LOCAL EXCHANGE 2013 SERVER'S REGISTRY

1. Download the Azure Rights Management Administration Tool from *http://go.microsoft.com/fwlink/?LinkId=257721*.

2. Install the Azure Rights Management Administration Tool.

3. Open PowerShell and run the **Import-Module AADRM** command.

4. Run the **Connect-AadrmService** command and authenticate with your Azure RMS administrative credentials.

5. Run the **Get-AadrmConfiguration** command. In the output, copy the value of the LicensingIntranetDistributionPointUrl property. Remove the /_wmcs/licensing portion of the URL. The remaining portion of the URL is your Azure RMS connector URI. The connector URI should be in the following form: *https://xxxxxxxx-xxxx-xxxx-xxxx-xxxxxxxxxxxx.rms.na.aadrm.com*.

6. At the PowerShell prompt, navigate to the location of the GenConnectorConfig.ps1 script.

7. Run the **.\GenConnectorConfig.ps1 –ConnectorUri http://<*YourConnectorUri*>
-SetExchange2013** command. Note that you can use the SetExchange2010,
SetExchange2013, SetSharePoint2010, SetSharePoint2013, and SetFCI2012 parameters
based on the type of server that you want to connect to Azure RMS.

CREATE REGISTRY FILES AND THEN RUN THEM ON SERVERS

The GenConnectorConfig.ps1 script can generate all of the necessary registry files automatically. You just need your connection URI to generate the registry files. See the previous steps
for information on how to obtain your connection URI. Then, perform the following steps:

1. Open a PowerShell prompt and ensure that the AADRM module is loaded. You can
run the **Get-Module AADRM** to see if it is loaded. If it is, you'll see the version and
cmdlets. If not, you can import the module by running the **Import-Module AADRM**
command.

2. Run the **.\GenConnectorConfig.ps1 –ConnectorUri <*YourConnectorUri*>
-CreateRegEditFiles** command.

3. After the command completes, you will have several .reg files in the folder where you
ran the command. The files update the registry to point servers to the location of your
Azure RMS instance. A sample registry file for Exchange 2013 is shown in in Figure
3-36. Note that portions of the URIs are blacked out for privacy.

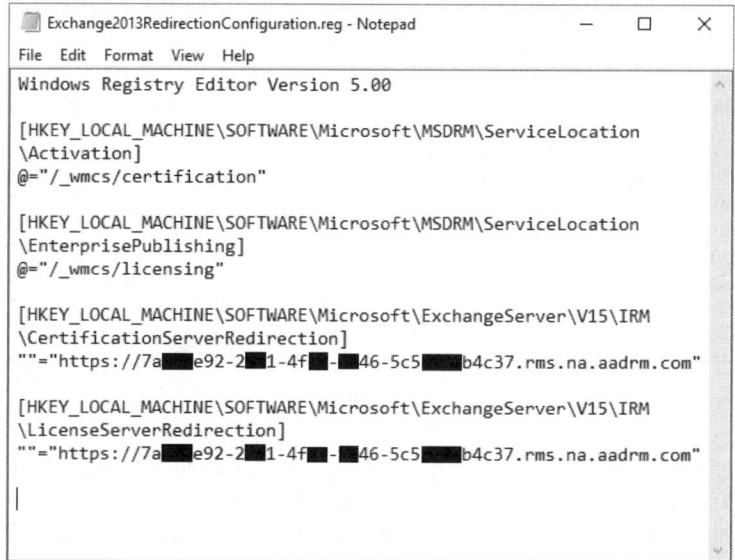

FIGURE 3-36 A registry file to update an Exchange 2013 server for Azure RMS

4. Double-click the appropriate registry file. For example, for Exchange 2013, you double-click Exchange2013RedirectionConfiguration.reg. In the Registry Editor message box warning you of the risk of modifying the registry, click Yes.

5. In the Registry Editor message box that says that the registry values have been successfully added to the registry, click OK.

CREATE GPOS FOR CONFIGURING SERVERS FOR AZURE RMS

Now you need to review the last option, which is to use GPOs to configure servers for Azure RMS. Perform the following steps:

1. Open a PowerShell prompt and ensure that the AADRM module is loaded. You can run the **Get-Module AADRM** to see if it is loaded. If it is, you'll see the version and cmdlets. If not, you can import the module by running the **Import-Module AADRM** command.

2. Run the **.\GenConnectorConfig.ps1 –ConnectorUri** *<YourConnectorUri>* **-CreateGPOScript** command.

3. In the folder where you ran the command, find the new script named CreateConnectorRedirectGPOs.ps1. Run the **.\CreateConnectorRedirectGPOs.ps1** command.

4. Open the Group Policy Management tool. Navigate to the GPOs and check to see if the new GPOs are present. Figure 3-36 shows the GPOs in a domain. All of the AAD RM GPOs were created from the script.

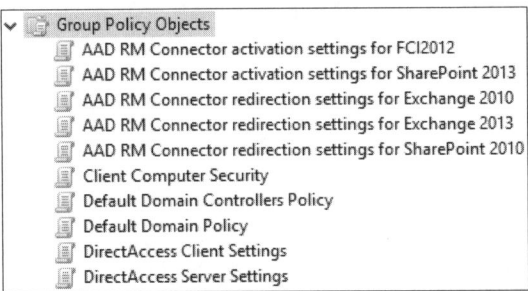

FIGURE 3-36 The automatically created Azure RMS GPOs

5. You should delete any GPOs that are not applicable to your environment. For example, if you plan to use only FCI with Azure RMS, you should delete all of the GPOs except for the GPO titled "AAD RM Connector activation settings for FCI2012." Besides the one GPO for FCI, there is one GPO for SharePoint 2010, one GPO for SharePoint 2013, one GPO for Exchange 2010, and one GPO for Exchange 2013.

6. The last step is to link the GPO to the OU, where the appropriate server objects are located. For example, if your Exchange 2013 servers are in an OU named "Exchange servers," you should link the GPO to that OU. By default, none of the GPOs are linked.

EXAM TIP

If you don't have access to Azure RMS, try going through the TechNet Labs. Look for a lab dedicated to Azure RMS which aligns with the 70-398 exam. See *https://technet.microsoft. com/virtuallabs/bb467605.aspx* and review the lab titled "On-Prem and Cloud App and data protection with Azure RMS."

Summary

- DAC provides dynamic access to files based on who your users are, which device they are using, and which data is being accessed. DAC reduces administrative overhead, reduces the use of AD DS security groups, and automates some resource access.

- DAC integrates with FCI to enhance your file services environment. With FCI, you can classify data automatically. Then, you can use DAC to dynamically provide access to users based on the data classification.

- The Web Application Proxy role service is the new role service that supersedes the AD FS Proxy role service. You can publish HTTP and HTTPS applications to the Internet so that remote clients can gain secure access to the applications.

- A Web Application Proxy server is usually deployed to a perimeter network to maximize security. This ensures that your servers on the LAN are not directly exposed to the Internet.

- Azure RMS is the cloud-based offering that provides AD RMS functionality for your on-premises environment, your cloud environment, or your hybrid environment.

- The key steps to begin using Azure RMS are: activate Azure RMS in the Azure portal, create Rights Policy templates, configure the templates so that people can access them, install the Azure RMS connector, integrate Azure RMS with Exchange, Share-Point, or FCI, and configure your applications to use Azure RMS.

Thought experiment

In this thought experiment, demonstrate your skills and knowledge of the topics covered in this chapter. You can find answers to this thought experiment in the next section.

You are a system administrator for Alpine Ski House, a luxury mountain sports provider of mountain lodging, recreational activities, and special events facilities and services. The current IT environment consists of a single-domain AD DS forest. All servers run Windows Server 2012 R2 and all client computers run Windows 10. All member servers are located in the Servers OU. All client computers are located in the Client Computers OU. The company operates 14 locations worldwide. Each location has a local server room. The main site, located in Jackson, Wyoming, hosts the company's primary datacenter. The company recently moved its disaster recovery site to Microsoft Azure. Currently, the company has 2,200 employees and 350 contractors. The following pain points have been identified:

- **Personal devices are being used often.**   Employees and contractors have been bringing their personal devices to work and have been asking to connect them to the company networks. This has resulted in a rapidly expanding supported device list and many security exceptions because the company does not have many written policies for personal device use.

- **Lost and stolen devices are reported too often.**   During the busy season, employees are reporting lost and stolen devices on a regular basis. The company is concerned about data loss.

- **Internal email messages are being leaked to news organizations.**   Alpine Ski House opened two new resorts in the last year. In both cases, details about the resorts leaked to the public, which impacted the marketing and opening of the resorts. Marketing sends out the information in internal email messages but sometimes attaches Microsoft Word files with the information. Marketing suspects that employees forwarded the email or sent the attachments outside of the company.

The company has recently decided to start a project to implement enhanced security company-wide. The following requirements have been identified:

- All company devices must implement a solution so that company data is inaccessible to unauthorized users, such as malicious users that steal company devices. The solution must provide a good user experience.

- The security team needs to monitor data transfers to portable storage devices to protect against data leakage by employees.

- The marketing team needs a solution to minimize the chances of their internal-only marketing information being seen by people outside of the company.

You need to design a solution to protect company data on company client computers, provide the security team with logs of data transfers to portable storage devices, and minimize the chances of internal marketing information from being leaked to news organizations. What should you do?

Thought experiment answer

This section contains the solution to the thought experiment.

To protect company data on client computers, you should use BitLocker with Network Unlock because it is the best solution to the problem and aligns with company goals. BitLocker with Unlock encrypts the entire volume and provides a good user experience because users do not need to type a PIN at startup when they are connected to a company network.

To provide the security team with logs of data transfer to portable storage devices, you should use advanced audit policies, specifically the removable storage devices auditing, because it is the best way to meet the goal of providing the security team with information about data being copied to removal storage devices. Because the security team wants to know about data copies, you must use the Success setting, which captures actions that occur (such as a copy) and not actions that do not occur (such as a user attempting to read a file that they do not have access to).

To minimize the chances of internal marketing information from being leaked to news organizations, you should use Azure RMS. You need to deploy Azure RMS and integrate it with Microsoft Exchange so that you can protect email messages and attachments in a way that meets the company requirements. Additionally, you need to create a Rights Management template and configure it so that employees are limited to viewing information. This prohibits printing, copying, and forwarding. While it doesn't protect against every potential misuse (such as using a smartphone to take screen captures of the information and then sending them via SMS message), it meets the requirement to minimize data leakage.

CHAPTER 4

Design for remote access

This chapter reviews the two design aspects for remote access, including how to plan for the remote connectivity, which consists of the connection method that users and clients use, as well as the authentication method that is used. The second design aspect for remote access is the number of mobility options available to users today. With many organizations implementing a Bring Your Own Device (BYOD) model, managing how users and devices can connect to a corporate network has become increasingly important.

Skills covered in this chapter:

- Plan for remote connectivity
- Plan for mobility options

Skill 4.1: Plan for remote connectivity

With a BYOD model, people are working from multiple devices and multiple networks. In some scenarios, users might expect to have full functionality from their device, as if they were working in a corporate office. As a system administrator, you need to provide the solutions to enable these users to work efficiently and effectively, so that regardless of where they are connecting from, or what device they are using, they can access the resources they need. This section reviews the design considerations and requirements for remote connectivity with clients and devices in an enterprise environment. Later in the section are discussions about the different remote connectivity options through Remote Desktop Services (RDS) and Virtual Private Networks (VPNs).

> **This section to:**
> - Design remote authentication
> - Configure Remote Desktop settings
> - Design VPN connections and authentication
> - Enable VPN reconnect
> - Configure broadband tethering

Design remote authentication

You can categorize remote authentication for clients and devices with Windows Server 2016 into different methods. Each method has different requirements and considerations for planning and designing the implementation of the technology. These authentication methods include:

- **DirectAccess** The Microsoft flagship remote connectivity solution, DirectAccess provides remote users with always-on connectivity to the corporate network. It enhances the user experience because the remote connection occurs automatically, without user intervention. Compared to other technologies, DirectAccess offers the best security and user experience.

- **Virtual Private Networks** VPNs are the most common method of remote connectivity, as they are easy to implement and manage. You can configure VPNs with high security, but doing so doesn't always translate to an easy experience for the user. With VPNs, manual intervention is typically necessary for a user or device to connect to the network, which can cause additional support or help desk requirements.

- **Remote Desktop Services** RDS is one of the easiest methods of providing remote connectivity, but it isn't considered the most secure. Users must still be authenticated, but if a user account has set up a weak password, RDS is the easiest method of connectivity to compromise. However, RDS provides a consistent method of access that is easy for users, and RDS ensures that all users can have access to the applications and corporate resources that they need to function efficiently.

DirectAccess

To plan and design a successful DirectAccess implementation you need to account for the following technologies (all of which are critical in the operation of DirectAccess), before you attempt to configure DirectAccess:

- Network topology and configuration
- Firewall design and implementation
- Certificate requirements
- DNS requirements
- Network location server
- Management servers
- Active Directory Domain Services (AD DS)
- Group Policy Objects (GPOs)

These technologies are critical in the operation of DirectAccess, and must be designed accordingly for a successful implementation.

Network topology and configuration

You can implement DirectAccess using either one or two network adapters, depending on the environment in which the DirectAccess server is being installed. Figure 4-1 displays the basic topology when using a DirectAccess server with two network adapters.

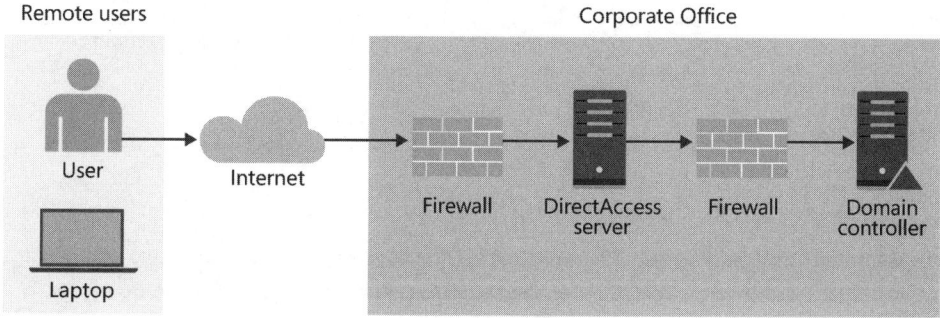

FIGURE 4-1 A DirectAccess server with two network adapters

In Figure 4-1 the Network Address Translation (NAT) device is represented by the edge firewall. If you use two network adapters, you can connect one network adapter to the internal LAN, while the other adapter is connected directly to the Internet.

Alternatively, one network adapter can be internal, while the other network adapter is in a perimeter network. With a single network adapter, the server is connected only to the private network, and is located behind a NAT device. Figure 4-2 displays a basic topology when using DirectAccess with a single network adapter.

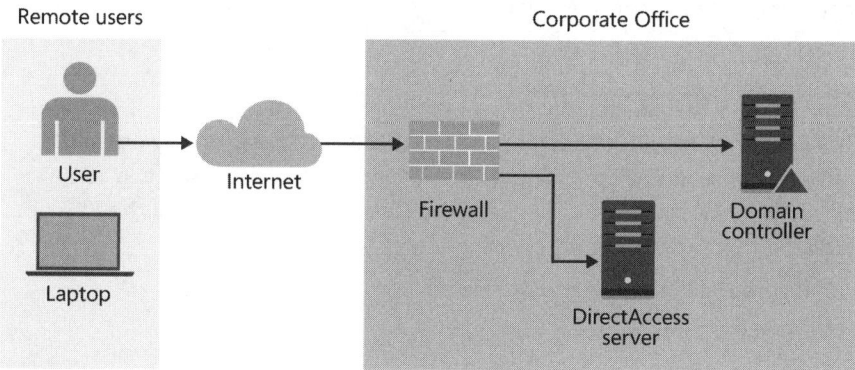

FIGURE 4-2 A DirectAccess server with one network adapter

In a scenario that uses two network adapters, they can be configured as: both as IPv4, both as IPv6, or one as IPv4 and one as IPv6. However, you should be aware of the specific configuration depending on the version types that are used. To manage remote DirectAccess clients, IPv6 must be used. Table 4-1 lists the basic requirements for the different types of connections that you can use with DirectAccess.

TABLE 4-1 Network adapter requirements

	External network adapter	**Internal network adapter**
IPv4 Internet and Intranet	■ Two public IP addresses ■ Default gateway of firewall or ISP	■ Internal IPv4 address ■ DNS suffix of the Intranet ■ No gateway
IPv6 Internet and Intranet	■ Use configuration provided by ISP ■ Ensure that an IPv6 default route exists	If necessary, modify the default preference levels for the Intranet traffic
IPv4 Internet and IPv6 Intranet	■ Use a Microsoft 6to4 adapter	N/A

> **IMPORTANT** **IPV4 WITH TEREDO**
>
> When using only IPv4 addresses with Teredo, the DirectAccess server requires two consecutive public IP addresses. This enables the DirectAccess server to act as a Teredo server, so Windows clients can detect the type of NAT devices used.

Another network consideration for DirectAccess is the use of force tunneling. When used with IPv6 and Name Resolution Policy Table, DirectAccess clients separate the traffic destined for the internal network and the Internet. Force tunneling sends all traffic, regardless of the destination, through the Intranet tunnel. In this scenario, the clients use IP over Secure HTTP (IP-HTTPS) to connect to the DirectAccess server.

> **EXAM TIP**
>
> When using force tunneling, you should configure at least two tunnels. Force tunneling in a single tunnel configuration is not supported.

Firewall design and implementation

Determining the number of network adapters and the IP version to use is only one part of planning for network changes. The components that you choose to use also need to be permitted through any network firewalls between the DirectAccess server and the Internet. This could include the following traffic when using IPv4:

- Teredo, UDP port 3544 inbound and outbound
- 6to4 traffic, IP port 41 inbound and outbound
- IP-HTTPS, TCP port 443 inbound and outbound

When using IPv6, you should plan for the following remote access ports:

- IP protocol ID 50
- UDP port 500 inbound and outbound
- ICMPv6 inbound and outbound when using Teredo

Certificate requirements

After designing and planning the network, you should plan for the use of certificates for DirectAccess. Consider the following three scenarios when using certificates with DirectAccess:

- **IPsec**   IPsec is a secure protocol that ensures confidentiality, integrity, and authentication.
- **IP-HTTPS**   IP-HTTPS is used with DirectAccess to provide remote connectivity to a client computer.
- **Network location server**   The network location server is how clients determine whether they are already connected to a corporate Intranet, or if they are on a remote network.

When using IPsec authentication, you must use an internal enterprise Certification Authority (CA . The CA issues computer certificates to the DirectAccess servers and clients when Kerberos is not being used. The best method of installing the certificates is to use a Group Policy Object (GPO) to configure automatic enrollment. This ensures that all client computers in the domain receive the computer certificate from the enterprise CA. Figure 4-3 displays a sample certificate that has been issued to web.tailspintoys.com.

FIGURE 4-3 A sample certificate issued to web.tailspintoys.com

When using IP-HTTPS, you might also use an internal CA to issue the IP-HTTPS certificate. An additional requirement is that the Certificate Revocation List (CRL) distribution point must be available from the Internet. IP-HTTPS also provides the ability to use self-signed certificates, or certificates issued by a public CA. The CRL for a self-signed certificate must also be available externally. Certificates issued by public CAs ensure that this requirement is met.

For the network location server, you can also use an internal enterprise CA. The CRL distribution point should be configured with redundancy. The certificate for both IP-HTTPS and the network location server must have *Subject Name* defined in the certificate. Without this value defined, the Remote Access Wizard cannot identify the certificate as valid.

EXAM TIP

You cannot use a self-signed certificate in a multisite DirectAccess deployment.

DNS requirements

Name resolution is also an important aspect of planning a DirectAccess implementation. DirectAccess clients use DNS to communicate with the network location server, which determines whether the client is internal or external. If the connection to the network location server is successful, the client is determined to be on the internal network. In this case, the client would not use DirectAccess to connect to the network. If the network location server is unavailable, the client would use DirectAccess and the Name Resolution Policy Table (NRPT) to determine the appropriate DNS server to use. The NRPT is a name resolution table that is stored in the registry and determines how a client issues queries based on the name that needs to be identified. Before sending a DNS query, a client uses the NRPT to determine how or where the request is sent.

You can configure DirectAccess clients to use DNS64 to resolve names, or configure it to use an internal DNS server. When a DirectAccess client needs to resolve a name, it first uses the NRPT to determine how to handle the query. If the DNS name for the query is located in the NRPT, the client uses the specified DNS server. If the query is in the NRPT, but does not have an associated DNS server, then the default DNS server is used. The default DNS server on an internal network that uses IPv4 is the DNS64 address of the DirectAccess server. If the internal network uses only IPv6, the default server is the DNS server in the internal network.

Network location server

The network location server is the component of a DirectAccess implementation that determines how a client connects to the network. In short, if a client can communicate with the network location server, the client must be on the corporate network. If the client can communicate with the network location server, DirectAccess is not used. Therefore, it is vital that the network location server is not accessible from the Internet. If the client cannot communicate with the network location server, the client must be outside of the network. In this scenario, DirectAccess would be used to connect to the Intranet. In a best practice design, the network location server should meet the following requirements:

- The network location server's web service uses an HTTPS certificate. This provides identity verification for both the client and the server. The web service responds to clients using HTTPS through the web service.

- The DirectAccess clients trust the CA that issues the certificates to the network location server. If the CA is not trusted, the certificate that has been issued won't be trusted. This can cause connectivity issues when communicating with the network location server.

- The DirectAccess clients can resolve the fully qualified DNS name (FQDN) of the network location server. By resolving the FQDN, the client can use the NRPT to determine how it issues the DNS query, and whether or not it is on the internal network.

- The network location server should be configured with redundancy. Because the network location server is critical to determining the location of the client, if a single server were to be unavailable it could cause all clients to begin using DirectAccess to connect, even if they were already on the Intranet.

- The network location server must not be accessible from the Internet by DirectAccess clients. For the same reason, if the clients are always able to communicate with the network location server, they never use DirectAccess to connect to the Intranet.

- The presented certificate must be validated against the CRL. The certificate that is presented by the network location server must be valid. If the CA revokes the certificate, the clients will not trust the server with which it is communicating.

Management servers

Management servers assist in managing the DirectAccess clients after deployment. The Remote Access management console identifies management servers, such as AD DS domain controllers and System Center Configuration Manager servers. These servers are necessary to manage clients and perform actions like software and hardware inventory assessments. After DirectAccess is implemented, these management server types are discovered automatically. For a management server to function properly, it must be available over the DirectAccess infrastructure tunnel. Additionally, for a management server to initiate contact to DirectAccess clients, the clients must fully support IPv6.

Active Directory Domain Services

From a domain perspective, the DirectAccess server must be a member of the domain to which it is providing service. However, the DirectAccess server cannot also be a domain controller, and the external adapter that is used cannot communicate with a domain controller. Use Kerberos and NTLMv2 to authenticate the DirectAccess clients against the AD DS domain.

Group Policy Objects

Finally, GPOs ensure that the DirectAccess policies are applied to the client computers in the domain. By default, the Remote Access Wizard can create the required GPOs, which are:

- **DirectAccess client GPO**   This GPO contains the client settings, including the IPv6 information, NRPT, and Windows Firewall rules. The GPO applies to the security groups that contain the client computers that are configured to use DirectAccess.

- **DirectAccess server GPO**   This GPO contains the configuration settings used for any DirectAccess server that needs to be configured. Other settings in the GPO include the Windows Firewall settings that need to be in place for DirectAccess.

Once GPOs have been set, you cannot configure DirectAccess to use a different set of GPOs. To create the GPO automatically, a remote access administrator must have the permission level to create GPOs. Optionally, you can create these GPOs manually. If you choose to create them manually, they should exist before configuring them within DirectAccess. In this scenario, the remote access administrator must have the permission level to edit the existing GPOs.

Configuring the GPOs manually enables you to easily customize and control the settings used for DirectAccess clients. The GPOs must specify a security group that contains the client computers to be used. For the DirectAccess server GPO, a server computer object must exist in the Organizational Unit (OU) that is in use. If an object does not exist in the OU, the GPO is linked to the root domain. You can modify the GPO links after running the DirectAccess Wizard, if necessary. You can also link the GPO to a specific OU, and remove the link to the domain, if necessary.

It is also possible that the remote access server does not see the newly created GPOs. In this case, wait or force AD DS replication on the network to obtain the latest GPOs.

Virtual Private Networks

This section provides a review of Virtual Private Networks (VPNs) as they relate to remote authentication. Later in this chapter you should also review the discussion about designing VPNs. VPNs provide secure connections between two devices. This could be a site-to-site VPN, where two networks are connected together, as shown in Figure 4-4.

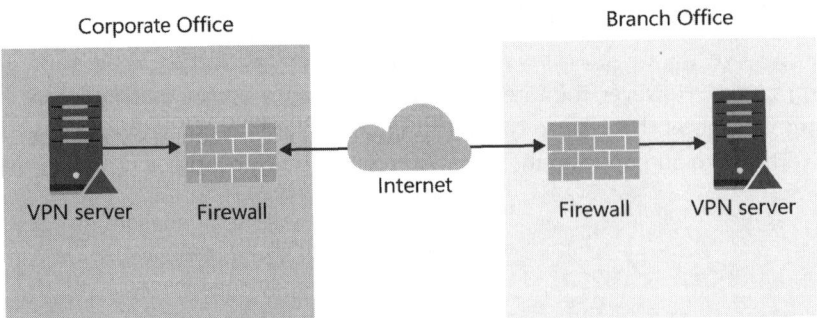

FIGURE 4-4 Two networks connected by a site-to-site VPN tunnel

Another use case for a remote access VPN is where a client can access an intranet remotely. You can choose from a few different protocols to establish a VPN:

- **Point-to-Point Tunneling Protocol (PPTP)** You can use PPTP to establish both site-to-site and remote access VPNs. However, PPTP is not considered a secure connection because the encryption method does not provide any data integrity during transit. Therefore, traffic could be modified before it reached the destination without detection.

- **Layer Two Tunneling Protocol/Inter Protocol security (L2TP/IPSec)** The L2TP/IPSec protocol can also be used for both site-to-site and remote access VPNs. When you use IPSec, you must install a certificate on the remote access servers that are used to establish the VPN connections. By using IPSec, you ensure that data in transit remains confidential, maintains integrity, and is authenticated.

- **IPSec tunnel** The IPSec tunnel mode is primarily used for site-to-site VPNs, and can be used to connect remote access servers to network devices, such as firewalls or routers. The IPSec tunnel mode also requires the use of a certificate on the remote access servers.

- **Secure Socket Tunneling Protocol (SSTP)** You can use the SSTP protocol to establish remote access VPNs from devices to a corporate network. The tunnel is established by using a Secure Sockets Layer (SSL) tunnel, and assists with the ability of VPN traffic to cross firewalls and proxy servers.

Table 4-2 compares these protocols in terms of their level of security and the associated port number used, as well as the typical use for each protocol.

TABLE 4-2 Comparing VPN protocols

Protocol	Level of security	Port number	Typical use
PPTP	128-bit	TCP 1723	Legacy
L2TP/IPSec	256-bit	UDP 500/4500 ESP 50	Remote access or site to site
IPSec	Certificate-based	TCP 50/51 UDP 500/4500	Site-to-site tunnels
SSTP	Certificate-based	443	Remote client access

When you are using remote access, you can configure authentication for the protocols in the previous table either using Remote Authentication Dial in User Service (RADIUS) or Windows authentication. RADIUS uses an external server that users are authenticated against.

- **Extensible authentication protocol (EAP)**   EAP provides an authentication framework for point-to-point connections. EAP is enabled by default when using a remote access VPN server. EAP should also be used when a Network Access Protection (NAP) server is used. EAP is also commonly used with 802.1X wired and wireless policies.

- **Microsoft encrypted authentication version 2 (MS-CHAP v2)**   MS-CHAP v2 is another default option that you can select when you configure a VPN tunnel with remote access. MS-CHAP v2 can be used with PPTP, or as an authentication option with RADIUS servers. MS-CHAP v2 is the primary authentication method when using Protected EAP (PEAP). MS-CHAP v2 provides password change and authentication retry options, and defines failures within the packet that is transferred.

- **Challenge-Handshake Authentication Protocol (CHAP) or encrypted authentication**   CHAP is another available method of authentication when using remote access. CHAP requires the use of a plain-text secret that both the client and the server have predefined when configuring the tunnel, even though the secret key is never exchanged over the network.

- **The Password Authentication Protocol (PAP) or unencrypted password**   PAP is a point-to-point protocol that uses unencrypted passwords to establish a connection. Because the password is never encrypted when exchanged, it is considered very insecure. PAP should only be used as an emergency or last chance method of establishing a remote connection.

- **Machine certificate authentication for IKEv2**   The Internet Key Exchange (IKE) protocol is used to establish a connection when using the IP security (IPSec) protocol. IKE uses X.509 certificates to validate and authenticate both the client and the server in a remote connection. IKE tunnels create a shared session secret each time a tunnel is configured.

- **Anonymous (unauthenticated) access**   As the name implies, anonymous access does not require any type of authentication to establish a connection. Ideally, this type of connection should never be used in a production environment. It might be useful in a development or test environment, where simply having the remote connection is required. In this environment you would not need to be worried about authentication or encryption.

Windows authentication can use a combination of the following options, as shown in Figure 4-5.

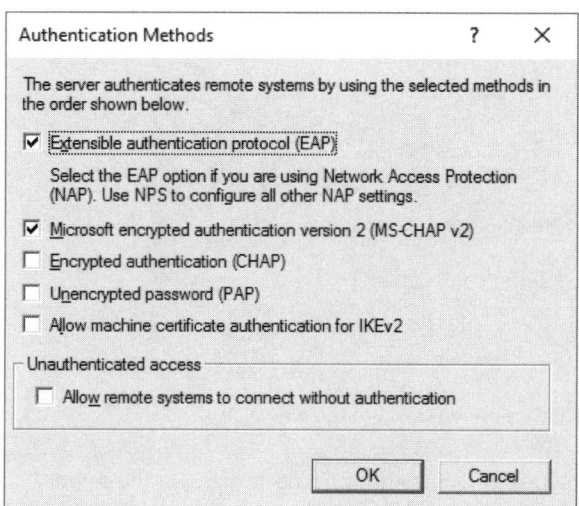

FIGURE 4-5 The Authentication Methods dialog box with the default options selected

In Figure 4-5, EAP and MS-CHAP v2 are enabled by default as the Windows authentication methods for a VPN connection. It is also possible to combine VPN services with Network Access Protection (NAP) to control the ability of clients to connect to the network. For example, clients with outdated antivirus definitions or Windows Updates can be denied remote connectivity until their health status is remedied.

Remote Desktop Services

Remote Desktop Services (RDS) provides a method of connecting to a server remotely by using a graphical interface. By default, the connection process occurs on TCP port 3389. You provide your credentials, and a domain controller authenticates you, as if you were logging on locally. You are granted a Kerberos ticket to log onto the RDS server. Figure 4-6 illustrates an example of a remote user connecting to an RD Session Host through a firewall. The RDS server then communicates with the domain controller to authenticate the user, and if successful, the domain controller issues a Kerberos ticket.

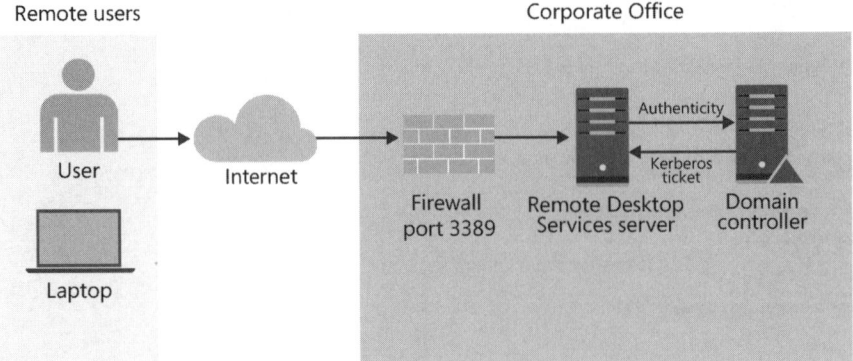

FIGURE 4-6 A remote user makes an RDS authentication request through a firewall

You can enhance RDS authentication by using certificates and smartcards. On the server side, using a certificate with RDS ensures that server identity is trusted before providing credentials. This assists in preventing man-in-the-middle attacks by verifying the server's identity.

To use certificates with RDS, they must be placed in the local computer's *Personal* certificate store. The certificate that is used must also have a corresponding private key. To use Enhanced Key Usage, the corresponding field of the certificate must have a value that matches "Server Authentication" or "Remote Desktop Authentication."

Figure 4-7 illustrates a remote user connecting to an RD Session Host server indicated by step 1. The RDS server presents the installed certificate to the client, identified by step 2. The client then verifies that it trusts the certificate, and that the certificate has not been revoked by looking up the Certification Revocation List (CRL) on the Certification Authority, identified by step 3. Finally, if the certificate is valid and trusted, the typical authentication process occurs, identified by step 4.

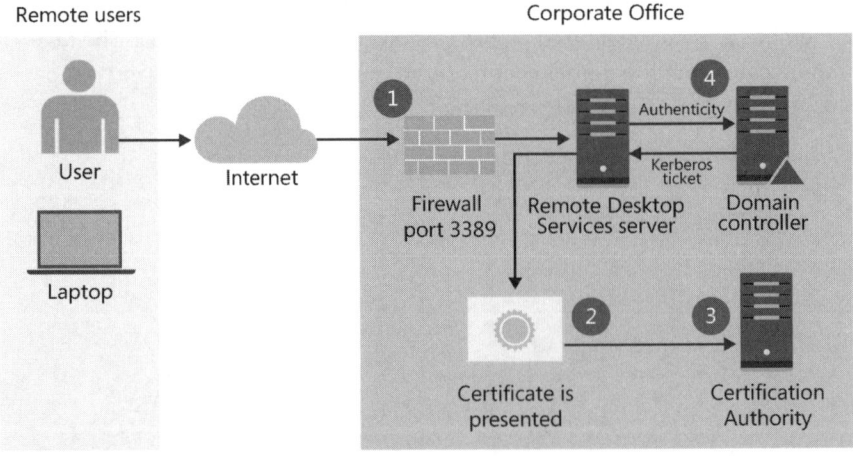

FIGURE 4-7 An RDS authentication request through a firewall with certificate

A common deployment method of RDS includes configuring a Remote Desktop Gateway (RD Gateway) to act as the broker between Internet requests and the RDS session hosts. If you don't use an RD Gateway server, clients must be able to communicate directly with the RD Session Host server. This isn't a good practice because it exposes the internal domain to the Internet. In this scenario, the certificate would need to be installed on the RD Gateway. However, the certificate must also include the name of the session hosts in the Subject Alternative Name (SAN) of the certificate. For example, Table 4-3 identifies the RDS servers that are being deployed.

TABLE 4-3 Tailspin Toys RDS servers

Computer name	RDS role
web.tailspintoys.com	RD Gateway
rds1.tailspintoys.com	RD Session Host
rds2.tailspintoys.com	RD Session Host
rds3.tailspintoys.com	RD Session Host

Each server listed in Table 4-3 must have a certificate installed to identify each component. The certificate can be configured in one of two ways:

- List each FQDN in the SAN
- Use a wildcard certificate

Listing each FQDN in the SAN means that the primary name of the certificate would be web.tailspintoys.com, and the SAN field would include rds1.tailspintoys.com, rds2.tailspintoys.com, and rds3.tailspintoys.com. This design would enable the clients to verify the identity of the servers, but it does not provide an easy growth strategy. If you need to add another RDS Session Host, the certificate must be regenerated, and reinstalled on each server.

Obtaining a wildcard certificate is one easy method of using certificates. The primary name of the certificate could still be web.tailspintoys.com, but the SAN would be simply: *.tailspintoys.com. Therefore, any host in the tailspintoys.com domain could use the certificate, and be accepted as an alternative. This makes it easy to grow and add additional servers without requiring any certificate changes. Simply import the wildcard certificate, and it responds and is trusted. However, this ease of management does not come without risk. A single certificate can be problematic if the private key is compromised, or if the certificate is used elsewhere.

As with DirectAccess, you can use either an internal CA to issue the certificate, or use a public CA. An internal CA provides more control over the certificate and the enrollment process, but is more complex to manage. A certificate issued by a public CA would ensure that the certificate is trusted, and has an externally available CRL.

Configure Remote Desktop settings

In this section, your review includes configuration of Remote Desktop settings to control and configure remote access. When using RDS for remote connectivity, you need to configure several settings on both the server and the client. The client settings relate mostly to the user experience, such as screen size, local redirects, and graphic performance. Server settings typically define security requirements, the session environment, and other variables. In this section, you need to review both server-side and client-side settings.

RDS server-side settings

The following authentication and security options can be configured on an RD Session Host server, as shown in Figure 4-8:

- **Server authentication certificate template**   When configured, this setting defines the certificate that is presented to clients when they connect.

- **Client connection encryption level**   The client connection encryption-level setting defines the level of encryption that clients are required to use when connecting to the remote server.

- **Always prompt for password upon connection**   This setting requires the user's credentials to be entered after a secure connection has been established.

- **Require secure RPC communication**   This setting requires that clients use secure RPC communication when establishing a connection to the remote server.

- **Require use of specific security layer for remote connections**   A specific security layer can be defined to require clients to connect at certain levels of encryption. This setting is used in conjunction with the connection encryption-level setting.

- **Require user authentication for remote connections by using Network Level Authentication (NLA)**   Requiring NLA ensures that only clients with the latest versions of an RDP client can connect to the remote server. NLA also requires authentication to occur earlier in the connection process.

Setting	State
Server authentication certificate template	Not configured
Set client connection encryption level	Not configured
Always prompt for password upon connection	Not configured
Require secure RPC communication	Not configured
Require use of specific security layer for remote (RDP) conn...	Not configured
Do not allow local administrators to customize permissions	Not configured
Require user authentication for remote connections by usin...	Not configured

FIGURE 4-8 Group Policy Object portion shows the Security options available with RDS

Server authentication certificate template

The server authentication certificate template can easily create certificates that can be used for authentication. Figure 4-9 displays a screen shot of the Certificate Template's management console.

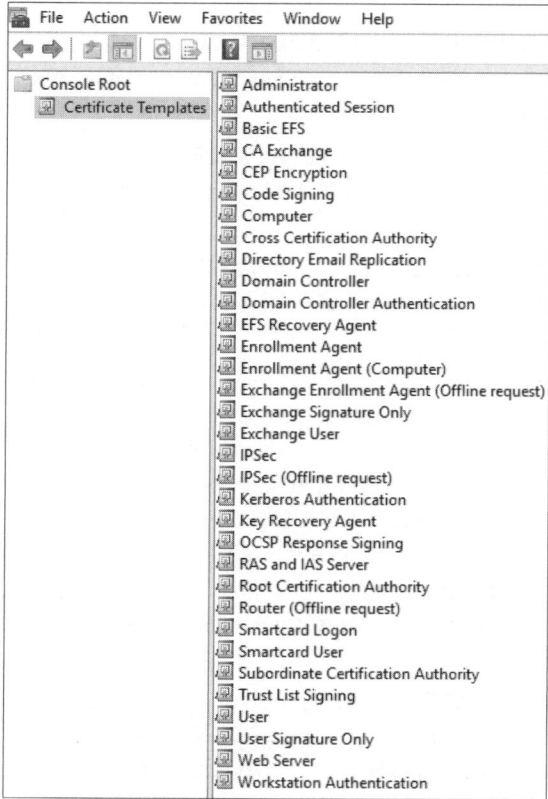

FIGURE 4-9 The available certificate templates

A certificate is required to authenticate a host when TLS is used to secure the traffic between the client and the server. Only certificates that were created based on the template can be specified when configuring this option. If a template is not selected, the RD Session Host server automatically selects a certificate. If a certificate cannot be found that was created based on the template, the server generates a certificate enrollment request.

EXAM TIP

If more than one certificate exists, the server uses the certificate that matches the server name and has the expiration date that is dated furthest out.

Client connection encryption level

You can specify the client connection encryption level for clients that are connecting to an RD Session Host. This setting is not necessary if you are using SSL encryption, but can be set if using native RDP encryption.

You have three options you can set for the encryption level:

- **Client compatible** This setting encrypts the communication between the client and the server at the highest level that both the client and the server support. The level is determined automatically during the connection process.

- **High level** This setting encrypts the connection data by using a 128-bit key. If the client that is trying to connect does not support using 128-bit keys, it cannot connect to the server.

- **Low level** This setting encrypts the connection data by using a 56-bit key. If the client that is trying to connect does not support using 56-bit keys, it cannot connect to the server.

By default, the encryption level is set to High. This setting encrypts the data to the server by using a 128-bit key. Therefore, if the client requesting a connection to the RD Session Host does not support using a 128-bit key, that client would be unable to connect. The official Remote Desktop Client supports this level of encryption, but when using third-party applications or devices, this setting might not be supported. The Low settings use a 56-bit encryption key between the client and the RD Session Host. If you are setting up a mix of clients, the client-compatible settings can be used to enable the client to configure the highest setting that it can achieve. This option is used in conjunction with the security layer setting, which is discussed later in this section.

A Federal Information Processing Standard (FIPS) 140 level of encryption can also be configured by using the Security Options tree of a GPO. By enabling FIPS-level encryption for the system, RDP can also use FIPS 140 encryption algorithms for remote connectivity. This setting should be enabled in environments that need the highest level of encryption.

By using FIPS 140 for encryption, the server and client use a range of algorithms that are FIPS 140 approved. The highest level of encryption that FIPS 140 supports is 256-bit Advanced Encryption Standard (AES). For hashing, FIPS 140 supports up to 512-bit Secure Hash Algorithm (SHA).

Always prompt for password upon connection

You can configure a GPO to require the RD Session Host to prompt for a password upon connection. This can be configured to ignore any passwords that are saved in the RDP client, and cause the authentication to occur after the encrypted connection has been established. By default, if a client stores the credentials for connection on the client, the server uses those credentials to log on the client.

Require use of secure RPC communication

You can configure a GPO to require the use of secure RPC communication. RPC is used with Remote Desktop to administer and configure RDS. By default, unsecure communication is allowed by the Remote Desktop service and RPC. By enabling this configuration, RDS accepts requests only from clients that support secure requests. Remote Desktop does not accept requests from clients that do not support secure connections.

Require the use of a specific security layer

You can configure the specific security layer used for RDP connections by using a GPO. This setting can be configured to specify what security layer must be used to secure the traffic between an RDP client and the RD Session Host. The three available options for the security layer are:

- **Negotiate**   The negotiate setting enables clients to automatically connect at the highest level of security that both the client and the server support.
- **RDP**   The RDP setting forces the use of the native RDP encryption methods to encrypt the data that is transmitted between the client and the server.
- **SSL**   The SSL setting forces the use of the highest level of encryption that is supported. If the client does not support SSL, it cannot connect to the server.

If the security layer is set to Negotiate, the RD Session Host and the RDP client can connect at the highest security level supported between the two. If available, Transport Layer Security (TLS) 1.0 can be used to authenticate the connection. If TLS is not available, native RDP encryption can be used, and the RD Session Host cannot be authenticated.

If the RDP security layer is configured, the connection uses native RDP encryption to secure the connection between the client and the RD Session Host. When this option is configured, it can be used in conjunction with the encryption level setting discussed earlier in this section to determine the level of encryption that the connection can use. Typically, native RDP encryption is not recommended for use.

The SSL security layer setting requires TLS 1.0 to authenticate the RD Session Host. However, if the client does not support TLS, the connection will fail. This is the recommended method to configure Remote Desktop authentication.

Figure 4-10 displays the collection settings of an RD Session Host. In this example, the local settings are configured to use SSL (TLS 1.0) as the security layer, and the encryption level is set to FIPS Compliant. The Allow Connection Only From Computers Running Remote Desktop With Network Level Authentication setting is also enabled.

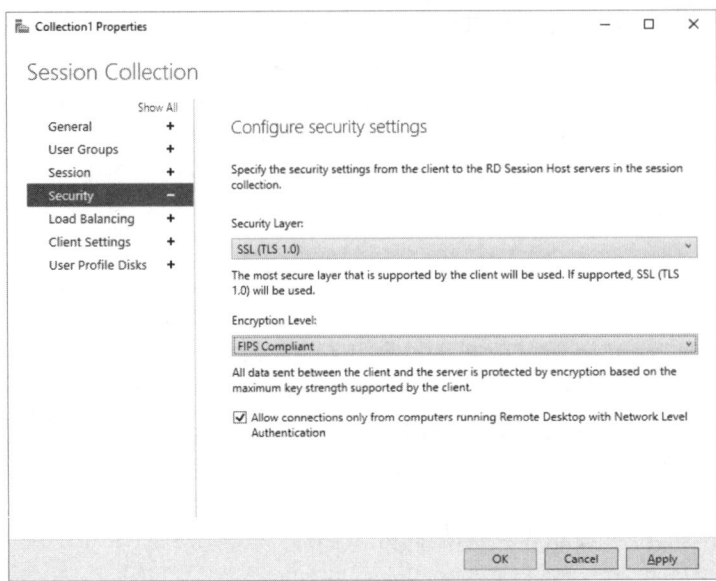

FIGURE 4-10 The Session Collection properties dialog box displays the configured security settings

RDS session settings

As shown in Figure 4-11, you can configure several settings on the server side to control individual sessions. These session settings can be configured on the RD Session Host after an RDS collection has been created.

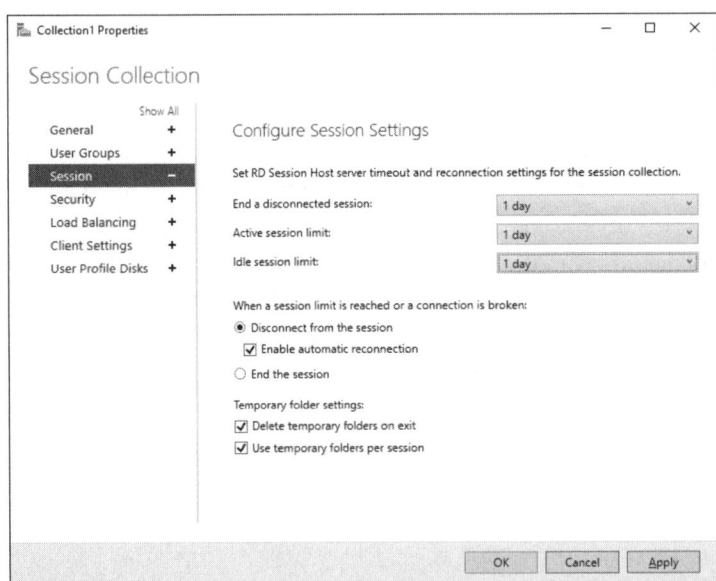

FIGURE 4-11 The Session Collection properties shows configured session settings

Figure 4-11 shows the following available settings:

- **End a disconnected session**   This setting enables you to force a session that has been disconnected, but not logged off, to end. This frees up resources on the server by ending the session that is not currently in use by a client. The available configuration time ranges from one minute to five days.

- **Active session limit**   This setting controls how long a session can remain active before the client is disconnected. This setting is also useful to free up resources when a client leaves a session connected. This can typically happen if a session is left open overnight. The available configuration time ranges from one minute to five days.

- **Idle session limit**   This setting controls how long a session can remain open without being used. This setting disconnects the client if the server does not detect any activity, such as a mouse or keyboard input, in the specified time. The available configuration time ranges from one minute to five days.

- **Session limit or broken connections**   This setting enables you to control how the server responds if users have reached the session limit, or if a connection is unexpectedly terminated. You can configure the server to disconnect the session, with automatic reconnection. You can also configure the server to simply terminate the session.

- **Temporary folders**   Sessions can also use temporary folders for each user session. You can also configure the server to delete the temporary folders after each session ends, to ensure that the server resources are not wasted.

Client settings

The client settings can be defined on the client, which you can review later in this skill, but the overall devices and resources that a client can use with an RD Session Host can also be configured on the server side. Figure 4-12 shows the Configure Client Settings page.

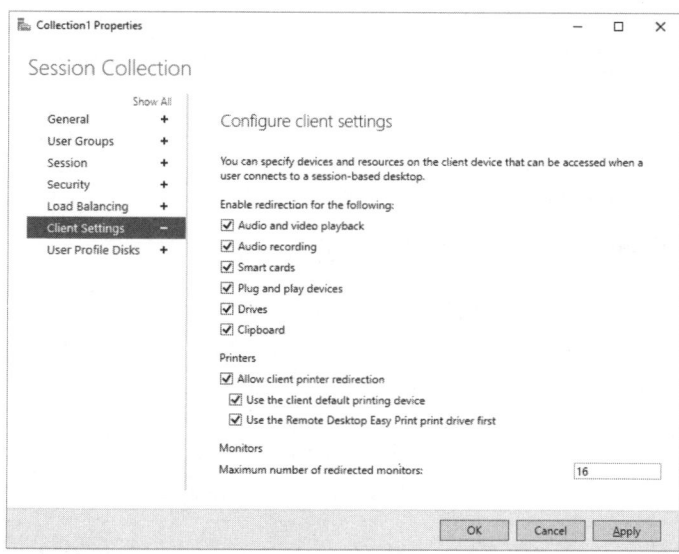

FIGURE 4-12 Session Collection properties are displayed showing the client settings

Figure 4-12 shows the following options that can be configured on the server to enhance or restrict the client experience.

- **Audio and video playback**   This enables audio and video playback through the Remote Desktop Protocol on the client device.
- **Audio recording**   This enables a local microphone to be used on the remote server.
- **Smart cards**   This enables smart card redirection for authentication purposes.
- **Plug and play devices**   This enables local devices, such as USB devices, to be redirected and used on the remote server.
- **Drives**   This enables local fixed and removable disk drives to be accessed through the remote connection.
- **Clipboard**   This enables the user to be able to copy and paste text, graphics, and files between the local and remote session.
- **Printers**   Remote servers can use locally defined printers by using printer redirection. Additional options for redirected printers are whether or not to set it as the default printing device, and the type of driver that should be used to print to the device.

Require user authentication for remote connections by using NLA

Finally, the last Remote Desktop security setting that you need to review is the use of Network Level Authentication (NLA) for remote connections. This GPO setting enables you to require NLA, which requires the client to authenticate earlier in the connection process to the RD Session Host. This means that the user or computer must provide credentials before a connection can be established. This ensures that only users with appropriate permissions can begin a session on the remote server. With this setting enabled, only clients that support NLA are allowed to authenticate to the RD Session Host. By default, this setting is enforced on Windows Server 2012 and later. You should not disable this setting because it causes more resources to be used in the event an unauthorized user makes repeated attempts to log on to the remote server. Table 4-4 displays the remote desktop client versions and corresponding operating system that supports using NLA when connecting remotely.

TABLE 4-4 Remote Desktop clients that support NLA

Remote Desktop Connection client version	Supports NLA
RDC 6.1 (Windows XP SP3, Vista SP1)	No
RDC 7.0 (Windows 7)	Yes
RDC 8.0 (Windows 8)	Yes
RDC 9.0 (Windows 8.1)	Yes
RDC 10.0 (Windows 10)	Yes

RDS client-side settings

Many of the client-side settings relate to the usability and experience of the client. Figure 4-13 displays the local resource options for an RDP client. These settings include:

- **Remote audio**   This controls how audio and recording devices can be used with the remote session. By default, remote audio is played back through the speakers on the local computer. By default, recording is not passed through the remote session.

- **Keyboard**   If the RDP session is in full screen mode, the typical Windows keyboard shortcuts can be sent to the remote computer instead of locally. For example, using ALT+TAB changes windows on the remote session. Similarly, CLT+ALT+DEL would bring up the management options for the remote computer.

- **Local devices and resources**   By default, locally defined printers, the clipboard, and smartcards are passed through to the remote session. Additional options that can be enabled include ports, disk drives, and other supported plug and play devices.

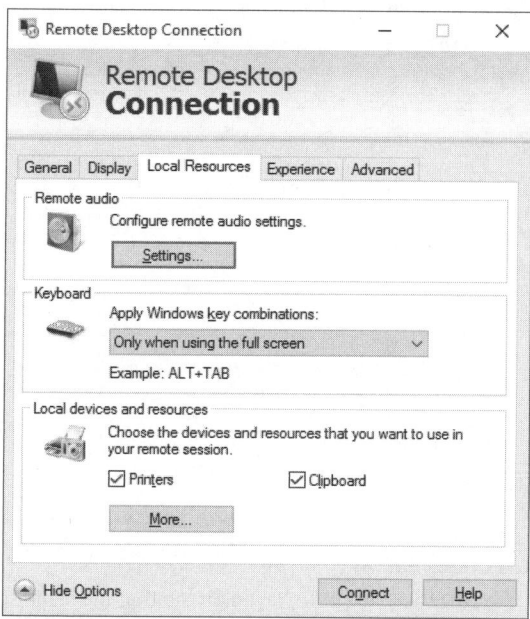

FIGURE 4-13 Local Resources tab for Remote Desktop Connections

Design VPN connections and authentication

As with DirectAccess, you need to be aware of several networking aspects when using VPNs. One of the first actions is to determine the server, and the network interfaces on that server to use. You also need to determine how the client that is connecting can communicate with the network. Should the client receive an IP address from the VPN server, or a DHCP server? If you choose to lease IP addresses from a DHCP server, you need to ensure that there is one available on the subnet, or if the DHCP server is on a different network, that a DHCP relay agent is configured between the networks. When using a DHCP server, the VPN server can lease up to 10 addresses at a time from the DHCP server to issue to clients. Otherwise, you can plan to have the VPN server issue IP addresses from a pre-defined pool as the clients connect.

You must also ensure that the user accounts in AD DS are enabled for VPN services. This can be performed through AD DS per user account, or controlled by a Network Policy Server (NPS) Network Policy. Figure 4-14 displays the Properties dialog box for the user account named User1. The Dial-in tab of the account object controls the network access rules that the user account must follow. In this scenario, using a NPS Network Policy controls remote access.

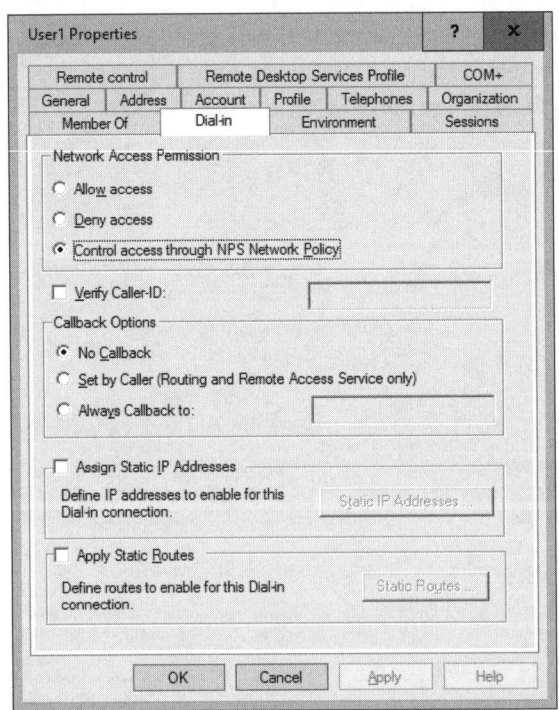

FIGURE 4-14 Properties dialog box for the User1 user account is displayed on the Dial-in tab, illustrating that an NPS Network Policy controls remote access

An NPS server is also useful for controlling the authentication methods, configure IP settings and addresses, and restrict the bandwidth a connection might use. Network policies can be used to grant or deny access based on the policy, which can also ignore the dial-in settings that are configured in the user account's properties in AD DS. Figure 4-15 illustrates the basic settings of a network policy that are configured. Notice that the Ignore User Account Dial-In Properties check box is selected.

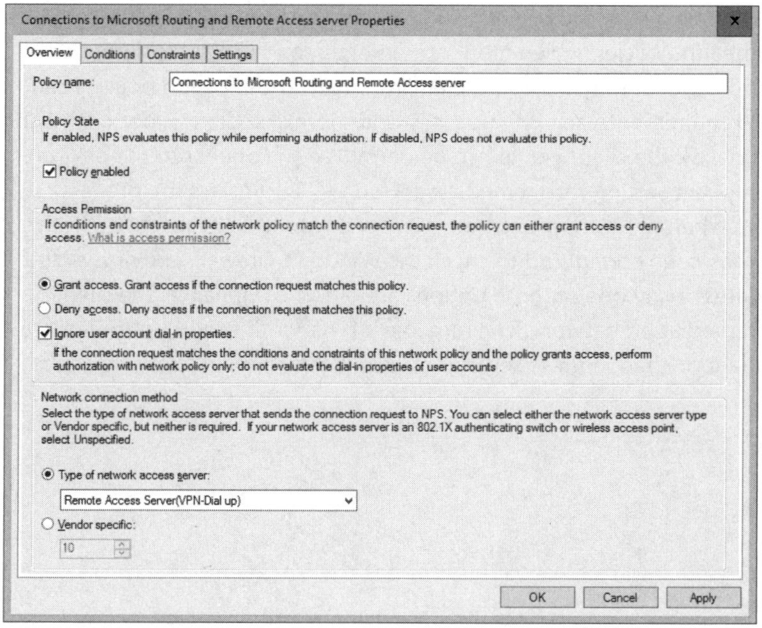

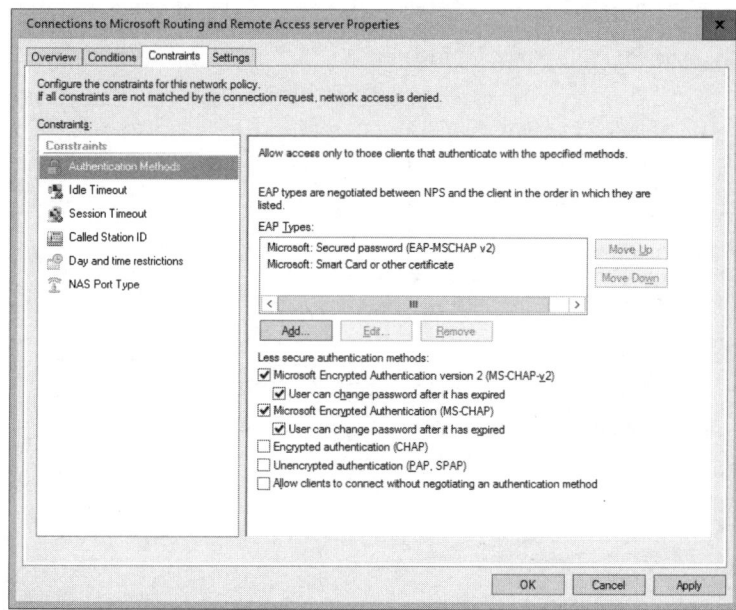

FIGURE 4-15 The Overview tab

A network policy can also be used to configure the authentication methods that were discussed earlier in this chapter. Figure 4-16 displays the Constraints tab of a network policy that includes the authentication method, session time, and other settings for remote access.

FIGURE 4-16 The Constraints tab

Another aspect of designing a VPN is using Network Access Protection (NAP). NAP enables administrators to create health policies, which monitors, enforces, and can remediate clients before they are allowed to connect to the corporate environment. These health policies can include specific software requirements, security updates, and more. If a client is not compliant with the required health policy, the client would not be permitted to connect to the environment until the out of compliant settings are remediated. Figure 4-17 displays the process that a client goes through when requesting to connect to the corporate environment. In this scenario, a health policy has been configured to check the Windows Firewall, antivirus definitions, and Windows Updates. If any one of these options are out of compliance, the client must only connect to a remediation network. The remediation network would be configured to allow the client to update or change these settings so that it can connect to the corporate network.

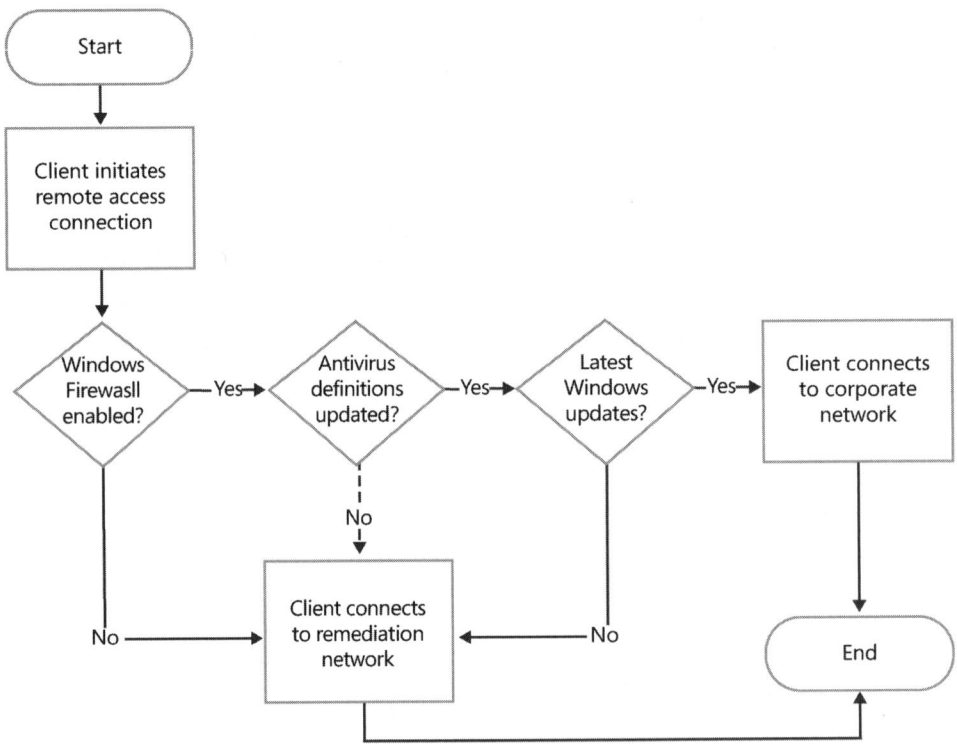

FIGURE 4-17 An illustration depicting the policy validation when using Network Access Protection

Enable VPN reconnect

VPN reconnect was introduced with Windows 7 and Windows Server 2008 R2 to automatically reconnect a VPN. This feature was intended to be used by mobile clients that used cellular hotspots for Internet access, or had another type of unreliable connection. However, this feature was removed in later versions of Windows. Because it is still listed as a skill that is measured, you should be aware of how the technology is used. A similar functionality can still be used today by using scheduled tasks to initiate an auto-redial.

VPN reconnect uses the IKEv2 standard to ensure reliable connectivity from a client to a remote environment. If a client temporarily loses Internet connectivity, then the VPN automatically reconnects to the environment. Because VPN reconnect relies on IKEv2, either computer-based certificates or EAP-based authentication can be used.

To enable VPN reconnect, the client must be allowed to communicate on UDP ports 500 and 4500 for IKE traffic, and IP Protocol ID 50 for Encapsulating Security Protocol (ESP) traffic. When the VPN service is configured on the server, these firewall rules should be configured automatically. Any internet-facing firewalls must also be configured to allow this traffic to the VPN server.

Other than using IKEv2 with certificates or EAP, there is no specific server configuration to enable VPN reconnect. VPN reconnect is a client-based configuration that can be enabled after the VPN object has been created.

Configure broadband tethering

Broadband tethering enables you to use a mobile data connection, such as a cellular hotspot, to connect a device to the Internet. By default, tethered and mobile connections through Windows devices are metered. By default, apps and updates don't download over the broadband connection. On individual devices, metering can be enabled per Wi-Fi profile.

You can configure a Windows device to tether additional devices by sharing the connection. When sharing an existing Wi-Fi connection, the properties to create a network name and password are automatically generated. If necessary, these credentials can be changed for distribution. Up to 10 devices can be tethered off of the device hosting the connection.

To configure broadband tethering, you must have a local wireless adapter. First, create a hosted network by using the **netsh** command. For example, to create a hosted network named TT-WiFi-Tether, use the following command.

```
netsh wlan set hostednetwork mode=allow ssid=TT-WiFi-Tether key=12345678
```

Figure 4-18 displays a newly created hosted network adapter.

```
C:\WINDOWS\system32>netsh wlan show hostednetwork

Hosted network settings
-----------------------
    Mode                    : Allowed
    SSID name               : "TT-WiFi-Tether"
    Max number of clients   : 100
    Authentication          : WPA2-Personal
    Cipher                  : CCMP

Hosted network status
-----------------------
    Status                  : Started
    BSSID                   : d2:7e:35:93:8d:25
    Radio type              : 802.11n
    Channel                 : 153
    Number of clients       : 1
        80:be:05:37:cc:dd         Authenticated
```

FIGURE 4-18 The CLI shows a hosted network configuration

After the hosted network has been configured, you can share the connection so that it can be used with other devices. To share the connection, go to the Sharing tab of the Wi-Fi Properties dialog box. Figure 4-19 displays the Sharing tab with the configured settings.

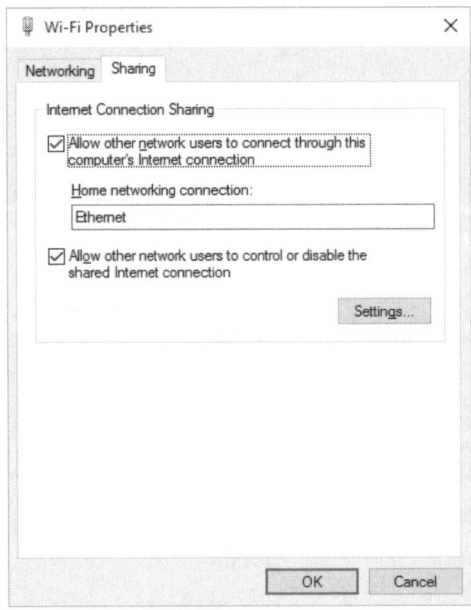

FIGURE 4-19 The Sharing tab for Wi-Fi Properties

Finally, once the sharing has been configured, the adapter displays as shared in the network connections menu. This verifies that the adapter is shared and is available for use by nearby devices, as shown in Figure 4-20.

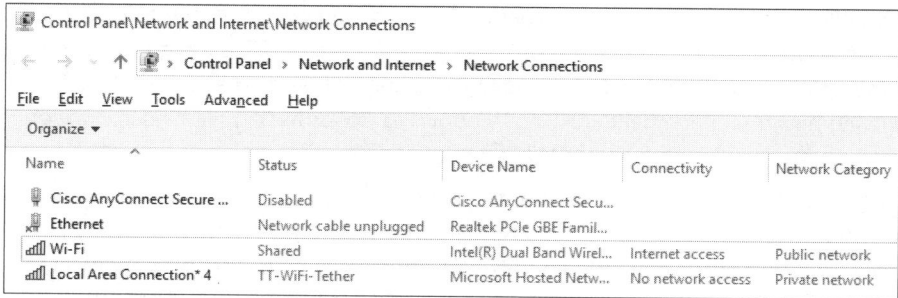

FIGURE 4-20 Available network connections

Summary

- DirectAccess, VPNs, and RDS can be used to establish remote connectivity for clients. These technologies provide different scenarios for authenticated users to have remote access to a corporate network.

- DirectAccess can be used with either one or two network adapters. However, IPv6 is required to manage the DirectAccess clients remotely.

- VPNs are used to configure site-to-site and remote access connections. SSTP is a good protocol to use for enhanced remote access security.

- RDS provides an easy method of accessing a remote environment. A good practice when using RDS is to implement an RD Gateway, which enhances the security of the implementation.

- RDS can be configured to use AES 256-bit encryption with FIPS Compliant mode.

- VPN reconnect was introduced with Windows 7 and was used to automatically reconnect clients with unreliable Internet connections.

- Broadband tethering enables up to ten devices to share an Internet connection. You can configure this easily by using the command line to set up an ad hoc (as needed) network.

Skill 4.2: Plan for mobility options

Computing on the go has become a requirement for many organizations. Companies that still require employees to use a desktop computer are a rare breed. Employers like the idea that their staff can telecommute, or continue working after hours, providing a higher quality of service to their customers. Desktop computers still have their place in the business world, but the demand for ultrabooks and tablets is far greater.

In the first part of this chapter you review how to provide mobile users with remote connectivity. To do this, you need to take a closer look at the user's mobile device and how to manage the various mobility options that come with Windows 10. Working outside of the corporate network can introduce a number of obstacles. The goal for this skill is to identify those obstacles and deliver the appropriate solutions. This includes: designing policies that address offline file needs, improving power management, utilizing Windows to Go, providing file synchronization options, and understanding the benefits of Wi-Fi direct.

When you think about managing devices in the enterprise, consider the fact that many of those devices might not always be connected to the corporate network, and the times that they are connected typically involve using a VPN. With this in mind, you need to understand some of the fundamental mobility options that Microsoft has included in Windows 10.

This section covers how to:
- Design for offline file policies
- Design for power policies
- Design for Windows to Go
- Design for sync options
- Design for Wi-Fi Direct

Design for offline file policies

Even with the abundance of wireless hotspots and cellular access points, working offline is sometimes necessary. In a situation where users cannot connect to the corporate network, how would your finance team continue working on those critical year-end spreadsheets that are stored on the corporate file server? One solution is to implement offline file policies. Offline files, also referred to as client side caching (CCS), enables users to access files on the network in cases where the network might be slow or unreliable. It accomplishes this by caching a copy of the shared resource to the user's local machine. File shares that are marked for offline availability are synchronized in the background based on the settings you configure, whether those are local or through Group Policy. The following list provides some real-world use cases for implementing offline files.

- **Slow network**   You can use offline files to improve your users' experience when they are connected to a slow or unreliable network, providing seamless access, whether or not the network is running smoothly. For example, if a user is editing a large document on a file server and you have enabled offline availability, the user does not experience any work disruption regardless of any hiccups with the file server or network. This is ideal in situations where a branch office might have a slow wide area network (WAN) link to the main office, and in extreme cases where that WAN link fails completely.

- **Unavailable network**   You can use offline Files to provide users with access to important files without being connected to the corporate network, enabling them to continue working offline. For example, if a user is leaving the office early but needs to continue editing a PowerPoint presentation that is stored on the corporate file server, Offline Files allows that user to open that document offline, make edits, and then synchronize it with the file server the next time the user is connected. This is ideal in travel situations, such as when users do not have wireless access on an airplane or train.

■ **Centralized data** You can use offline files to assist administrators in centralizing corporate data. For example, your human resources department could have multiple shared folders across the organization and each folder might contain employee information. You decide to centralize that data on a secure file share, and allow the HR team to access it regardless of their current location. Offline files can maintain synchronization with the HR team's computers, while reducing the file share footprint and improving the security of the data.

As is the case with most Windows features, users can control offline files locally. For the purpose of this exam you need to focus on using Group Policy to manage offline files. However, it is important that you are familiar with the UI elements on the client computer so you understand how these policies affect the user experience. Start your review by taking a look at the available options on the client computer.

To manage offline files in Windows 10, you need to open the Sync Center, as shown in Figure 4-21.

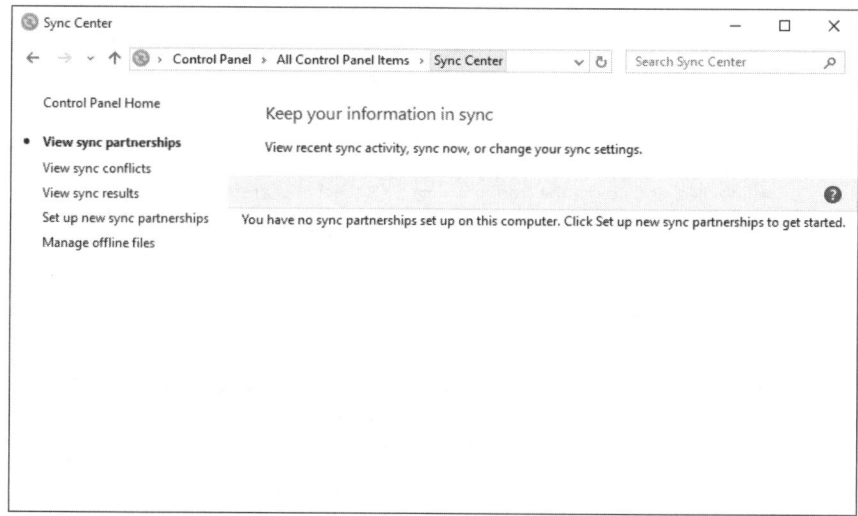

FIGURE 4-21 The Sync Center in Windows 10

You can access the Sync Center in a few ways:

■ Open Control Panel and change the view to Small Icons, as Sync Center is not available under Category view. Click Sync Center to access the Management page.

■ Click Start and type **Sync Center** in the search box and press Enter.

■ Right-click the Start button and click Run. On the Run dialog box, type **mobsync.exe** and click OK.

The Sync Center management page includes multiple options on the task list. This is where offline files are set up on a client computer, but it is important to note that Sync Center is also used to sync content from mobile devices. In a later section you can review this in more detail. The available tasks in Sync Center are outlined in Table 4-5.

TABLE 4-5 The Sync Center tasks list

Task	Description	Important elements
View Sync Partnerships	Displays your active sync partnerships, sync activity and progress, and enables you to perform a forced sync on one or more partnerships.	Offline files are listed under a single partnership. If you have multiple folders set up for offline files, they are listed under the same Sync Partnership.
View Sync Conflicts	Displays file synchronization conflicts that have occurred while you were offline. Provides you the option to resolve conflicts by choosing a preferred version, or keeping both by renaming one of the files.	The user deals with sync conflicts. If a file has multiple changes by multiple users, this can be problematic. It is recommended that users choose to keep both versions and compare the results before overwriting.
View Sync Results	Displays the results of your last file synchronization.	This view shows the last successful sync attempt and any failed attempts. For additional troubleshooting refer to the event logs for offline files.
Set Up New Sync Partnerships	Enables you to set up new sync partnerships.	This page is focused on detecting connected mobile devices and is not used for offline files.
Manage Offline Files	Opens the Offline Files Management window. This window gives you the option to enable or disable offline files, configure your local storage limits for offline files, delete temporary offline files, enable encryption of offline files, and adjust the frequency for network connection tests.	The Delete Temporary Files button is not tied to the offline files feature. It deals with deleting cached copies of documents outside of offline files. For encryption, a full disk solution like BitLocker is preferred, but the option to encrypt the cached offline files is available.

After you review the available tasks in the Sync Center, review the enabling the offline files feature, and the process for making network resources available offline. In the following example, imagine you are an employee for Tailspin Toys. You have a few projects started that you expect to evolve over the next several weeks. The content for these projects is stored on the corporate file server. You need to have access to these files constantly, regardless of your connection to the office. To accomplish this you need to use offline files. To begin, you must first enable offline files because they are disabled by default in this example.

1. Open Sync Center.
2. On the Sync Center page, click Manage Offline Files.
3. On the Offline Files properties window, click the Enable Offline Files button, as shown in Figure 4-22. When prompted, reboot your computer for the changes to take effect. Proceed with the reboot.

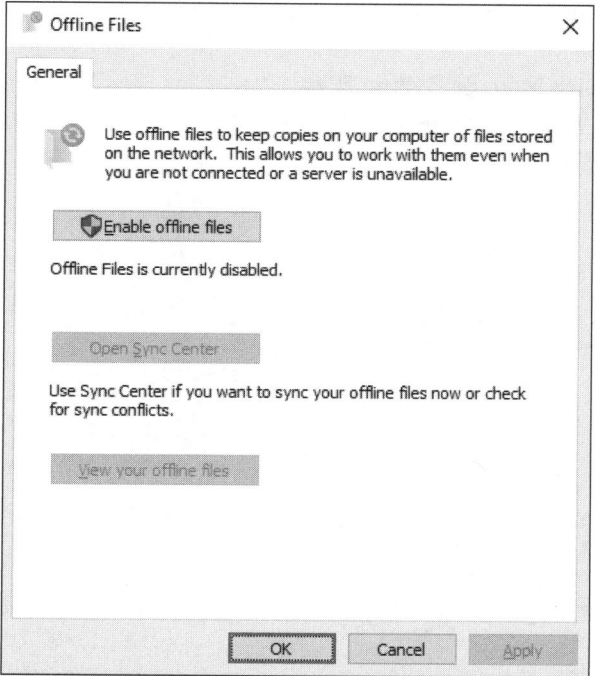

FIGURE 4-22 The Offline Files Properties page

Once your computer has rebooted you can confirm that Offline Files have been successfully enabled in one of three ways:

- **Open File Explorer and browse to a network share**   Right-click to a shared file or folder and look for an option titled Always Available Offline, as shown in Figure 4-23.

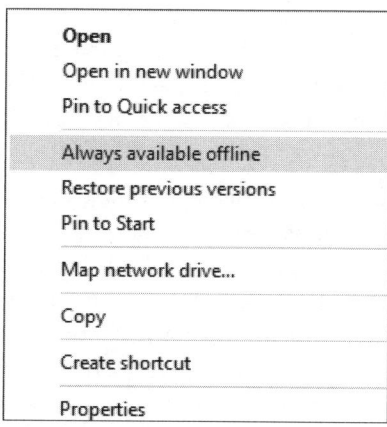

FIGURE 4-23 Right-click from a file share to view the context menu

- **Open the Services management console**   Look for the service called Offline Files. It should now be in a running state.

- **Open Sync Center and click Manage Offline Files**   The Offline Files Properties window should now have four tabs across the top, as shown in Figure 4-24. These tabs include: General, Disk Usage, Encryption, and Network.

FIGURE 4-24 The Offline Files Properties page and corresponding options once you have enabled the Offline Files features

After you have confirmed that the offline files feature is enabled you need to be aware of a few other items before marking files and folders for offline use, as these can impact the user experience and the security of your data. Each of these components can be managed from the Offline Files properties window.

- **Disk usage**   Remember, any shared resources that you enable for offline usage are cached to your local system. On the Disk Usage tab of the Offline Files Properties page you can adjust the maximum amount of disk space made available for offline file usage by clicking the Change Limits button. The default value is a percentage of disk space based on the size of the primary drive, as shown in Figure 4-25. Cached content is stored in a protected directory structure under %SystemRoot%\CSC. This directory cannot be relocated, so it is important to monitor how much disk space is allocated to offline files.

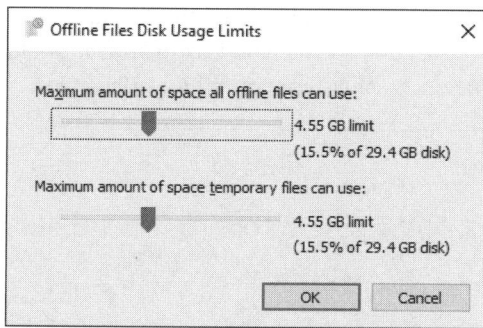

FIGURE 4-25 The Offline Files Disk Usage Limits dialog box

■ **Offline files versus temporary files**   The settings associated with temporary files are not related to offline files. Temporary files refer to Windows automatic caching. For example, if you open a document on a file share Windows automatically caches a local copy in case there are connection problems. The temporary files feature does not handle synchronization and is not network aware. This feature is meant to protect files that are opened remotely.

■ **Encryption**   Data security is a big concern underlying the implementation of offline files. Putting on your security hat, you realize that with offline files enabled, your users could potentially be synchronizing large amounts of corporate data to their portable devices (devices that are vulnerable to theft and carelessness). Offline Files includes an option to encrypt the file database that contains all synchronized data. Using a standard EFS machine certificate, you can set up encryption, which can be maintained by an internal certificate authority, if available. Otherwise the system generates a local certificate. You can enable encryption on offline files by clicking the Encryption tab, then clicking the Encrypt button. Likewise, you can decrypt files by clicking the Unencrypt button.

■ **Network performance**   The offline files feature regularly checks your active network connection for performance issues or loss of connectivity. You have the option to adjust the interval at which this check is performed. The default value is five minutes, but in cases where you regularly work offline you might choose to set this to 30 or 60 minutes. You can adjust a slow connection by clicking the Network tab and updating the value for Check For A Slow Connection every x Minutes.

Now that you have a better understanding of the options available for offline files, you can proceed with setting up a few file shares for offline use.

1. Open File Explorer and browse to your organization's file server.

2. Identify a directory that makes sense for offline usage. When making this decision be aware of the file or folder size and content.

3. Right-click the folder you want to make available offline. From the menu, click Always Available Offline, as shown in Figure 4-26. Look for a dialog box to appear briefly that confirms whether or not the folder was successfully configured for offline files.

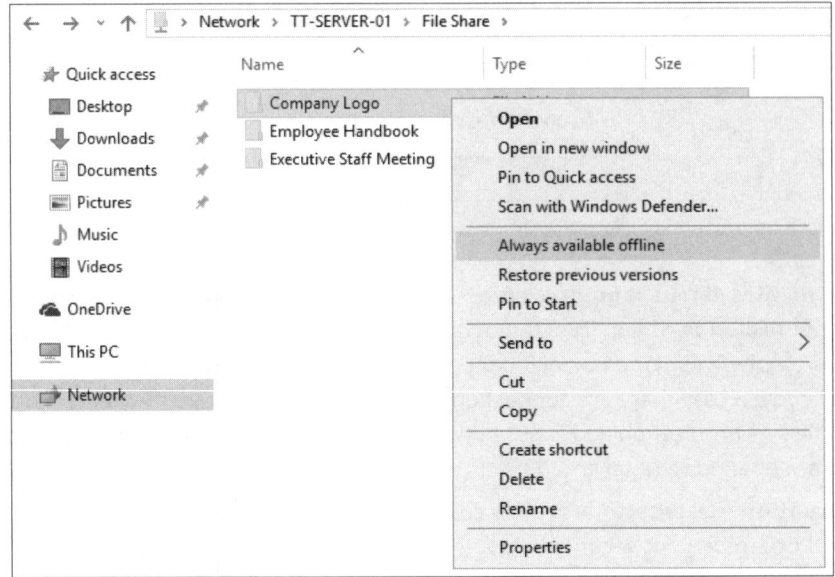

FIGURE 4-26 The Windows 10 right-click options for offline files

4. Open Sync Center.

5. On the View Sync Partnerships page, right-click the Offline Files partnership and select Open. This reveals the Sync Partnership Details page, which displays the list of file shares that have associated content enabled for offline use, as shown in Figure 4-27. A separate pointer is visible for each file server you connect to.

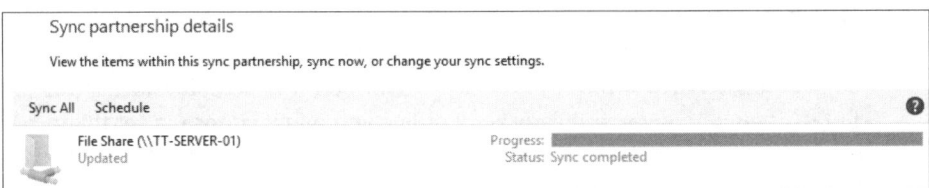

FIGURE 4-27 The Sync Center Sync Partnership Details page for Offline Files

At this point you have enabled offline files, reviewed the default options, and set up a few file shares for offline file access. Your last step in this example walks you through accessing these files when you are disconnected from the network.

1. Open Sync Center.

2. On the Sync Center page, click Manage Offline Files.

3. On the General tab of the Offline Files property window, click View your offline files. This opens up File Explorer, pointing at Offline Files Folder. To simplify future navigation, you can pin this directory or any of the subdirectories to your Quick Access list in Explorer.

4. Double-click Computers, followed by the name of the file server, and finally the File Share folder. The final directory listing should show all of the folders and/or files that you enabled for offline access. Figure 4-28 is an example directory with a few documents that have been synchronized for offline use. You might notice the Offline Status column reads Offline (not connected). Once this computer is reconnected to the network this status updates automatically and any changes are synchronized.

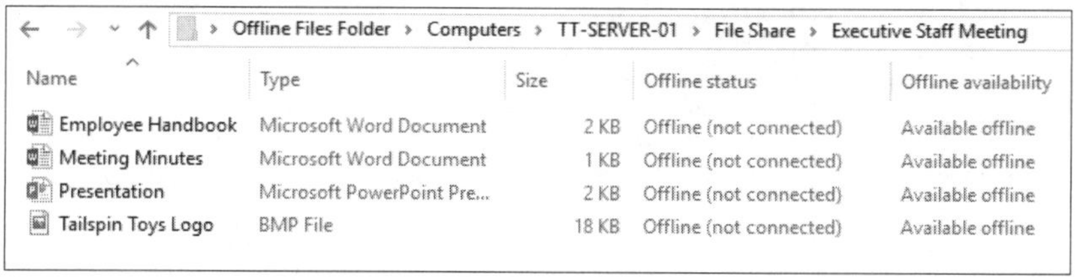

Name	Type	Size	Offline status	Offline availability
Employee Handbook	Microsoft Word Document	2 KB	Offline (not connected)	Available offline
Meeting Minutes	Microsoft Word Document	1 KB	Offline (not connected)	Available offline
Presentation	Microsoft PowerPoint Pre...	2 KB	Offline (not connected)	Available offline
Tailspin Toys Logo	BMP File	18 KB	Offline (not connected)	Available offline

FIGURE 4-28 An Explorer window pointing at the Offline Files Folder path with various offline documents listed

The final piece to review for this section covers the Group Policy options for offline files. You have taken an in-depth look at how offline files function, and the various options that are present to the user. Now you can leverage Group Policy to manage these settings and provide a consistent user experience that meets the guidelines of your organization.

You can find the Group Policy settings that control Offline Files under Computer Configuration\Administrative Templates\Network\Offline Files. Figure 4-29 shows the Group Policy Management editor with the available settings for Offline Files.

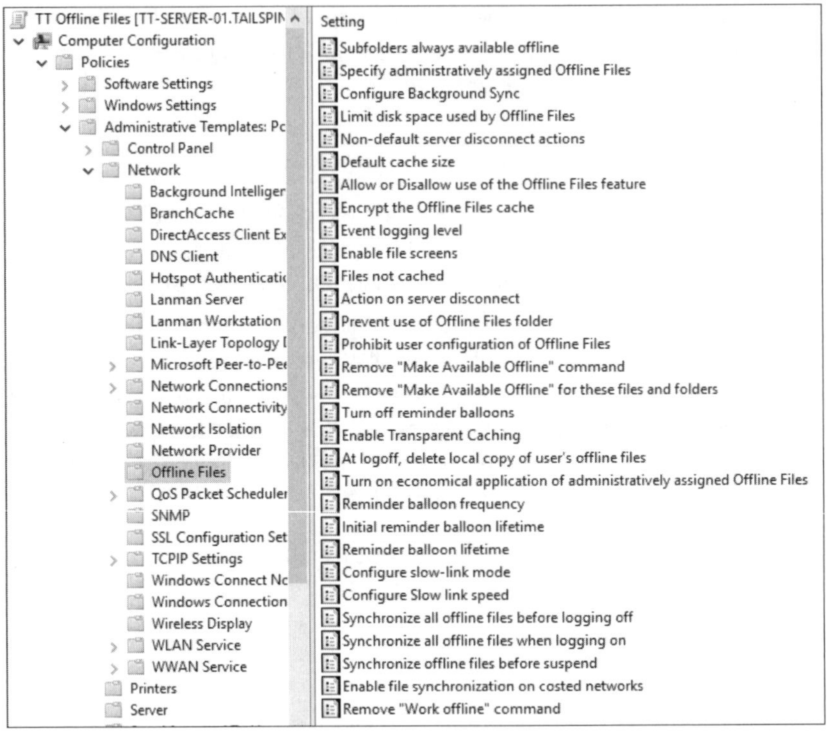

FIGURE 4-29 The Group Policy Management editor showing the Offline Files policy settings

Among the list of available settings shown in Figure 4-29, you should know about a few that stand out from an exam perspective. In a scenario where you plan to use offline files and manage them with Group Policy, these settings would be used to control several of the configuration options you reviewed earlier. Refer to Table 4-6 for more information about these settings.

TABLE 4-6 Commonly used Group Policy settings for Offline Files

Setting	Description	Important elements
Allow or Disallow the use of the Offline Files feature	This policy setting gives you the ability to enable or disable the offline files feature.	Use of this setting is important to ensure client computers have offline files enabled, so they can begin using it when needed.
Configure Background Sync	This policy setting gives you the ability to manage the sync interval, sync variance, and block out time frame for background synchronization when on a slow link.	The default sync interval for Windows 7 is 360 minutes. For Windows 8, 8.1, and 10, the default is 120 minutes. If you plan to pre-assign offline files you should consider assigning a set value, along with a variance to avoid a heavy amount of synchronization traffic at any given time.
Configure Slow Link speed	This policy setting gives you the ability to assign the bandwidth threshold before a network connection is considered slow.	The default slow link threshold is 64,000 bps. If you are using Offline Files for a branch office, assign a value that coincides with the link speed to that office.
Encrypt the Offline Files cache	This policy setting gives you the ability to enforce encryption of the offline file cache.	This option should be closely considered, especially if your client computers do not have full-disk encryption.
Limit disk space used by Offline Files	This policy setting gives you the ability to assign a limit to the amount of disk space that offline files can use.	The default limit for offline files is 25 % of the total space on the system drive. If you run into disk space issues you should consider assigning a set value.
Prohibit user configuration of Offline Files	This policy setting gives you the ability to restrict changes to the offline files configuration.	If you are managing offline files for your users, you should consider enabling this setting to avoid users making changes to the settings you apply.
Specify administratively assigned Offline Files	This policy setting gives you the ability to pre-assign offline files that users can have enabled by default.	In situations where you know users require offline access to specific file shares, this setting enables you to pre-configure those paths so users don't need to configure anything.

> *NEED MORE REVIEW?* **CONFIGURING OFFLINE FILES**
>
> To learn more about ways to configure and manage offline files, visit *https://technet.microsoft.com/library/ff633429(v=ws.10).aspx*.

Design for power policies

The battery life for many modern-day ultra-portables and tablets has surpassed the average eight-hour workday, which is a great benefit for those who are working on the go. However, power management is still something that should be configured to avoid unnecessary battery consumption. Simple considerations like screen brightness can make a noticeable difference on overall usage. In this section, you need to review the client-side options for configuring power settings, and then address how to manage those setting using power policies.

Windows 10 provides multiple ways to configure power settings. The most common method is to create or modify a Power Plan, also referred to as power options or power schemes in previous versions of Windows. In Figure 4-30 you are looking at the Power Options in Control Panel. Windows 10 includes three built-in Power Plans.

- **Balanced (recommended)**   This is the default Power Plan and offers a midpoint between performance and power saver.
- **High performance**   This Power Plan focuses on maximum performance, disabling features like sleep and increasing the default screen brightness.
- **Power saver**   This Power Plan focuses on maximum battery life, utilizing very short thresholds before dimming the screen or going into sleep mode.

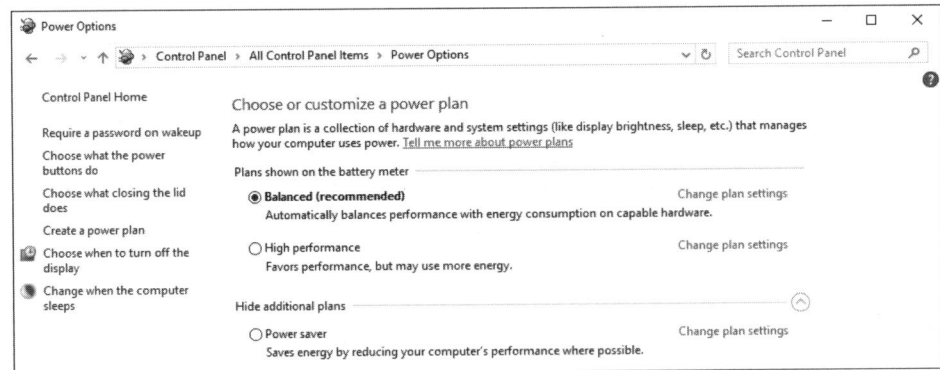

FIGURE 4-30 The Power Options item in Control Panel displaying the available Power Plans

In your organization, if the built-in Power Plans do not meet your needs you have the option to modify any of them, or create your own. To create a custom Power Plan, follow these steps:

1. On the Power Options Control Panel item, click Create A Power Plan in the tasks list.

2. On the Create A Power Plan page, select a built-in Power Plan on which you would like to base your new plan. Enter a name for your custom Power Plan and click Next.

3. On the Edit Plan Settings page, configure the desired display, sleep, and brightness settings for both the battery and plugged in scenarios, as shown in Figure 4-31. Click Create to complete the wizard. Your custom Power Plan is selected and listed in the Power Options.

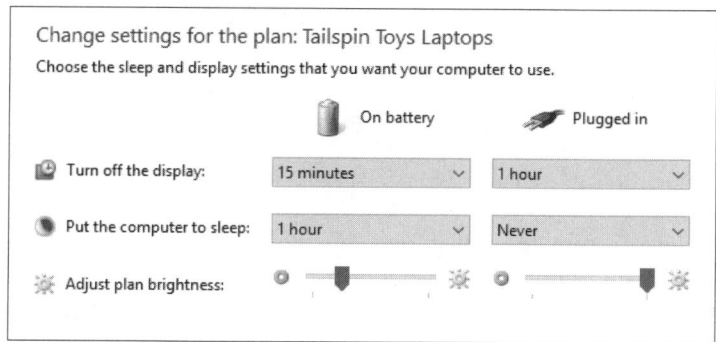

FIGURE 4-31 The Edit Plan Settings page for a custom Power Plan

4. On the Power Options Control Panel for your custom Power Plan, click Change Plan Settings.

5. On the Edit Plan Settings page, click Change Advanced Settings.

6. On the Power Options Advanced Settings page you are presented with additional power settings, as shown in Figure 4-32. Some of these settings are specific to the hardware, such as the dedicated graphics card. To change the settings for an individual component, click the plus symbol to view the details for battery and powered scenarios. Proceed with any additional updates and click OK to save your changes.

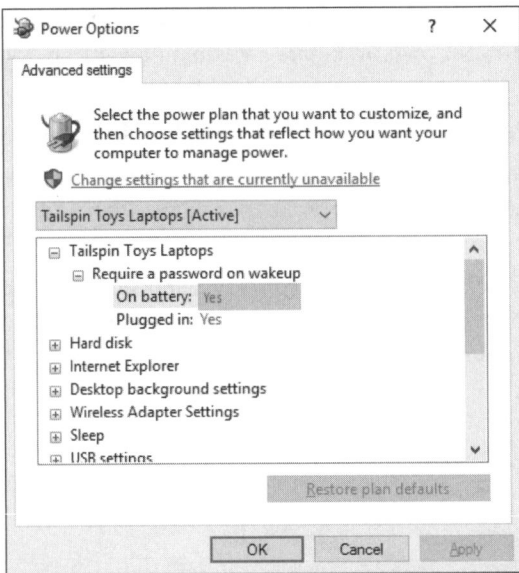

FIGURE 4-32 The Power Options Advanced Settings page displaying a custom Power Plan

Once you have created a custom Power Plan, you can administer it locally using the powercfg command-line utility. This utility provides an assortment of options for managing the Power Plans on your computer, including the ability to import, export, delete, and activate them. To use the powercfg utility you need to open a command prompt with administrative privileges. The following actions demonstrate how to administer your custom Power Plans using the powercfg utility.

- **List Power Plans** Each Power Plan has a globally unique identifier (GUID) assigned to it. In order to export a Power Plan you need to know the GUID of the plan that you are exporting. Figure 4-33 shows how to retrieve a list of the existing Power Plans, along with their GUID, using the powercfg /LIST command. Note that the active Power Plan has an asterisks next to its name.

```
C:\>powercfg /LIST

Existing Power Schemes (* Active)
-----------------------------------
Power Scheme GUID: 381b4222-f694-41f0-9685-ff5bb260df2e  (Balanced) *
Power Scheme GUID: 6f3afc9a-fa42-4b84-96e6-50ca7ed4791e  (TT Sales Team Power Plan)
Power Scheme GUID: 8c5e7fda-e8bf-4a96-9a85-a6e23a8c635c  (High performance)
Power Scheme GUID: a1841308-3541-4fab-bc81-f71556f20b4a  (Power saver)
```

FIGURE 4-33 The powercfg /LIST command

- **Export a Power Plan** To export a Power Plan, you need a path that you can save the file to, along with the assigned GUID. Figure 4-34 shows the export of the Tailspin Toys sales team Power Plan using the powercfg /EXPORT command.

```
C:\>powercfg /EXPORT C:\Temp\tt_sales_pp.pow 6f3afc9a-fa42-4b84-96e6-50ca7ed4791e
```

FIGURE 4-34 The powercfg /EXPORT command

- **Import a Power Plan** To import a Power Plan, you need to know the path to the .pow file. Figure 4-35 shows the import of the Tailspin Toys engineering team Power Plan using the powercfg /IMPORT command.

```
C:\>powercfg /IMPORT C:\Temp\tt_eng_pp.pow
Imported Power Scheme Successfully. GUID: e7ac7350-21d3-4af1-8212-cb0d5b325835
```

FIGURE 4-35 The powercfg /IMPORT command

- **Set an active Power Plan** After a new plan is imported, you need to activate it before the settings take effect. Figure 4-36 shows how to activate the Tailspin Toys engineering team Power Plan using the powercfg /SETACTIVE command.

```
C:\>powercfg /SETACTIVE e7ac7350-21d3-4af1-8212-cb0d5b325835
```

FIGURE 4-36 The powercfg /SETACTIVE command

- **Delete a Power Plan** To delete a Power Plan, you need to know the assigned GUID. Figure 4-37 shows how to delete the Tailspin Toys sales team Power Plan using the powercfg /DELETE command.

```
C:\>powercfg /DELETE 6f3afc9a-fa42-4b84-96e6-50ca7ed4791e
```

FIGURE 4-37 The powercfg /DELETE command

The previous process is useful for applying custom power settings in situations where Group Policy is not available, such as a workgroup environment, or in imaging workflows used for operating system deployment. For the exam you should be familiar with the Power Plan UI elements on the client computer, as well as how to manage power settings using Group Policy.

There are multiple locations in Group Policy where you can manage the power settings for your client computers. Review the following options:

- **Power Options** These policy settings have been a part of Group Policy dating back to Windows XP. They enable you to configure Power Plan options for both users and computers. When deployed, these settings are applied, but the user can still alter the settings or change to active Power Plan. Figure 4-38 shows how to create a new Power Plan that is compatible with Windows 7, Windows 8, Windows 8.1, and Windows 10. Once created, the configurable settings mimic the UI settings you see on the client computer when looking at the advanced settings window for a Power Plan. You can create Power plans in the following locations:
  - Computer Configuration\Preferences\Control Panel Settings\Power Options
  - User Configuration\Preferences\Control Panel Settings\Power Options

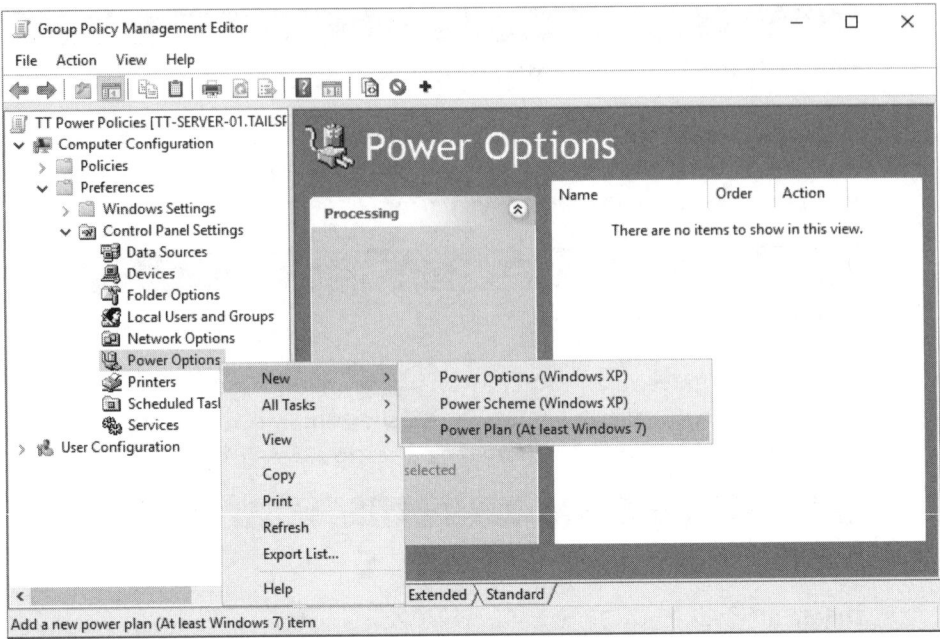

FIGURE 4-38 Creating a new Power Plan by using Group Policy

- **Power Management** These power settings offer a more restrictive approach to power management, compared with the power options reviewed previously. Policy settings that you enforce under power management cannot be altered by the user and are applied regardless of the Power Plan that is selected. For example, Figure 4-39 shows the configuring of the lid close action for a laptop computer. Once this setting is applied, all targeted laptops take no action when the lid on their computer is closed while it is running on battery. This is the case whether the laptops are using the high performance Power Plan or the power saver Power Plan. You can configure Power Management options in the following locations:

  - Computer Configuration\Policies\Administrative Templates\System\Power Management

  - User Configuration\Policies\Administrative Templates\System\Power Management

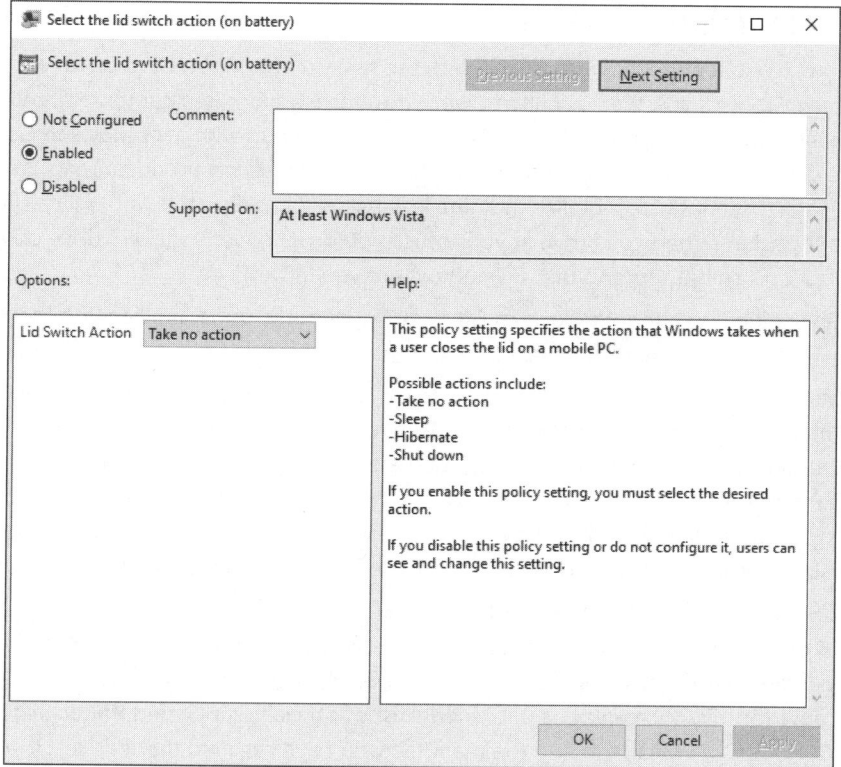

FIGURE 4-39 Group Policy setting for the lid switch action

For the exam, be prepared to answer questions that ask where a particular power setting would be configured in Group Policy. You should be familiar with the differences between applying power plans with a GPO, and enforcing power settings with a GPO.

 **Quick check**

- As a system administrator you are implementing some default Power Management settings to help improve the experience of your mobile users. You need to implement a power policy that can apply the company defaults, but allows users to adjust as needed. How would you accomplish this?

Quick check answer

- Under the Power Options section, create a GPO that assigns a custom Power Plan and sets it as the active plan on client computers. On the commons settings for the GPO, select the check box that applies the GPO once and does not reapply it.

Design for Windows to Go

The Windows to Go feature was introduced with the release of Windows 8. From a mobility perspective, it gives you the flexibility to work from nearly any client computer by simply booting from a pre-configured USB drive. The USB drive is set up with a Windows to Go workspace that contains all of your settings, applications, and important documents. In many ways this feature defines true mobility. Imagine leaving on a business trip with only a USB drive in your pocket. When you arrive at your office you simply boot your USB drive on any of the organization's computers and pick up right where you left off.

To begin working with the Windows to Go feature, you need to be familiar with the prerequisites. These include the following:

- **Enterprise only**   To create a Windows to Go USB drive, you must be using Windows 8 Enterprise, Windows 8.1 Enterprise, or Windows 10 Enterprise. Other versions of Windows do not have the Windows to Go feature available. Note that the correct operating system version is only required to create the Windows to Go workspace. The host computer that you boot the USB drive from does not need to have an enterprise version of Windows installed. The host could even be a computer that does not run Windows, as long as the hardware meets the requirements.

- **USB drive**   Windows to Go is a disk-intensive feature, and not all USB drives are able to handle the excessive number of reads and writes that occur with average usage. To account for this, Microsoft has published a list of officially supported Windows to Go USB drives. Both Microsoft and the manufacturer have certified these drives. If you connect a USB drive that is not certified and Windows recognizes it as a removable drive, Windows to Go does not allow you to use it. If you connect a USB drive that is not certified and is not recognized as removable drive, such as a USB external hard drive. Windows to Go works, but you are given a warning message stating that the device is not certified.

- **Host computers**   The host computer that you boot Windows to Go on must have hardware compatibility with Windows 7 at a minimum. The computer needs to have a USB 2.0 port, preferably a 3.0 port for optimal throughput, and a BIOS that has USB boot enabled. External USB hubs are not supported. The USB drive must be connected directly to the host device.

- **Hardware drivers**   The Windows to Go workspace is built from a .WIM file (Windows imaging file format). A .WIM file is a file-based disk image that Microsoft introduced with Windows Vista. The .WIM file contains the Windows operating system, which includes basic hardware drivers certified by Microsoft. There is a very good chance that this basic set of hardware drivers does not cover the various types of hardware that you need to boot from. To resolve any driver conflicts you need to inject the necessary drivers into the .WIM file before creating a Windows to Go USB drive. This can be done using the deployment image servicing and management (DISM) command-line tool.

- **Domain membership** A generic Windows to Go drive is not bound to your organization's domain by default. Domain membership is supported with Windows to Go.

Now that you have a better understanding of the Windows to Go requirements, you need to review the process for building a Windows to Go USB drive. First, retrieve a copy of the Windows 10 Enterprise .WIM file used for your corporate image. If you do not have one available, you can copy the Install.wim file from the Windows 10 Enterprise media, found in the sources directory. Copy the .WIM file to a client computer in your organization that has Windows 10 Enterprise installed on it.

1. Sign-in to a client computer running Windows 10 Enterprise. Make sure to have the Windows 10 .WIM file saved to the local system.

2. Connect your Microsoft-certified Windows to Go USB drive to the client computer.

3. Type **Windows to Go** in the search box and press Enter. This tool requires administrative rights, so you might be prompted with a User Account Control dialog box. Click Yes to approve.

4. On the Create a Windows To Go Workspace Wizard, under Choose The Drive You Want To Use, select the USB drive to which you wish to install, as shown in Figure 4-40. Note that in this example you are using a USB drive that has not been certified by Microsoft, but is compatible with Windows to Go. Click Next.

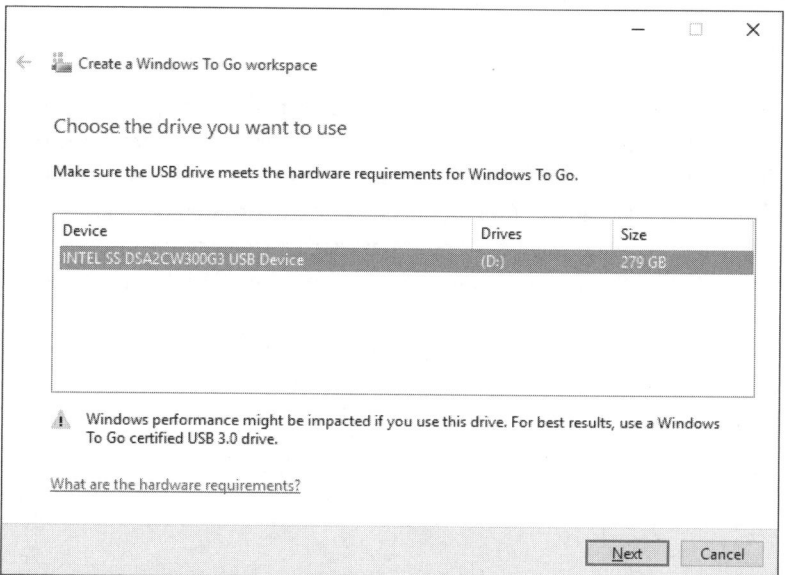

FIGURE 4-40 Windows to Go Wizard, device selection page

5. On the Choose A Windows 10 Image page, select the .WIM file you are using to create the Windows to Go workspace, as shown in Figure 4-41. If the file did not appear in the list, click Add Search Location and browse to the desired path. Click Next.

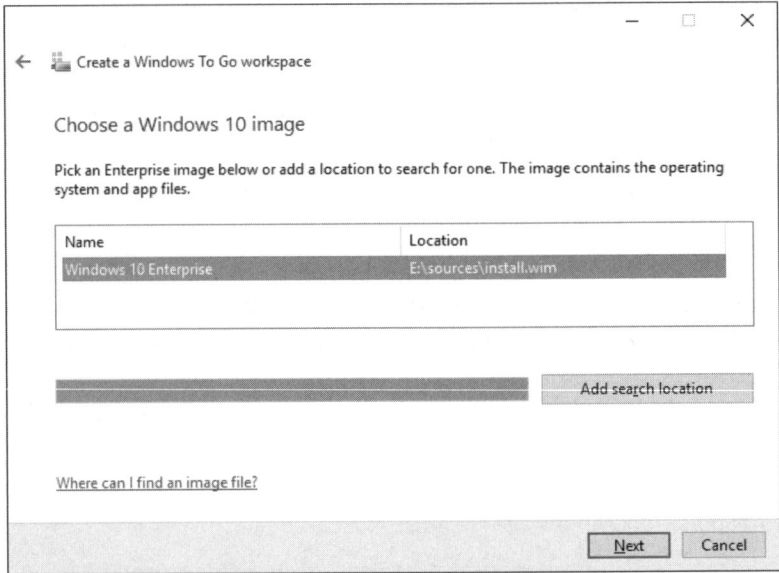

FIGURE 4-41 Windows to Go Wizard, image selection page

6. On the Set a BitLocker Password (Optional) page, select the check box to use BitLocker and enter a BitLocker password to unlock the drive, as shown in Figure 4-42. This action encrypts the USB drive, requiring you to enter the BitLocker password each time you boot the Windows to Go drive. Note that because this is a roaming install of Windows, BitLocker uses a pre-operating system boot password instead of the onboard trusted platform module (TPM), which is tied to the host computer. Click Next.

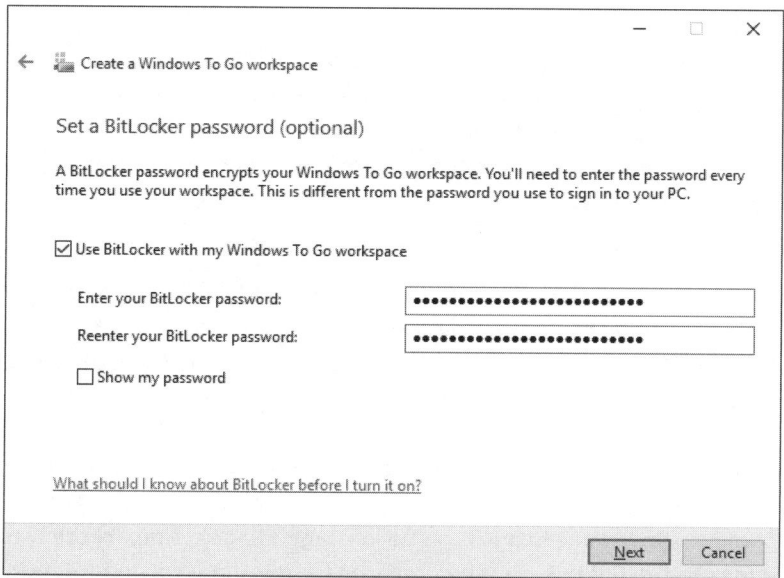

FIGURE 4-42 Windows to Go Wizard, showing the BitLocker setup option

7. On the Ready To Create Your Windows to Go workspace page, confirm the settings you have chosen and click Create. Note that the contents of the targeted USB drive are erased as part of this process.

8. On the Choose a Boot Option page, select Yes. This is the final page in the wizard and enables you to boot directly to the USB drive at next reboot. Click Save and Restart. When your computer restarts it now boots from the new Windows to Go drive.

9. Upon booting from the Windows to Go drive you should be prompted for your BitLocker password. Type your password and press Enter. Now Windows 10 Enterprise runs directly from the USB drive.

The first time Windows to Go loads it installs drivers and sets up the operating system, similar to a first-time install of Windows. When Windows comes up for the first time you are prompted to select your language and location, accept the Microsoft license agreement, and choose the Windows 10 configuration settings that come with every new install. The computer reboots at this stage. The next time you boot your Windows to Go drive it prompts you to join a domain. You have two options: Azure AD or a local domain. If you select Azure AD it prompts you to connect. If you select to join a local domain you need to create a local user account and join the domain after logon. Once complete, you automatically are logged in to your desktop.

The setup process for Windows to Go is minimal overall, and becomes increasingly easier to manage if you keep this feature in mind when designing your operating system deployment workflow. Those same images can be leveraged to help simplify the setup process for Windows to Go. For the exam be prepared for questions related to the Windows to Go prerequisites and system requirements. Make sure you understand the benefits of Windows to Go and how it can be used.

> *NEED MORE REVIEW?* **DEPLOYMENT CONSIDERATIONS FOR WINDOWS TO GO**
>
> To learn more about Windows to Go and the deployment considerations, visit *https://technet.microsoft.com/library/jj592685.aspx*.

Design for sync options

With Windows 10 devices you can synchronize common user settings between devices. This is ideal for truly mobile users. For example, imagine that you have both a Surface Pro and a Windows to Go USB drive. You can enable synchronization so that your user settings are automatically updated when you log in to either of your devices.

To enable this feature, you need to link your Windows 10 devices with a Microsoft account. To accomplish this task, follow these steps:

1. Sign in to your Windows 10 device.

2. In the search box, type **manage your account** and press Enter. This opens the Accounts page, as shown in Figure 4-43.

3. On the Accounts page, scroll to the bottom and click Add A Microsoft Account.

4. Enter your Microsoft account credentials and click Sign In.

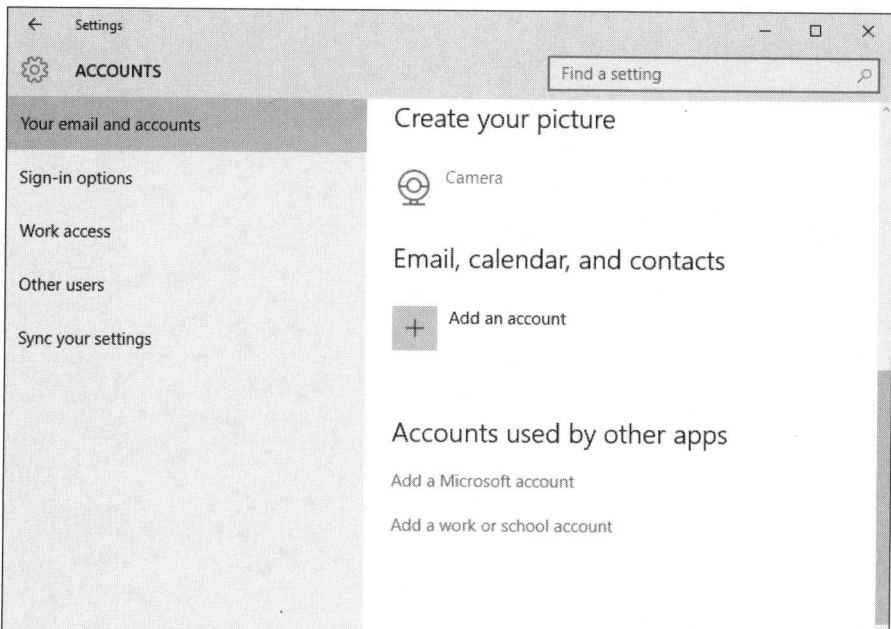

FIGURE 4-43 The Accounts page

Using a Microsoft account also enables you to synchronize personal settings between Windows 10 devices. To accomplish this, follow these steps:

1. Sign in to your Windows 10 device.

2. In the search box, type **Sync Your Settings**. This opens the Sync Your Settings page, as shown in Figure 4-44.

3. Enable or disable the available settings according to your needs.

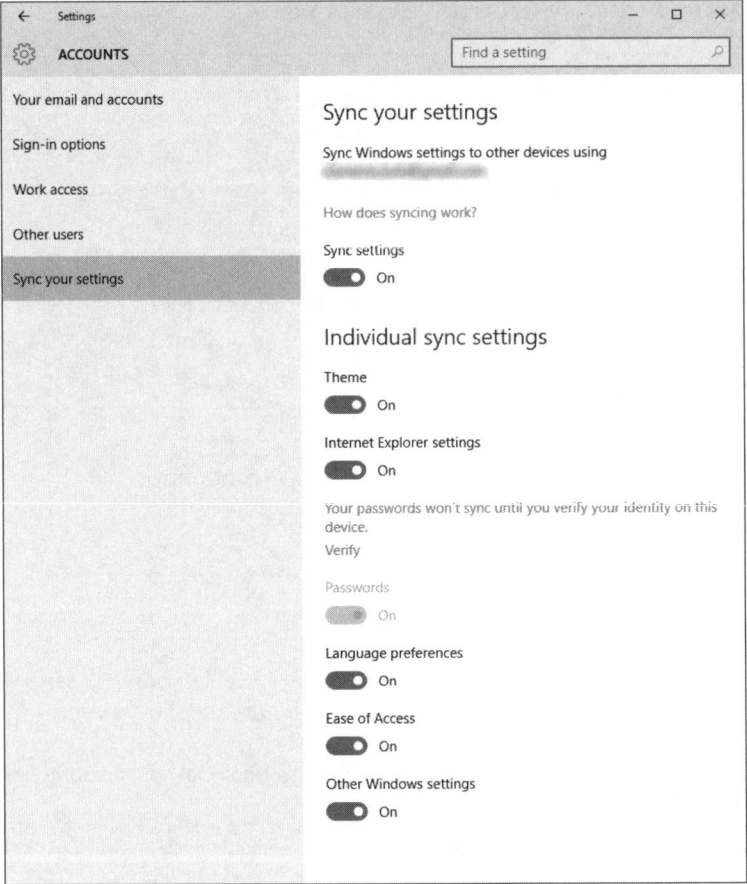

FIGURE 4-44 The Sync Your Settings page

Design for Wi-Fi Direct

Wi-Fi Direct is an up and coming technology that enables users and devices to connect and share data seamlessly. This technology mirrors the capabilities of a Bluetooth connection, in that the two connected devices are communicating over an isolated network. Likewise, a wireless ad hoc network accomplishes the same task, but with Wi-Fi Direct you do not need to configure static IP addresses or other custom IP settings.

Wi-Fi Direct is designed to simplify client communication without the need for additional network equipment. The diagram in Figure 4-45 demonstrates a very basic use case for Wi-Fi Direct. On the left you have a typical file transfer scenario, and on the right you see how Wi-Fi Direct can handle peer-to-peer communication. In terms of requirements, client computers that have a Wi-Fi Direct-capable application need a Wi-Fi Direct-compatible wireless network adapter. Other use cases include Wi-Fi Direct printing, which require no additional drivers or software, simply a peer-to-peer connection between your device and the printer.

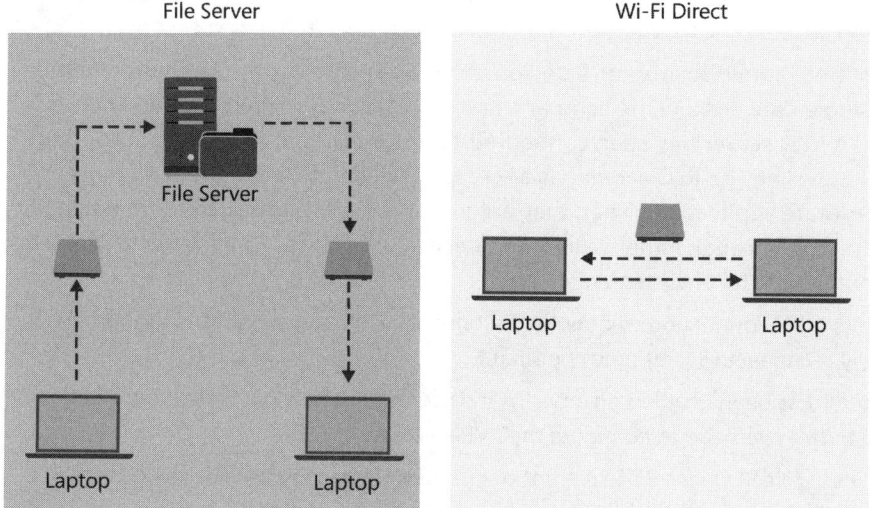

FIGURE 4-45 An illustration shows how diagramming as a basic use case for Wi-Fi Direct

The release of Windows 10 has expanded the capabilities of Wi-Fi Direct. Microsoft has incorporated several new APIs for software developers to leverage. Over time more apps will take advantage of the wireless standard.

Summary

In this section, you took a closer look at the management capabilities for client mobility options. For the exam you should be familiar with the following topics:

- Know the different use cases for Offline Files, including which Group Policy settings you should use for certain scenarios.
- Know how to resolve sync conflicts for Offline Files.
- Know how to create a power plan and apply it to multiple client computers.
- Know the difference between applying a power plan using a GPO and enforcing power options using a GPO.
- Know the requirements for Windows to Go, both for the administrator and the user.
- Know how to enable sync options and configure the sync settings.
- Know what Wi-Fi direct is and what it accomplishes for mobile users.

Thought experiment

In this thought experiment, demonstrate your skills and knowledge of the topics covered in this chapter. You can find the answer to this thought experiment in the next section.

You are a system administrator for Blue Yonder Airlines. The current IT environment consists of a single data center. The company uses AD DS for all authentication and authorization. A VPN server has been configured to allow pilots to access the corporate network when traveling. An RDS server has also been installed to ensure that all personnel can access corporate applications when they are not at the office. All aircraft are equipped with wireless Internet through either cellular or satellite connectivity. The following problems have been identified in the environment:

- The pilots' VPN connection frequently disconnects while traveling, forcing them to manually reconnect several times per flight.

- VPN access has been enabled on a per-user basis. Adding new pilots requires additional administrative overhead to configure their VPN access.

- Employees typically leave RDS sessions open, even when they are not using them.

- You need to identify a resolution for problems that have been reported in the environment. What should you do? (Choose all that apply.)

    A. Migrate pilot's remote access method to DirectAccess.

    B. Migrate pilot's remote access to RDS.

    C. Configure session policies on the RD Session Host.

    D. Configure security policies on the RD Session Host.

    E. Configure a Network Policy on an NPS server.

    F. Configure a Health Policy on a NAP server.

Thought experiment answer

This section contains the solution to the thought experiment. Each letter answer explains why the answer choice is correct or incorrect.

A. **Correct.** DirectAccess works better in this scenario because the pilots do not have to reconnect to a VPN. If the client loses Internet access, it automatically re-establishes when the client regains an Internet connection.

B. **Incorrect.** An RDS solution requires manual intervention on the user's part when the client loses an Internet connection. While it might be an alternate solution, it does not solve the problem that was identified.

C. **Correct.** Session policies control how long remote sessions can be maintained on the server while the client is disconnected. Modifying the session policy enables you to set a time limit for a disconnected session.

D. **Incorrect.** A security policy cannot control how long a disconnected session can be maintained on the RD Session Host server.

E. **Correct.** A Network Policy can be used to control how users connect to a VPN, including who has access to do so. Once configured, individual AD DS user accounts can be set to use the NPS server for the VPN policies.

F. **Incorrect.** A Health Policy is used with NAP to identify if clients need remediation for local settings and configurations, such as antivirus definitions and Windows updates.

CHAPTER 5

Plan for apps

M anaging applications in the enterprise, whether it is from a developmental or user perspective, is one of the many responsibilities that you face as a systems administrator. Over the years, this task has grown more complicated with the growth of the mobile device ecosystem. Administrators have a variety of device platforms, operating systems, and application types that must be supported on a day-to-day basis. Common factors, such as licensing and support, still carry the same level of importance, but many organizations are facing platform diversity and the evolution of the mobile application market.

Solutions like Configuration Manager and Microsoft Intune provide you with the tools to manage your application source files, track licensing compliance, and simplify distribution. These are staple technologies for system administrators. However, when a user has two, three, or possibly four devices connected to your organization, localized application support becomes tedious and unrealistic. The question becomes, how can you provide these same applications to your users while reducing costs and minimizing the overhead needed to support them? You can provide these solutions using virtualized applications and cloud-based services. The market for virtualization hasn't stopped growing since its conception, bringing an entirely new way of managing applications in the enterprise. The days of installing software directly on a user's device are diminished.

This chapter covers several planning skills for application management. These skills include the use of RemoteApp And Desktop Connections to provide hosted application services with Microsoft Azure, and you will review support and compatibility considerations as you design your application portfolio and hosted services.

Skills covered in this chapter:

- Manage RemoteApp
- Plan app support and compatibility

Skill 5.1: Manage RemoteApp

Imagine you are a systems administrator for a large software development company that places a large emphasis on mobile device application development. This focus on mobile devices requires many of the employees to use a combination of device types on a daily basis given that they test and develop their software. You are using a hybrid Microsoft Intune and Configuration Manager Infrastructure to manage these devices and provide applications to your users. Your boss has asked you to propose a solution that enables users to access productivity applications, and in some cases entire virtual desktops from any mobile device, without the need to install the software locally. To accomplish this task, you implement RemoteApp and Desktop Connections.

Remote Desktop Services (RDS) is the delivery mechanism used to provide RemoteApp and Desktop Connection services. The RemoteApp and Desktop Connections feature was announced with the release of Windows Server 2008 and has continued to evolve throughout the Windows Server and Microsoft Azure product lifecycle. RemoteApp focuses on the provisioning of hosted Windows-based applications, enabling users to connect through a Remote Desktop Protocol (RDP) session and stream applications to any device. Desktop Connections is based around Microsoft Virtual Desktop Infrastructure (Microsoft VDI), providing users with remote access to a virtualized desktop.

> **This section covers how to:**
> - Design RemoteApp and Desktop Connections settings
> - Configure Group Policy Objects (GPOs) for signed packages
> - Subscribe to the Azure RemoteApp and Desktop Connections feeds
> - Export and import Azure RemoteApp configurations
> - Support iOS and Android
> - Configure Remote Desktop Web Access for Azure RemoteApp distribution

Design RemoteApp and Desktop Connections settings

An RDS deployment is relatively straight forward. With Microsoft Azure, your deployment can quickly be up and running. Once deployed, there are several settings available for configuring RemoteApp and Desktop Connections. It is important to understand the architecture that RDS uses to host services like RemoteApp and Desktop Connections. Start by taking a look at the three available deployment scenarios and their requirements.

1. **On-premises**   An on-premises deployment of RemoteApp uses your on-premises Active Directory, network, and server infrastructure. This deployment scenario requires you to create and manage the RDS server infrastructure, which depending on the scale of the deployment, can encompass multiple servers. The benefits for a deployment of this type include application authentication with your on-premises Active Directory, and granular control over your RDS infrastructure. Items to be aware of include scalability and support of the on-premises infrastructure.

2. **In the cloud**   Through the use of Azure RemoteApp you have the ability to quickly deploy RemoteApp and start using pre-configured applications like Microsoft Office. Hosting RemoteApp with Azure includes the following benefits:

   - **Simplicity**   Azure RemoteApp can be deployed to the Azure cloud in a few clicks, reducing your setup and deployment time.

   - **Scalability**   As your environment grows you can scale Azure RemoteApp without incurring expensive infrastructure costs.

   - **Accessibility**   Azure RemoteApp can be accessed from Windows, iOS, Mac OS X, and Android devices using the Microsoft desktop and mobile apps.

   - **Reliability**   Your RemoteApp deployment is hosted on the Azure server farm and its global infrastructure.

   The biggest limitation with a cloud-only deployment is the inability to authenticate line of business LOB applications against an on-premises Active Directory. This is supported through hybrid deployment, which is covered later in this chapter.

3. **Hybrid**   A hybrid deployment connects your Azure RemoteApp collection with your on-premises Active Directory through a virtual network (VNET). To accomplish a hybrid deployment, use Azure Active Directory (Azure AD) in conjunction with your on-premises Active Directory environment. Implementing a hybrid solution includes the following benefits:

   - **Connectivity**   The hybrid solution provides you with full access to your on-premises network and resources, along with access to your Azure VNET and cloud-based services.

   - **Authentication**   The hybrid solution connects with your on-premises Active Directory environment. This enables you to include domain join access for applications and data that you need to provide through RemoteApp.

The configuration settings for RemoteApp and Desktop Connections vary based on the deployment scenario you choose. Review the variations for each of the three deployments in the following sections.

On-premises deployments

Hosting RemoteApp and Desktop Connections on-premises requires a lot of administration, starting with the deployment of the RDS infrastructure and the options available during setup. You must prepare to install RDS using the Add Roles And Features Wizard for Windows Server 2016, shown in Figure 5-1.

In this example, you have selected the Remote Desktop Services Installation option on the Select Installation Type page of the wizard. Selecting this option bypasses the standard list of roles and features, and instead provides you with specific deployment options for an RDS installation. This alternate set of options simplifies the task of deploying RDS by walking you through the setup for each of the mandatory roles. Along the way, you have the option to choose a Quick Start install, which initiates a single-server setup of RDS. This is beneficial in small environments or situations where you need to quickly provision an RDS instance for testing.

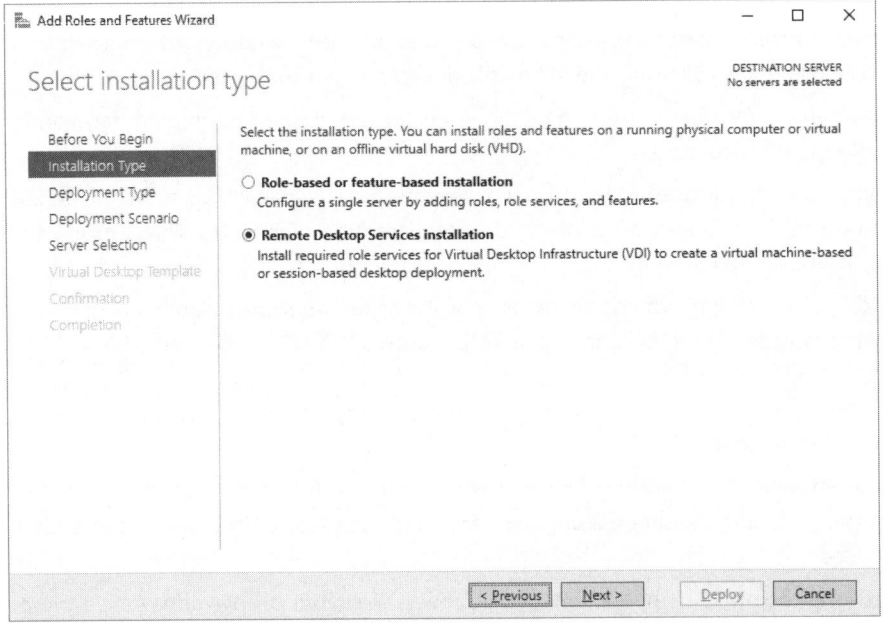

FIGURE 5-1 The Add Roles and Features Wizard with a dedicated RDS installation option

Each of the individual RDS server roles is identified in Table 5-1, along with the corresponding capabilities that they provide.

TABLE 5-1 Remote Desktop Services (RDS) roles

Server Role	Description	Important elements
Remote Desktop Connection Broker (RD Connection Broker)	This role allows users to reconnect to their existing virtual desktops, RemoteApp programs, and session-based desktops. It enables even load distribution across RD Session Host Servers in a session collection or across pooled virtual desktops in a pooled virtual desktop collection, and provides access to virtual desktops in a virtual desktop collection.	This role is mandatory and is installed by default when you deploy RDS. In instances where high availability is a requirement, you can deploy two RD Connection Brokers.
Remote Desktop Gateway (RD Gateway)	This role enables authorized users to connect to virtual desktops, RemoteApp programs, and session-based desktops on the corporate network or over the Internet.	This role is optional. In environments where remote access to RDS services is critical, this is an important role deployment. User authentication is regulated through the use of connection authorization policies. Client and server communication is encrypted through a mandatory SSL certificate.
Remote Desktop Licensing (RD Licensing)	This role manages the licenses required to connect to a Remote Desktop Session Host server or a virtual desktop. You can use RD Licensing to install, issue, and track the availability of licenses.	This role is mandatory for a production deployment, but is not included during the installation of RDS. The role can be installed after RDS is installed. For high availability requirements you can setup Windows clustering or deploy multiple RD Licensing servers.
Remote Desktop Session Host (RD Session Host)	This role enables a server to host RemoteApp programs or session-based desktops. Users can connect to RD Session Host servers in a session collection to run programs, save files, and use resources on those servers. Users can access an RD Session Host server by using the Remote Desktop Connection client or by using RemoteApp programs.	This role is optional depending on the deployment scenario. Multiple RD Session Host servers are pooled to create a session collection.
Remote Desktop Virtual Host (RD Virtualization Host)	This role enables users to connect to virtual desktops using RemoteApp And Desktop Connection.	This role is optional depending on the deployment scenario. It integrates with Hyper-V to provide personal virtual desktops to users.
Remote Desktop Web Access (RD Web Access)	This role enables users to access RemoteApp And Desktop Connection through the Start menu or through a web browser. RemoteApp And Desktop Connection provides users with a customized view of RemoteApp programs, session-based desktops, and virtual desktops.	This role is mandatory and is installed by default when you deploy RDS. The RD Web Access role handles client authentication for RemoteApp And Desktop Connection.

EXAM TIP

There are six roles that make up the RDS server infrastructure. Be prepared to identify which role is required to achieve a particular task. For example, in a scenario where users need to remotely access RemoteApp programs, know that a RD Gateway needs to be installed and configured with an SSL certificate.

After completing an on-premises installation of RDS, the configuration settings are available by opening Server Manager and clicking Remote Desktop Services. The RDS section of Server Manager is broken down into three subsections. Each subsection provides different configuration settings. The sections are broken down in the following order:

1. The Overview section gives you the ability to add additional servers to your RDS environment, remove existing servers, and modify the properties for your deployment. Adding additional servers is accomplished by clicking the Tasks drop-down next to Deployment Servers and selecting the server role that you want to install, as shown in Figure 5-2.

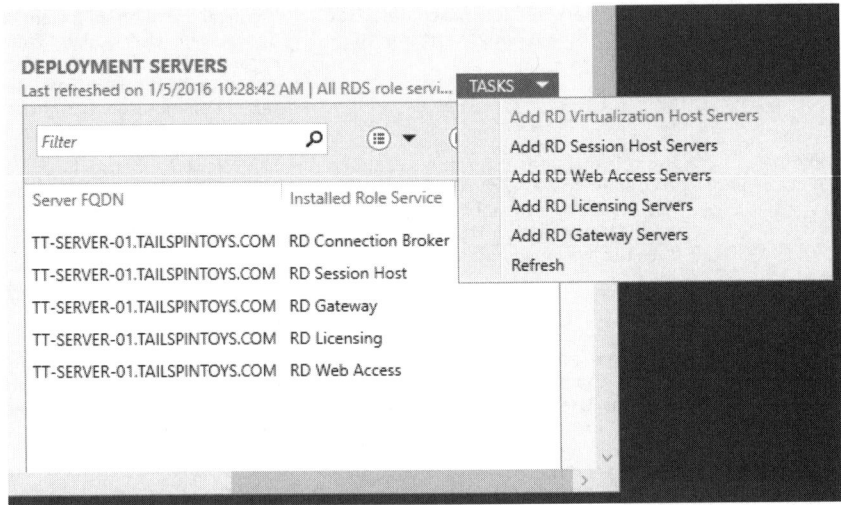

FIGURE 5-2 Administrators add additional RDS servers using the RDS Server Manager interface

2. Removing servers is accomplished by right-clicking the server role under Deployment Overview and clicking Remove RD Gateway Servers, as shown in Figure 5-3.

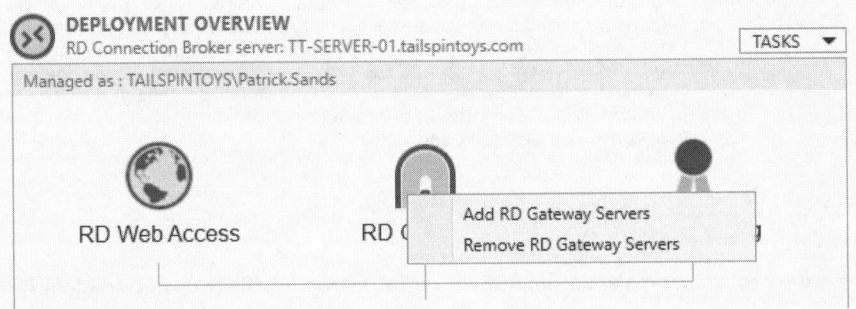

FIGURE 5-3 Administrators can remove RDS roles using the RDS Server Manager interface

3. **Editing the deployment properties** is accomplished by clicking the Tasks drop-down next to Deployment Overview and selecting Edit Deployment Properties. This displays the Deployment Properties editor for RDS, as shown in Figure 5-4.

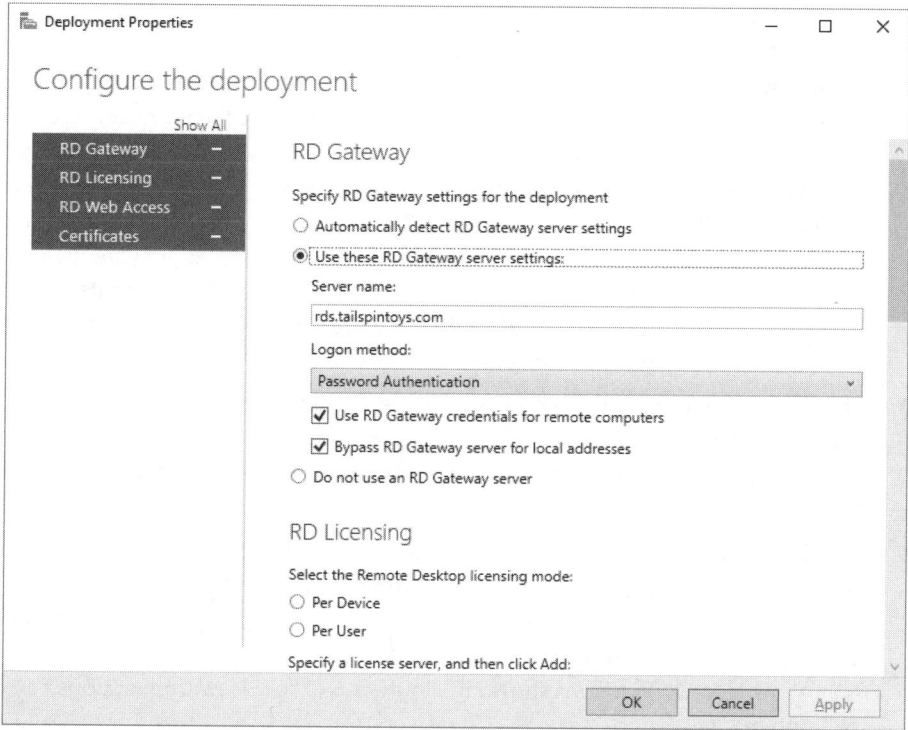

FIGURE 5-4 The RDS deployment includes multiple properties

The Deployment Properties editor contains the majority of your RDS settings. These settings are separated into the following groups:

- **High Availability (not shown)**   This group contains the values used to configure RDS for high availability (HA). This group is optional and is only displayed once HA has been configured. The values shown are read only and cannot be modified.

- **RD Gateway**   This group contains the settings associated with your RD Gateway server(s). You have the option to automatically detect RD Gateway server settings. For this to work you need to have the necessary configuration settings applied using a GPO. Alternatively, you can manually specify the server name and logon method used for client connections.

- **RD Licensing**   This group contains the settings associated with your RD Licensing server(s). You can set the licensing mode (user or device) based on the RDS client access licenses (RDS CALs) that you have assigned. You also have the ability to add additional licensing servers and order them based on priority.

- **RD Web Access** This group contains the URL assigned to your RD Web Access server(s). The values shown are read only and cannot be modified.

- **Certificates** This group contains the certificate settings for your RDS infrastructure. All certificate management for RDS is implemented from this page. There are three role services that use an SSL certificate. The first is the RD Connection Broker, used for both Single Sign-On (SSO) and publishing. For this you can use an internally signed certificate from a trusted CA. The second is RD Web Access, used for encrypting the web portal. The third is RD Gateway, used for encrypting inbound connections from the Internet. When configuring certificates, you are given the option to create a self-signed certificate or select an existing certificate. Creating a self-signed certificate isn't a recommended practice for production environments and should only be used for testing.

4. The Servers section mimics the All Servers section of Server Manager. From here you can manage each of the servers in your RDS infrastructure and accomplish common administrative tasks such as enabling performance counters or adding roles and features.

5. The Collections section is used for managing your collections. New collections can be created by clicking the Tasks drop-down next to Collections and selecting Create Session Collection. Blocking and allowing new connections to your host servers is accomplished by right-clicking the desired server under Host Servers and clicking Do Not Allow New Connections shown in Figure 5-5 (Allow New Connections can also be implemented).

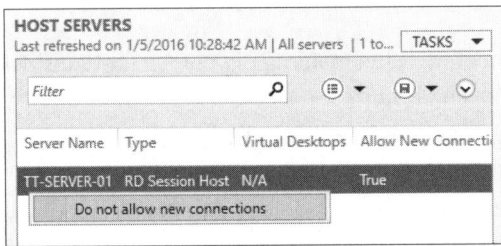

FIGURE 5-5 Control new connections on RD Session Hosts

Current connections are listed under Connections. From here you can disconnect users, send them a message, shadow their connection, and log them off, as shown in Figure 5-6.

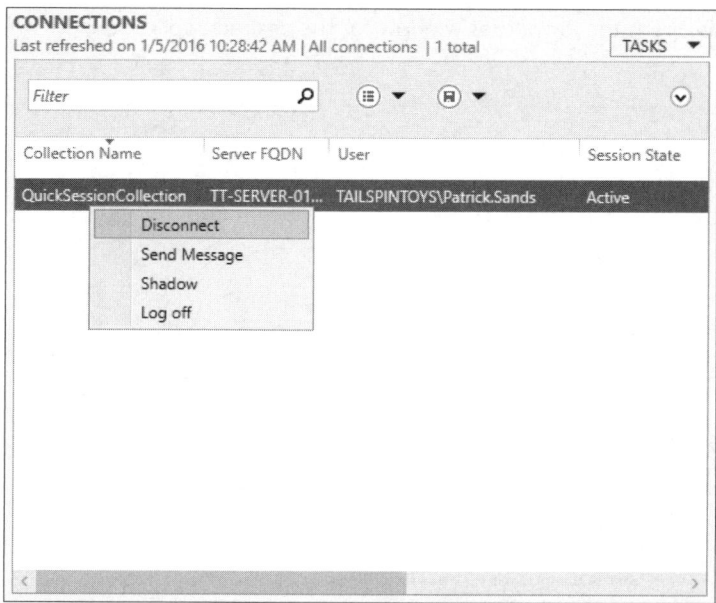

FIGURE 5-6 Interact with active RDS connections

After you create a new collection it appears under the Collections subsection. Click the collection to start publishing or unpublishing applications. Under RemoteApp Programs, Tasks has been clicked and the Publish RemoteApp Programs or Unpublish RemoteApp Programs can be selected, as shown in Figure 5-7.

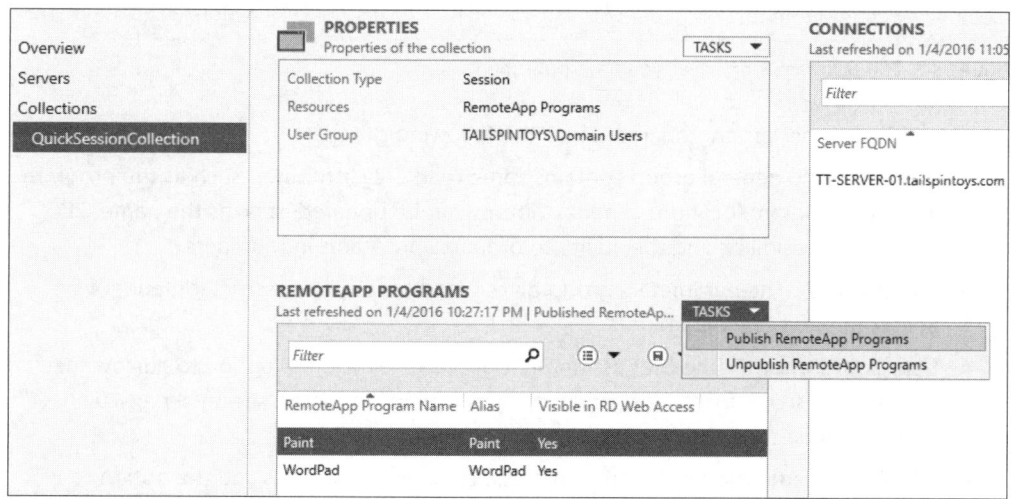

FIGURE 5-7 Additional programs can be published through Server Manager

Published applications include a variety of individual settings. You can access and modify the properties for any of your published applications by right-clicking the program and selecting Edit Properties. This opens the Properties window for the desired application, as shown in Figure 5-8.

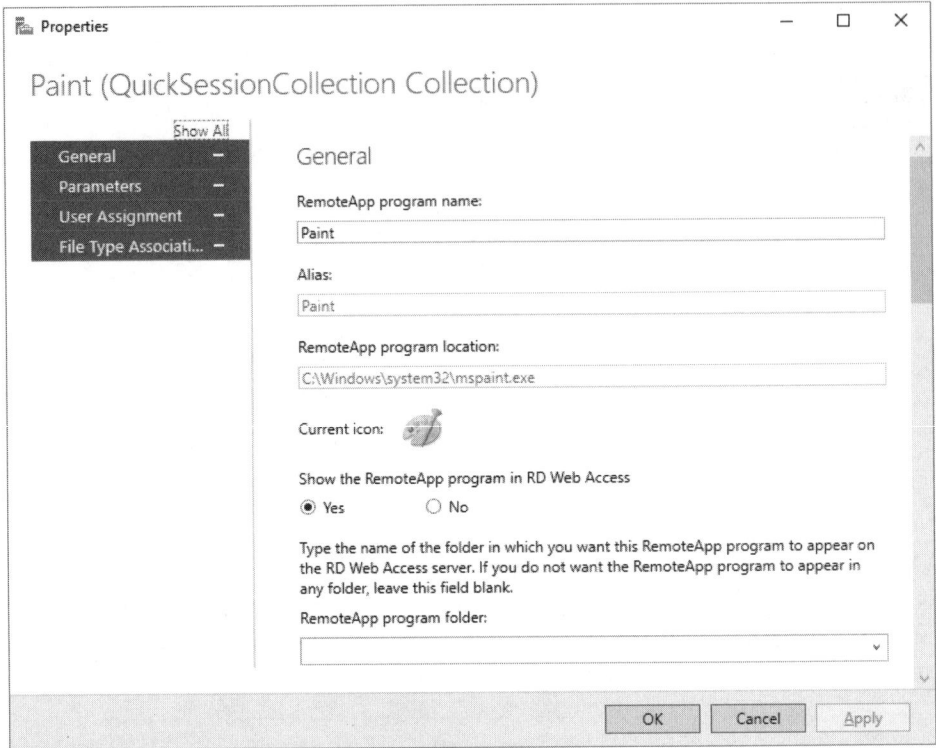

FIGURE 5-8 Edit program Properties you have published

Application properties are separated into the following groups:

- **General**   The general group contains some read only attributes, such as the program alias and program location. Other attributes can be updated, such as the name, RD Web Access visibility, and the ability to organize programs in to folders.

- **Parameters**   The parameters group gives you the option to restrict (default) or enable command-line parameters.

- **User Assignment**   The user assignment group gives you the option to narrow the deployment scope for the individual program, allowing you to specify a particular set of users and security groups.

- **File Type Associations**   The file type associations group gives you the option to adjust the default file types supported by the program. Note that these settings only apply to clients connecting through the RemoteApp And Desktop Connections Control Panel icons.

In the cloud deployments

Deploying RemoteApp with Microsoft Azure only requires a few clicks. The following example deploys RemoteApp for Tailspin Toys using the Microsoft Azure web portal.

1. Sign-in to the Microsoft Azure management portal at *https://manage.windowsazure.com*.

2. On the action pane, click New. Click App Services, RemoteApp, and then Quick Create. Using the web form, enter a name for the collection and select the target Region, Plan, and Template Image, as shown in Figure 5-9.

   - **Region** Azure spans across multiple data centers around the world. The region option enables you to deploy RemoteApp in a geographical location that meets the need of your organization. Choose this selection carefully, because the option cannot be changed without redeploying the RemoteApp collection.

   - **Plan** Azure RemoteApp is offered in four pricing tiers at the time of this writing. Each plan is targeted at certain application usage. For example, a basic plan provides enough performance to publish lightweight LOB applications. The premium plus plan supports heavy computing applications. The plan structure is flexible and can be changed at any time.

   - **Template Image** The template image contains the applications that can be published with RemoteApp. By default Microsoft provides a handful of images that contain different application models. Additionally, you have the option to upload a custom template image.

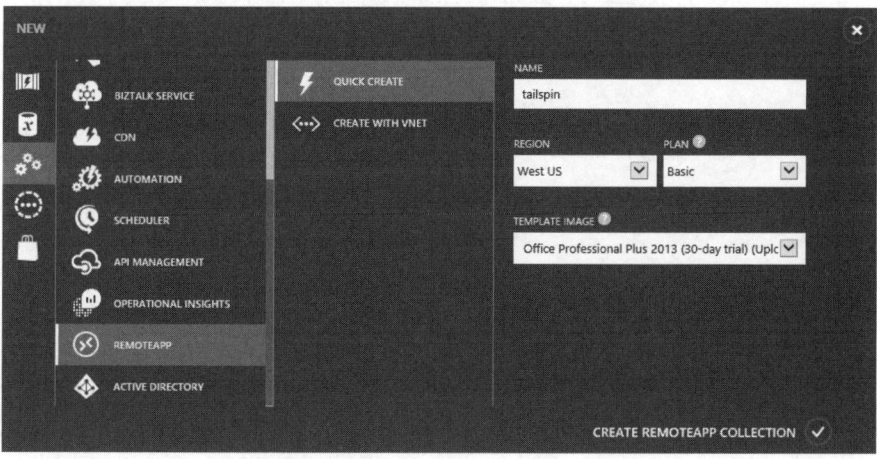

FIGURE 5-9 Deploy new Azure RemoteApp collection using the Azure Management Portal

3. Click Create RemoteApp Collection to initiate the provisioning task. Deployment times vary.

Once deployed, you can configure your RemoteApp collection through the Microsoft Azure management portal. To do so, sign-in to the portal, locate the RemoteApp collection under All Items, and click the name. This routes you to the configuration page. From here you can manage the collection, review statistics, and publish applications. Along the top of the RemoteApp collection page are six tabs. These tabs provide the following settings:

1. **Dashboard**   The Dashboard tab provides you with a basic overview of the collection and user usage. In Figure 5-10 you can see an example of the Dashboard tab. At the top of the page there is a graph that tracks your billed usage and overage data based on user activity. This data can be downloaded using the Download Usage link in the action pane at the bottom of the window. The lower half of the page includes a usage overview that displays usage on a per-user basis, including their total connected hours. The final section provides a Quick Glance that contains the basic settings for your collection.

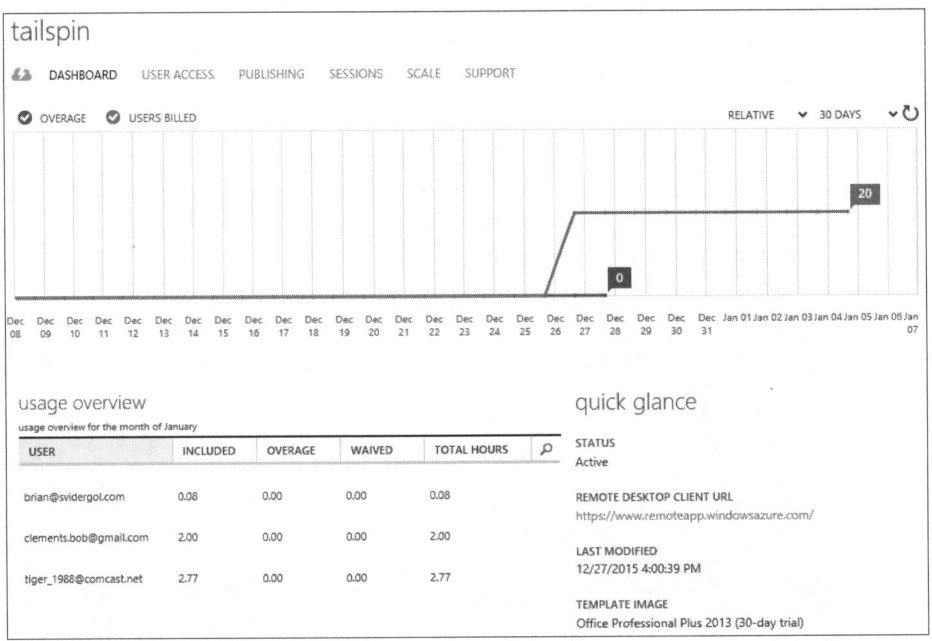

FIGURE 5-10 Azure RemoteApp usage seen from the Dashboard

2. **User Access**   The User Access tab provides a list of active user accounts associated with your RemoteApp collection. In Figure 5-11 you can see an example of the User Access tab. This page is used to remove accounts and add new accounts, either individually or using the Bulk Add Users link in the action pane. Azure can accept both Azure AD accounts and Microsoft accounts. Take note that every account you add is checked for resolution before being accepted, and any changes you make are not applied until you click the Save link in the action pane.

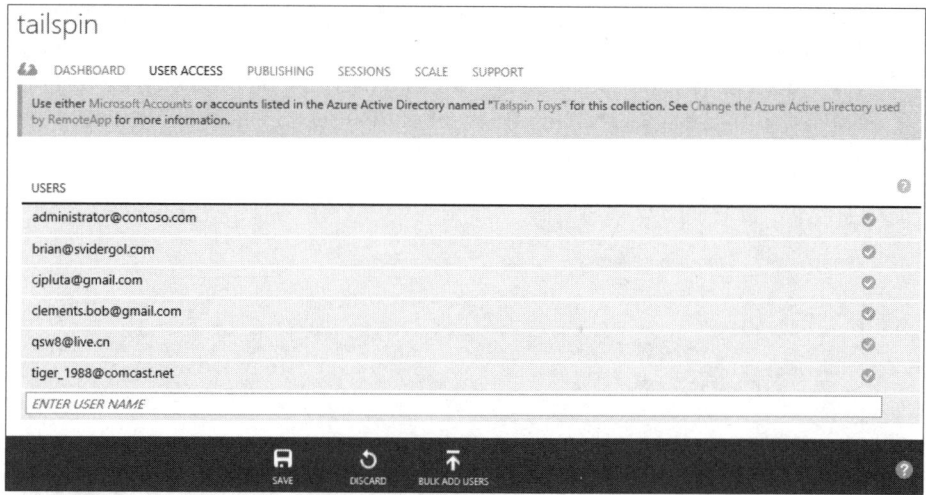

FIGURE 5-11 Azure RemoteApp user access managed from the User Access tab

3. **Publishing** The Publishing tab provides a list of currently published apps associated with your RemoteApp collection. In Figure 5-12 you can see an example of the Publishing tab. This page is used to publish new applications, edit existing versions, and unpublish those that aren't needed. The Edit link in the action pane provides you with two options: change the name of the app, and include a default command-line parameter.

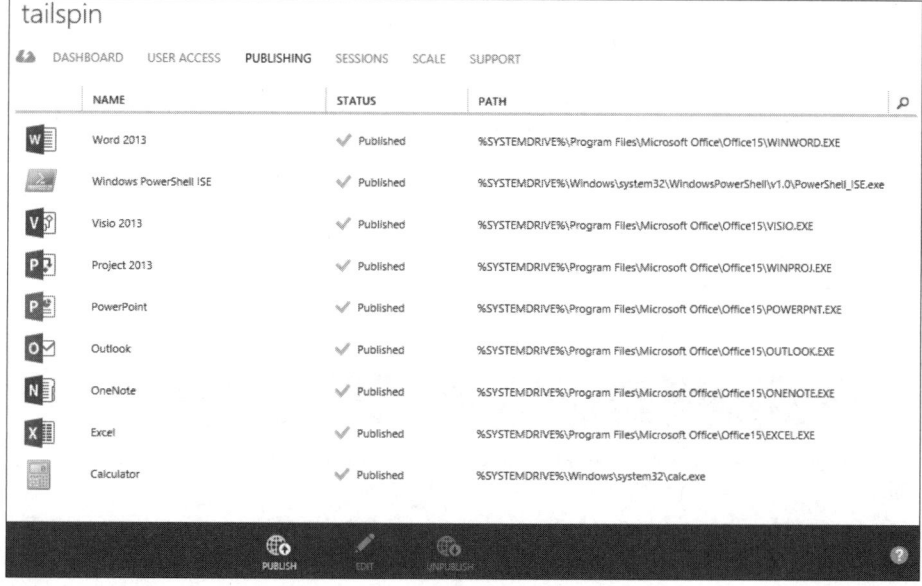

FIGURE 5-12 Azure RemoteApp programs published and modified from the Publishing tab

4. **Sessions** The Sessions tab provides you with a live view of the connections to your RemoteApp collection. In Figure 5-13 you can see an example of the Sessions tab. On the actions pane you have three options, and each can be used on a per-user basis or applied to all users. You can log users off of their session, disconnect users from their session, and send users a full screen message that must be accepted.

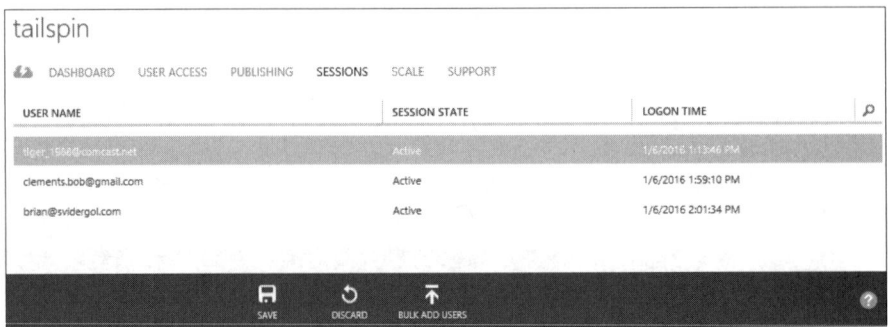

FIGURE 5-13 Active user sessions for Azure RemoteApp viewed from the Sessions tab

5. **Scale** The Scale tab displays your current service plan and billed user count. In Figure 5-14 you can see an example of the Scale tab. This page enables you to adjust your service plan on an as-needed basis. Take note that any changes you make are not applied until you click Save in the actions pane.

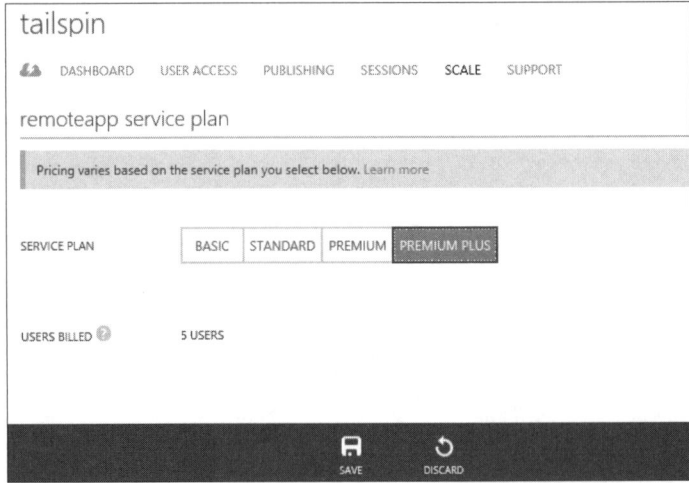

FIGURE 5-14 Azure RemoteApp service plan can be updated from the Scale tab

6. **Support** The Support tab provides links to documentation, discussion forums, feedback forums, customer service support, live chat, and the Microsoft RDS blog.

NEED MORE REVIEW? **CREATE A CLOUD COLLECTION FOR AZURE REMOTEAPP**

To review more about the requirements and steps for a cloud deployment of RemoteApp, visit *https://azure.microsoft.com/documentation/articles/remoteapp-create-cloud-deployment/*.

Hybrid deployments

The setup process for a hybrid RemoteApp collection requires a few additional steps beyond the cloud-based configuration that you reviewed in the previous section. The primary component revolves around interconnecting your on-premises network with Microsoft Azure. The following components are required to setup a hybrid collection for Azure RemoteApp.

- **Azure VNET**   You need to create an Azure VNET. This network is used by your RemoteApp collection to provide access to other Azure services and virtual machines (VMs). Once created, you can connect VNET with your on-premises network through the VNET configuration. This can interconnect your Azure VNET with your on-premises network, giving the RemoteApp collection to your local Active Directory environment.

- **Azure RemoteApp Collection**   You can create a new RemoteApp collection, similar to the steps you followed in the previous section, but instead of selecting Quick Create, you can choose Create With VNET. This enables you to select the RemoteApp VNET that you created previously, so you can choose to join your local domain. Take note that a new collection must be created for this deployment. There is not currently a method for converting a cloud-only collection into a hybrid collection.

- **Quick Start**   Once the new collection has been created, the status column reads Input Required. To complete the setup, click the new collection and access the Quick Start page. In Figure 5-15, you can see an example of the RemoteApp Quick Start page for a hybrid configuration. In this example, you have already setup the required VNET configuration and confirmed that VMs in the Azure cloud can communicate with your on-premises domain controller. You did this by creating a new VM, assigning it to your RemoteApp VNET, logging into the VM, and joining it to your on-premises Active Directory environment. There are two more steps unique to the hybrid deployment that needs to be completed.

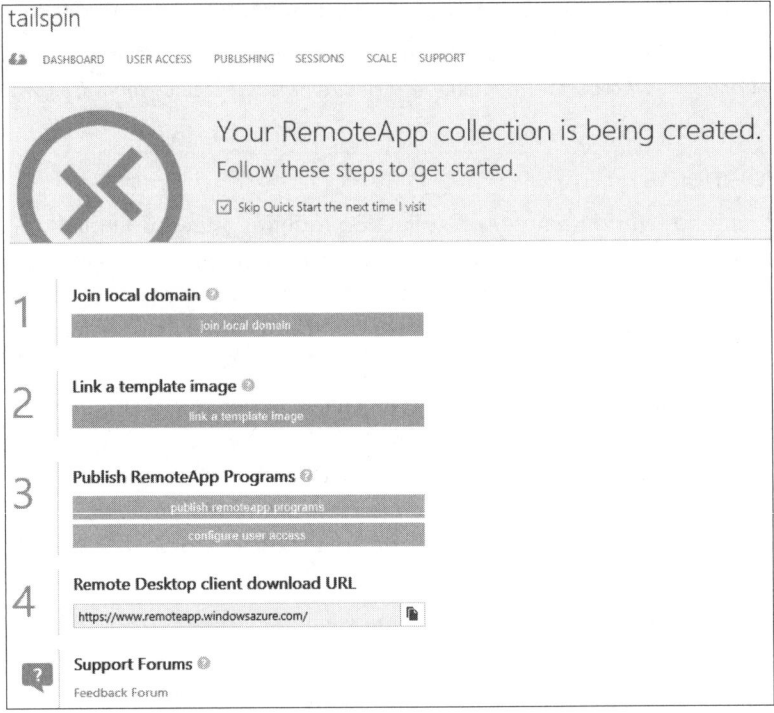

FIGURE 5-15 Azure RemoteApp hybrid deployment includes additional configuration requirements

- **Join Local Domain**   This RemoteApp instance is joined to your on-premises Active Directory. From the Quick Start page, click Join Local Domain and input the required criteria. This includes the domain name, organizational unit, service account name, and password.

- **Link A Template Image**   As discussed earlier, Azure RemoteApp enables you to use a custom image template. To assign an image template, click Link A Template Image. On the wizard, you have the option to import an image from your Azure VM library, upload a template, or select an existing template from the offerings that come preloaded with your Azure subscription. Take note that you have the ability to change the image template with this deployment, something that was not possible through the cloud-only configuration.

After completing these tasks, the management of the hybrid collection is very similar to the cloud-only deployment previously covered.

> ***NEED MORE REVIEW?*** **CREATE A HYBRID COLLECTION FOR AZURE REMOTEAPP**
>
> To review more about the requirements and steps for a hybrid deployment of RemoteApp, visit *https://azure.microsoft.com/documentation/articles/remoteapp-create-hybrid-deployment/.*

Configure Group Policy Objects for signed packages

Security is always a top priority for systems administrators. Before you implement a new service or technology it is important to understand how it impacts your security posture and what steps you need to take to protect the company resources involved. In the case of RemoteApp, Microsoft uses RDP to securely stream content to your devices. RDP is a mature protocol with security as a priority.

Depending on your deployment of RemoteApp, the applications you publish might already by digitally signed. For example, applications hosted through Azure RemoteApp are automatically digitally signed. Applications hosted on-premises should be digitally signed as well, but the certificates used to sign those apps are controlled by you. Figure 5-16 shows the certificate management for an on-premises install of RDS. The RD Connection Broker – Publishing certificate that you select is used to sign your RemoteApp packages.

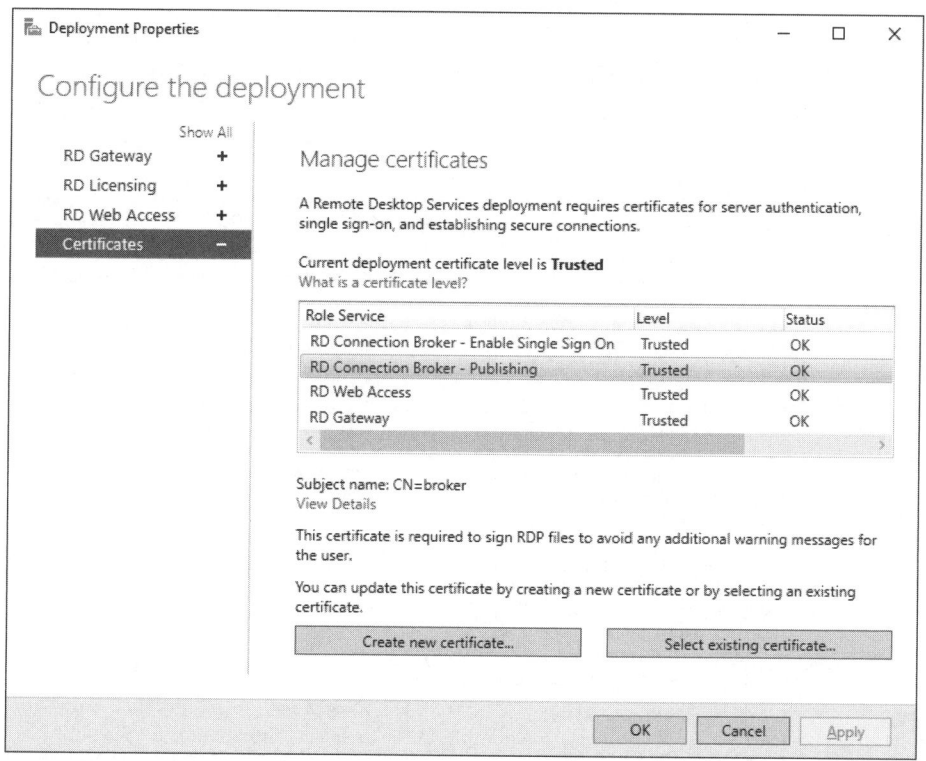

FIGURE 5-16 RemoteApp on-premises is encrypted using SSL certificates

Group Policy becomes an important tool to ensure that your devices trust the publisher providing the RDP files. To prevent any issues or unwanted security warnings, you can create a Group Policy Object (GPO) in your environment to control the client behavior when connecting to a digitally signed RDP session. This GPO needs to be applied to the clients connecting to RemoteApp. The policies to review are located in the following location (and shown in Figure 5-17): Computer Configuration\Policies\Administrative Templates\Windows Components\Remote Desktop Services\Remote Desktop Connection Client.

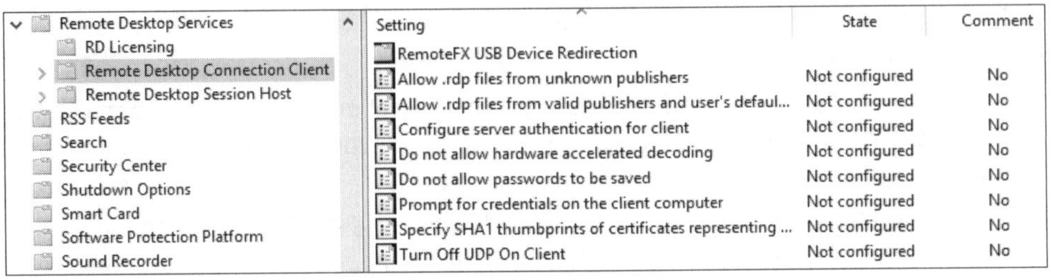

FIGURE 5-17 Access to perform RDP files from trusted publishers can be controlled by a GPO

There are three policies to examine for this requirement. Each policy deals directly with controlling how clients respond to digitally signed RDP files. Refer to Table 5-2 for a breakdown of these policies.

TABLE 5-2 GPOs for signed packages

Policy	Description	Important Elements
Allow .rdp files from unknown publishers.	This policy setting allows you to specify whether users can run unsigned Remote Desktop Protocol (.rdp) files and .rdp files from unknown publishers on the client computer.	By default, users can run unsigned .rdp files. This setting is generally only used when you need to restrict .rdp files to trusted publishers only. To accomplish this, you need to disable this policy.
Allow .rdp files from valid publishers and user's default .rdp settings.	This policy setting allows you to specify whether users can run .rdp files from a publisher that signed the file with a valid certificate. A valid certificate is one that is issued by an authority recognized by the client, such as the issuers in the client's Third-Party Root Certification Authorities certificate store. This policy setting also controls whether the user can start an RDP session by using default .rdp settings (for example, when a user directly opens the Remote Desktop Connection [RDC] client without specifying an .rdp file).	By default, users can run .rdp files that are digitally signed. This setting is generally enabled to help control the user experience and ensure that signed .rdp files are trusted. To accomplish this, you need to enable this policy.
Specify SHA1 thumbprints of certificates representing trusted .rdp publishers.	This policy setting allows you to specify a list of Secure Hash Algorithm 1 (SHA1) certificate thumbprints that represent trusted Remote Desktop Protocol (.rdp) file publishers.	This policy setting is used to directly specify the certificate thumbprints for the publishers you trust. Enabling this policy overrides the allow .rdp files from valid publishers and user's default .rdp settings. This policy is generally only used to control access to trusted content. To accomplish this, you need to enable this policy and enter the certificate thumbprint for every trusted publisher you want to allow.

For an Azure RemoeApp deployment you typically see the following GPO settings:

- **Disabled**   Allow .rdp files from unknown publishers.
- **Enabled**   Allow .rdp files from valid publishers and user's default .rdp settings.

For an on-premises RemoteApp deployment these GPO settings are relevant, unless you need to tightly restrict which publishers your devices can trust, in which case you use the following policy: Specify SHA1 thumbprints of certificates representing trusted .rdp publishers.

 Quick check

- You are in the final stages of rolling out Azure RemoteApp to your marketing department. As you begin your testing you receive reports that some users are receiving a trust warning when connecting to applications through RemoteApp. What GPO settings should you check to ensure that these warnings do not appear again?

Quick check answer

- You should confirm that the following policy setting is enabled and applied to the affected computers: Allow .rdp files from valid publishers and user's default .rdp is enabled. You should also confirm that the following policy setting is set to not configured or disabled on the affected computer: Specify SHA1 thumbprints of certificates representing trusted .rdp publishers.

Subscribe to the Azure RemoteApp and Desktop Connections feeds

With the introduction of Azure RemoteApp, connecting to hosted services has slightly changed. Before Azure, on-premises installs of RemoteApp were accessed using RemoteApp And Desktop Connections in the Control Panel, as shown in Figure 5-18. Users could manually connect using their credentials, or you had the option to create a client configuration file and distribute it to client devices.

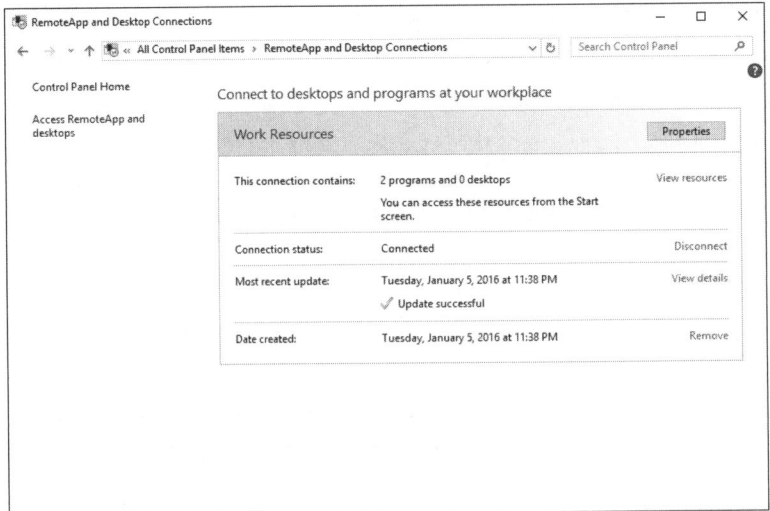

FIGURE 5-18 RemoteApp And Desktop Connections used to connect with on-premises RemoteApp deployments

Azure RemoteApp users need to have the Azure RemoteApp desktop client installed. This desktop client is required to access RemoteApp And Desktop Connection feeds through Azure. The following steps walk you through installing the new client and connecting.

1. Download the Windows Azure RemoteApp desktop client by navigating to *https://www.remoteapp.windowsazure.com/* and clicking Download.

2. You are prompted to install the desktop client, as shown in Figure 5-19. Click Run to proceed with the installation.

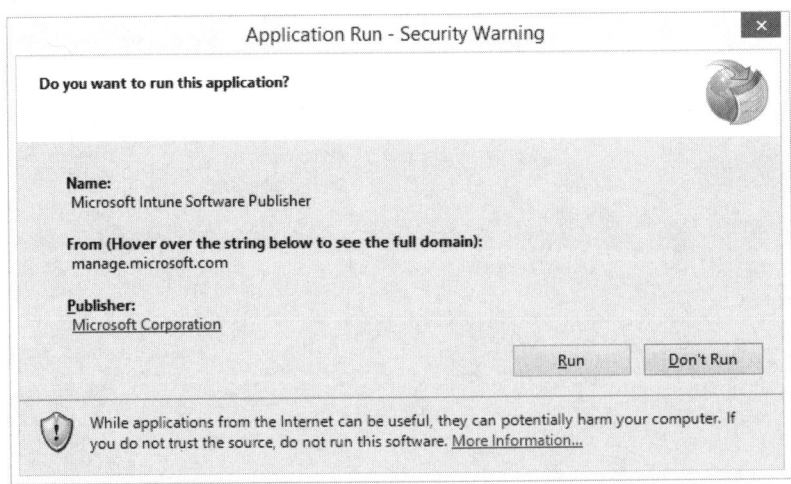

FIGURE 5-19 Azure RemoteApp requires a desktop client to connect

3. Once the installation completes, the Azure RemoteApp desktop client opens, as shown in Figure 5-20. Click Get Started on the Welcome page.

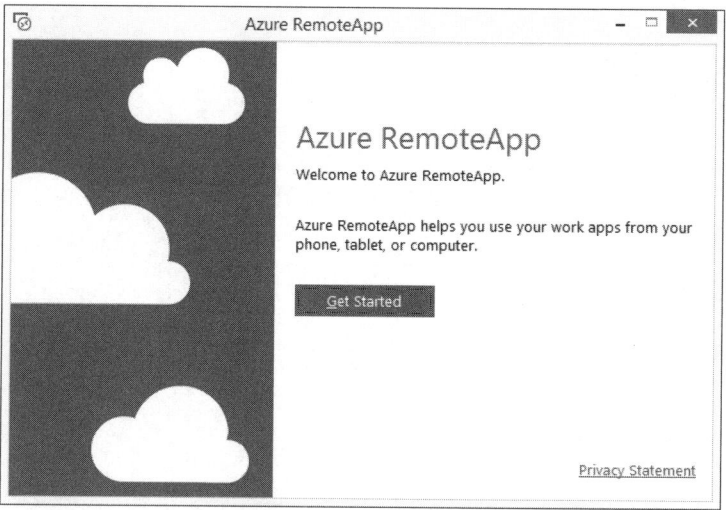

FIGURE 5-20 The Azure RemoteApp desktop client prompts for setup

4. Next, you are prompted to sign-in to Microsoft Azure, as shown in Figure 5-21. Enter your credentials and click Sign In.

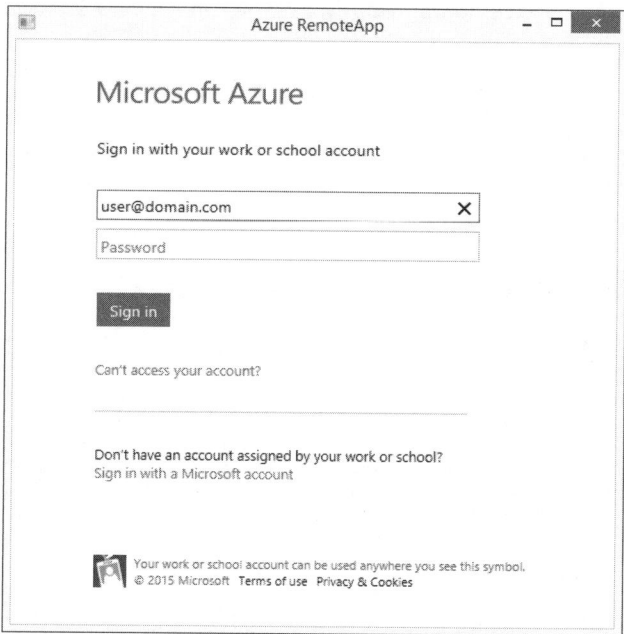

FIGURE 5-21 The Azure RemoteApp desktop client requires a sign in to connect

5. If this is the first time you have signed in, you are prompted to accept the RemoteApp invitation, as shown in Figure 5-22. Select the check box for the invitations you want to accept and click Done.

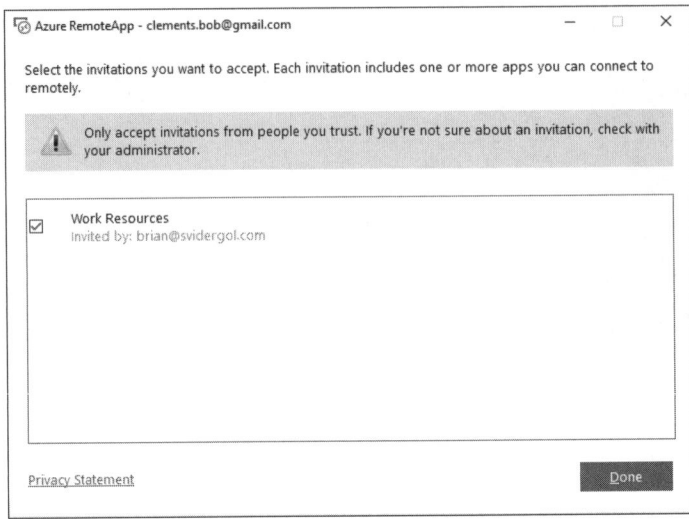

FIGURE 5-22 Azure RemoteApp displays the available invitiations

6. After completing the setup process, you are presented with the list of applications included in the invitations you accepted previously, as shown in Figure 5-23. Double-click any application to open it.

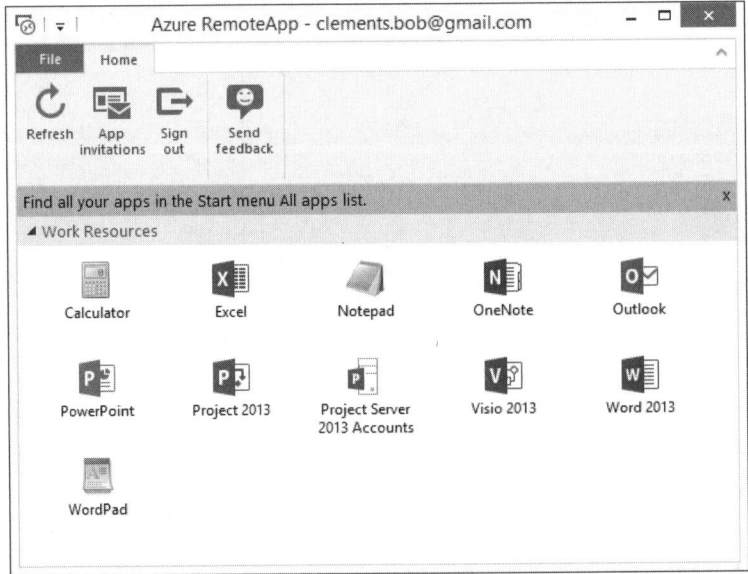

FIGURE 5-23 Azure RemoteApp desktop client displays all available published programs

Azure RemoteApp publishes the application shortcuts to the user's Start menu, blurring the line between local and remote resources. This eliminates the need for the user to repeatedly open the Azure RemoteApp desktop client. Instead they can continue launching applications from the Start menu just like they did before.

Export and import Azure RemoteApp configurations

The process of exporting and importing a RemoteApp configuration differs based on your deployment scenario. Exporting a RemoteApp configuration and importing it across multiple RD Session Host servers was first introduced for on-premises installations. This process was used in situations where a farm of RD Session Host servers were setup to host an identical set of RemoteApp programs, significantly simplifying the setup process. The alternative process focuses on Azure RemoteApp configurations, which is covered in this exam.

The catalog of applications that your organization supports might share similarities with other organizations, such as the use of Microsoft Office. However, often times there are LOB applications and custom configurations that are unique to your environment, and need to be included with your RemoteApp deployment. Custom configurations can include settings for a particular application, or changes to the underlining operating system, such as a series of registry entries.

Azure RemoteApp uses a system based on template images to provide programs to your devices. At its foundation, a template image is a customized Windows Server 2012 R2 installation that you create, export, and then import into Azure. Azure RemoteApp includes three template images out of the box to get you started. These options are shown in Figure 5-24 during the creation of a new RemoteApp collection.

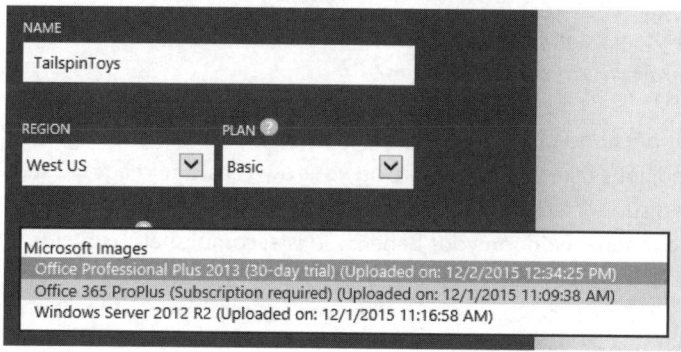

FIGURE 5-24 Azure RemoteApp includes three template images for a RemoteApp collection

Refer to Table 5-3 for a detailed breakdown on what Microsoft has included with their default template images.

TABLE 5-3 Azure RemoteApp Microsoft template images

Template Name	Available resources	Important Elements
Windows Server 2012 R2 "Vanilla Image"	.NET Framework 3.5, 3.5.1, and 4.5 Desktop Experience Ink and Handwriting Services Media Foundation Remote Desktop Session Host Windows PowerShell 4.0 Windows PowerShell ISE WoW64 Support Adobe Flash Player Microsoft Silverlight Microsoft System Center 2012 Endpoint Protection Microsoft Windows Media Player	This template image makes up the foundation for other images that Microsoft provides. The image is based on the Microsoft Windows Server 2012 R2 Datacenter operating system.
Office Professional Plus 2013 (30-day trial)	Extension of the Windows Server 2012 R2 image. Microsoft Office Professional Plus 2013: Access, Excel, Lync, OneNote, OneDrive for Business, Outlook, PowerPoint, Project, Visio, Word, and Microsoft Office Proofing Tools.	This template image is intended for evaluation purposes only.
Office 365 ProPlus (Subscription required)	Extension of the Windows Server 2012 R2 image. Office 365 ProPlus: Access, Excel, Lync, OneNote, OneDrive for Business, Outlook, PowerPoint, Publisher, Word, Microsoft Office Proofing Tools, Visio Professional, and Project Professional. SQL Native client ODBC Driver SQL Server Data Mining client Master Data Services client PowerQuery PowerMap	This template image requires an active Office 365 ProPlus subscription to use.

> *NEED MORE REVIEW?* **THE DEFAULT AZURE REMOTEAPP TEMPLATE IMAGES**
>
> To review more about the configuration of the default template images, visit *https://azure.microsoft.com/documentation/articles/remoteapp-images/*.

Now that you have a basic understanding of what a template image is and the configurations that Microsoft has included, let's take a look at building your own custom image. At the time of this writing there are two supported methods to create a custom image template. The first method requires a bit more manual work on your behalf and is accomplished by building an image from scratch using a local VM, then uploading it to Azure. The second method leverages a VM that you deploy in Azure, customize, and then import into RemoteApp. Both scenarios are covered in the next two sections.

Create a custom Azure RemoteApp template image locally

Before you begin building a local template image, there are several requirements that must be met in order for Azure to accept the image. These requirements include the following:

- **Virtual machine generation**   Azure requires that the VM be built using the generation 1 format. Generation 2 is not supported.

- **Virtual machine disk**   Azure requires the VM be installed on a Hyper-V virtual hard drive (VHD). VHDX files are not supported. Azure supports both fixed-size and dynamically expanding drives, although dynamically expanding is recommended to reduce the upload time. The VHD must be in multiples of megabytes (MBs). The size of the VHD must be between 60 GBs and 127 GBs. Anything smaller or larger is not accepted. The VHD must be initialized using the Master Boot Record (MBR) partitioning style. The GUID partition table (GPT) partitioning style is not supported.

- **Operating system**   Azure requires that the VHD contain a single installation of the Microsoft Windows Server 2012 R2 operating system. The RD Session Host role and Desktop Experience feature must be installed as part of the image. Encrypting File System (EFS) must be disabled. Azure requires that the image be sysprepped using the parameters /oobe /generalize /shutdown. The image must be fully patched using Windows Update.

Taking note of these requirements, you can start building a local template image. For this example, you are going to build a template image that includes the Microsoft Remote Desktop Connection Manager program.

1. Connect to a local Hyper-V instance and create a new VM using the New Virtual Machine Wizard. The VM can be created using Hyper-V Manager or Client Hyper-V. The VM should have the following configuration:

    - Generation 1

    - 60 GB VHD

    - Windows Server 2012 R2 ISO assigned to the DVD drive

2. Install the Windows Server 2012 R2 operating system image using the MBR partitioning style.

3. Install the RD Session Host role and Desktop Experience feature using PowerShell. To accomplish open a PowerShell window and type: **Install-WindowsFeature RDS-RD-Server,Desktop-Experience -IncludeAllSubFeature –Restart**. Press Enter to start the installation and automatically reboot the server.

4. Download and install Microsoft Remote Desktop Connection Manager, version 2.7 or later.

5. Confirm that a copy of the application shortcut has been placed in the all users Start Menu folder, as shown in Figure 5-25. The specific path is: %systemdrive%\ProgramData\Microsoft\Windows\Start Menu\Programs.

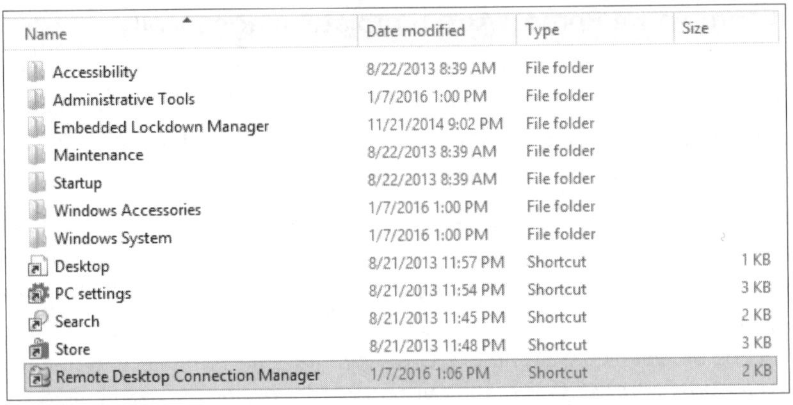

Name	Date modified	Type	Size
Accessibility	8/22/2013 8:39 AM	File folder	
Administrative Tools	1/7/2016 1:00 PM	File folder	
Embedded Lockdown Manager	11/21/2014 9:02 PM	File folder	
Maintenance	8/22/2013 8:39 AM	File folder	
Startup	8/22/2013 8:39 AM	File folder	
Windows Accessories	1/7/2016 1:00 PM	File folder	
Windows System	1/7/2016 1:00 PM	File folder	
Desktop	8/21/2013 11:57 PM	Shortcut	1 KB
PC settings	8/21/2013 11:54 PM	Shortcut	3 KB
Search	8/21/2013 11:45 PM	Shortcut	2 KB
Store	8/21/2013 11:48 PM	Shortcut	3 KB
Remote Desktop Connection Manager	1/7/2016 1:06 PM	Shortcut	2 KB

FIGURE 5-25 Program shortcuts must be in the Start Menu folder for publishing

6. Disable EFS. To accomplish this, open an elevated command window, run the following command, and then restart the VM to apply the change:

```
fsutil behavior set disableencryption 1
```

7. Open Windows Update and install the important updates. Note that this action can require multiple reboots and rechecks to confirm that all of the updates have been installed successfully.

8. Run Sysprep. To accomplish this, open an elevated command window and run the following command:

```
C:\Windows\System32\sysprep\sysprep.exe /generalize /oobe /shutdown
```

After completing these tasks, you are left with a VHD file that is ready to be imported. The next step involves importing that VHD file in to Azure using the Azure PowerShell module. To accomplish this, follow these steps:

1. Sign-in to the Microsoft Azure management portal at *https://manage.windowsazure.com*.

2. From the list of services on the left, click RemoteApp.

3. At the top of the RemoteApp page, click the Template Images tab.

4. On the Template Images page, click Add in the action pane.

5. On the Add RemoteApp Template Image Wizard, click Update A New Template Image.

6. On the RemoteApp Template Image Properties page, enter an appropriate name, and select a location relative to your organization. Click the arrow to continue.

7. On the Script Command page, as shown in Figure 5-26, save the Upload-AzureRemote-AppTemplateImage.ps1 script to the local system. Copy the PowerShell command that is provided and save it to a text file on the local system. Click the link to install the Azure PowerShell Module and follow the installation wizard. This requires a restart.

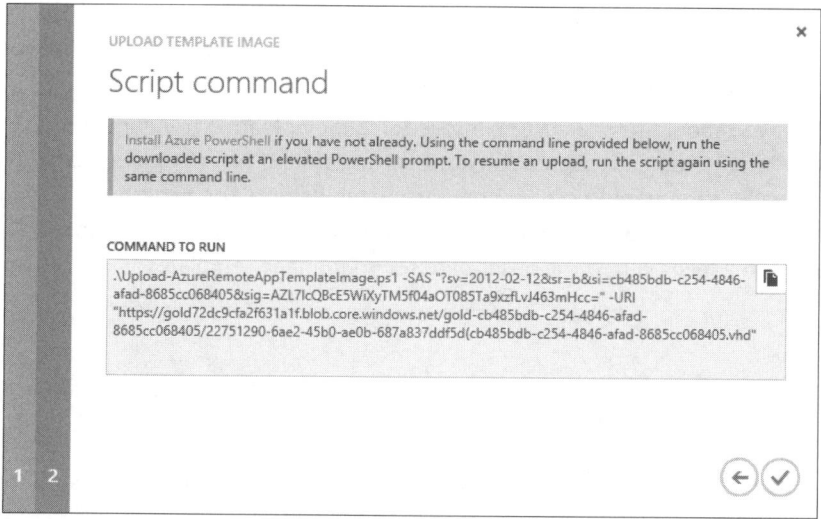

FIGURE 5-26 Uploading a custom template image requires the Azure PowerShell module

8. After your system has restarted, open an elevated PowerShell window. Run the following command to adjust your ExecutionPolicy:

```
Set-ExecutionPolicy RemoteSigned
```

9. Type **Y** to accept the change.

10. Open the text file that contains the PowerShell command that you previously saved . Copy the command to your clipboard.

11. In the elevated PowerShell window, change directories to the path where you saved the upload script. Paste the PowerShell command into the PowerShell window and press Enter. Type **R** to run the command. An example of these PowerShell steps is shown in Figure 5-27.

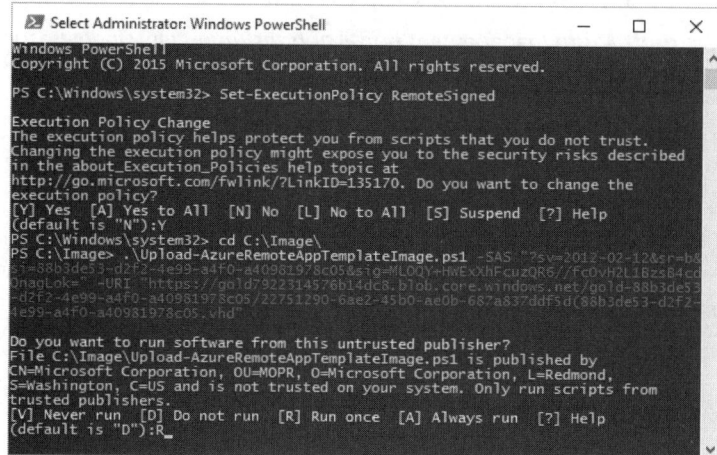

FIGURE 5-27 A pre-written PowerShell script uploads custom template images to Azure

12. As the implementation process proceeds you are prompted to locate and open your VHD file. Locate the file and click Open. Keep an eye on the PowerShell window as the validation and upload process proceeds. Any errors are recorded.

At this point you should have a custom image template uploaded to Azure and ready to assign to your RemoteApp collection. You can verify the state of the template image from within the Azure Management Portal by browsing to the RemoteApp service and clicking the Template Images tab, as shown in Figure 5-28. The image should have a status of ready. Clicking the image name takes you to the dashboard for your image templates.

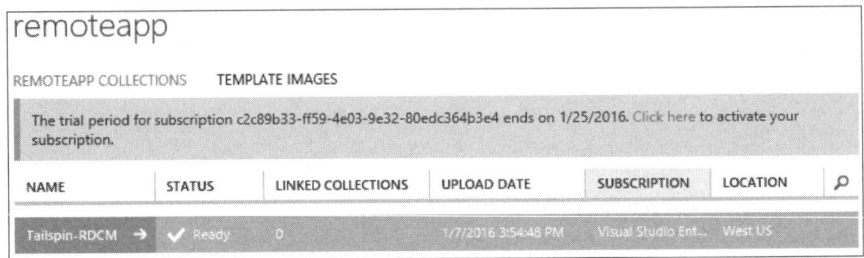

FIGURE 5-28 View an uploaded template image in the Azure Management Portal

> **NEED MORE REVIEW?** **LOCALLY CREATE A CUSTOM TEMPLATE IMAGE**
>
> To review more about requirements and steps for locally creating a custom template image, visit *https://azure.microsoft.com/documentation/articles/remoteapp-create-custom-image/.*

Create a custom Azure RemoteApp template image in Azure

Microsoft has extended the capabilities for template image creation. To help simplify the build process, RemoteApp can now accept a template image based on an Azure VM. Using this method greatly reduces your development time. The following steps walk you through provisioning the VM and linking it to your RemoteApp collection.

1. Sign-in to the Microsoft Azure management portal at *https://manage.windowsazure.com*.
2. On the portal homepage, click New in the action pane.
3. Click Compute, Virtual Machine, From Gallery.
4. On the Choose An Image page of the gallery, type **remote** into the search field. This leaves you with the Windows Server Remote Desktop Session Host image, as shown in Figure 5-29. This image meets all of the template formatting requirements, and has the RD Session Host role installed. Click to select the image. Click the arrow to continue, and proceed with the creation of this VM.

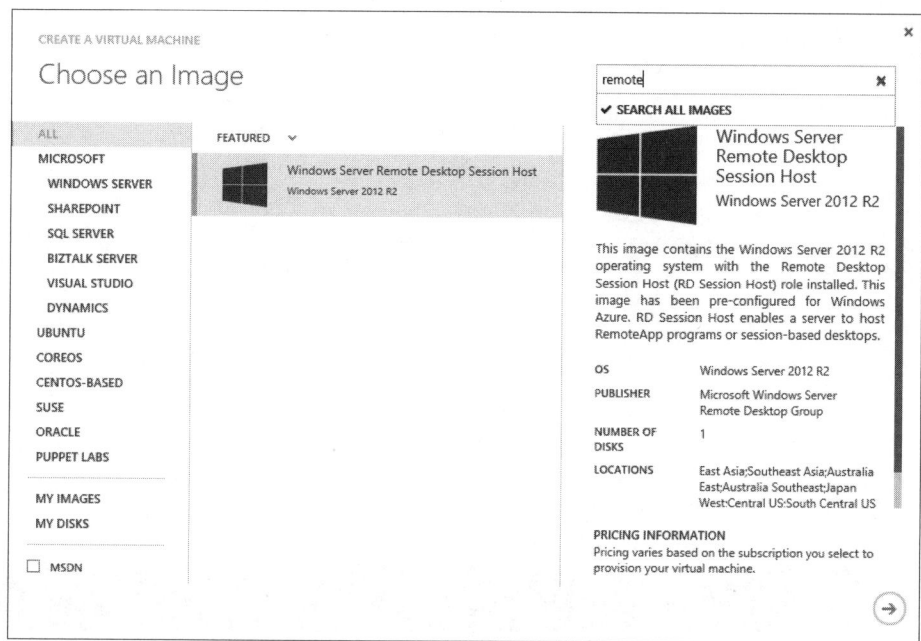

FIGURE 5-29 Create template images using an Azure VM gallery image

5. Connect to your new VM and start installing the programs you need available through RemoteApp. Complete all of your installs, configuration changes, and mandatory reboots.

6. On the desktop of your VM, double-click the ValidateRemoteAppImage PowerShell script. This script validates your RemoteApp template image and prompts you to run sysprep. Type **Y** to proceed with sysprep.

7. From the All Items page of the Azure management portal, as shown in Figure 5-30, confirm that your VM is now in a stopped state. With the VM selected, click Capture in the action pane.

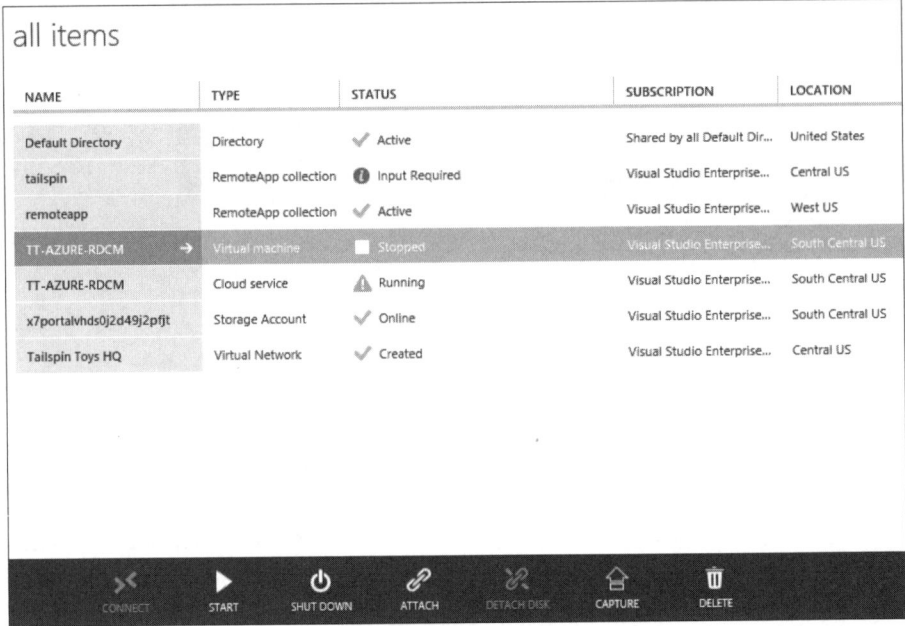

all items

NAME	TYPE	STATUS	SUBSCRIPTION	LOCATION
Default Directory	Directory	✓ Active	Shared by all Default Dir...	United States
tailspin	RemoteApp collection	ⓘ Input Required	Visual Studio Enterprise...	Central US
remoteapp	RemoteApp collection	✓ Active	Visual Studio Enterprise...	West US
TT-AZURE-RDCM →	Virtual machine	☐ Stopped	Visual Studio Enterprise...	South Central US
TT-AZURE-RDCM	Cloud service	⚠ Running	Visual Studio Enterprise...	South Central US
x7portalvhds0j2d49j2pfjt	Storage Account	✓ Online	Visual Studio Enterprise...	South Central US
Tailspin Toys HQ	Virtual Network	✓ Created	Visual Studio Enterprise...	Central US

CONNECT START SHUT DOWN ATTACH DETACH DISK CAPTURE DELETE

FIGURE 5-30 Azure VMs can be captured for use as template images

8. On the Capture The Virtual Machine page, enter the Image Description and select the I Have Run Sysprep On The Virtual Machine check box, as shown in Figure 5-31. Click the check mark to begin the capture process. After the image has been captured the virtual machine is deleted.

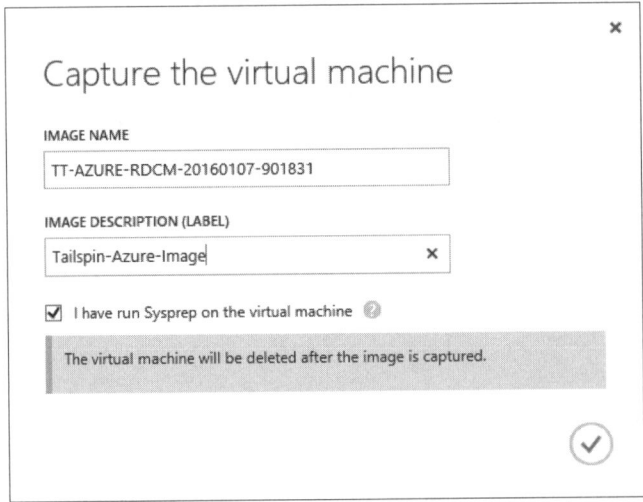

Capture the virtual machine

IMAGE NAME

TT-AZURE-RDCM-20160107-901831

IMAGE DESCRIPTION (LABEL)

Tailspin-Azure-Image

☑ I have run Sysprep on the virtual machine ⓘ

The virtual machine will be deleted after the image is captured.

FIGURE 5-31 Capture The Virtual Machine page

9. In the Azure Management Portal, click RemoteApp on the left.

10. On the RemoteApp page, click the Template Images tab.

11. On the Template Images page, click Add in the action pane.

12. On the Add RemoteApp Template Image page, click Import An Image From Your Virtual Machines Library (Recommended).

13. On the Select A Virtual Machine Image page, select the new image template from the drop-down box and select the I Confirm That I Followed These Steps To Create My Image check box. Click the arrow to continue.

14. On the RemoteApp Template Image Properties page, enter a name for the template image and select a location relative to your organization. Click the check mark to complete the wizard.

After completing these steps, you should see the new custom image template listed under the Template Images tab for RemoteApps, as shown in Figure 5-32. The status of the image reads Import In Progress until the image is ready for use, at which point you can create a new RemoteApp collection and link it to your custom image.

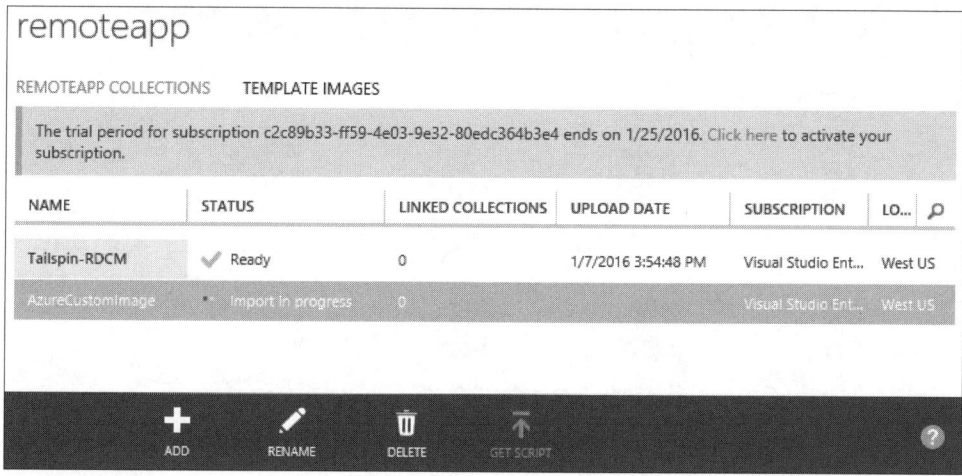

FIGURE 5-32 Captured VMs can be added to your available RemoteApp template images

Support iOS and Android

To expand support beyond the desktop, Microsoft has updated their Remote Desktop client and released it across multiple device platforms. Owners of iOS and Android devices can download the Microsoft Remote Desktop client free of charge. This mobile app includes full support for accessing Azure RemoteApp resources.

Remote Desktop app for iOS

For iOS devices, use the following steps to download, install, and configure the Remote Desktop app:

1. On your iOS device, open the App Store, search for the Microsoft Remote Desktop app, and install it.

2. Open the Microsoft Remote Desktop app.

3. In the Microsoft Remote Desktop app, tap the plus (+) icon in the upper-right corner of the app. Select Add Azure RemoteApp.

4. Tap Continue.

5. Enter the credentials associated with your Azure RemoteApp collection and tap Sign In.

6. On the Invitations page, select the invitations associated with your Azure RemoteApp collection and tap Done.

7. The final page includes the list of apps associated with your Azure RemoteApp collection, as shown in Figure 5-33. Tap on any of the apps to connect.

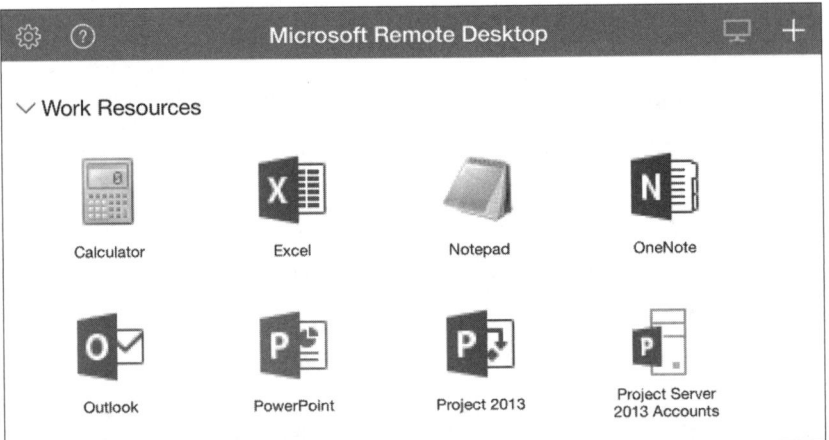

FIGURE 5-33 The Microsoft Remote Desktop app is available for iOS

Remote Desktop app for Android

For Android devices, use the following steps to download, install, and configure the Remote Desktop app:

1. On your Android device, open the Google Play store, search for the Microsoft Remote Desktop app, and install it.

2. Open the Microsoft Remote Desktop app.

3. In the Microsoft Remote Desktop app, tap the plus (+) icon in the upper-right corner of the app. Select Add Azure RemoteApp.

4. Tap Continue.

5. Enter the credentials associated with your Azure RemoteApp collection and tap Sign in.

6. On the Invitations page, select the invitations associated with your Azure RemoteApp collection and tap Done.

7. The final page includes the list of apps associated with your Azure RemoteApp collection, as shown in Figure 5-34. Tap on any of the apps to connect.

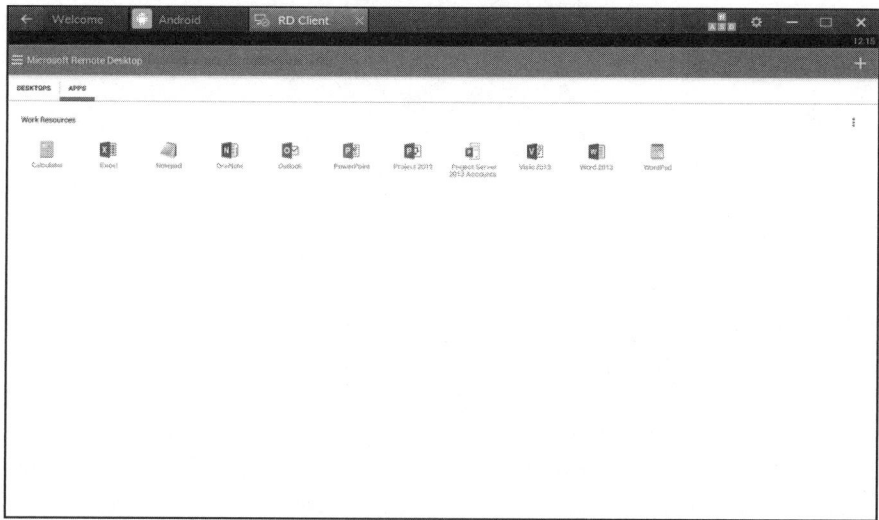

FIGURE 5-34 The Microsoft Remote Desktop app is available for Android

Configure Remote Desktop Web Access for Azure RemoteApp distribution

At the time of this writing, accessing Azure RemoteApp using RD Web Access is on the future feature availability list, with a targeted implementation date ranging between January 2016 and March 2016. In this section you review the RD Web Access role service as it exists today, as part of the on-premises deployment scenario.

When deploying your RDS infrastructure, take note that the RD Web Access role service is a mandatory component. The Remote Desktop Services installation Wizard includes this role service as part of the required configuration. RD Web Access provides a few critical functions.

1. **Web portal** The RD Web Access role service includes a web portal, providing users with a web-based frontend to your RDS services. This includes connectivity to RemoteApp programs and remote desktops. In Figure 5-35 you can see an example of the RD Web Access portal. The web site is encrypted using an SSL certificate and users authenticate with their Active Directory credentials to connect.

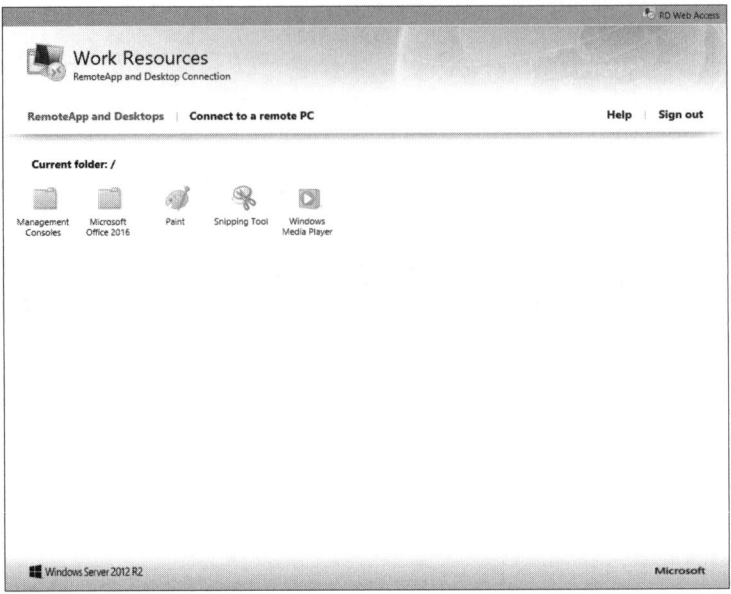

FIGURE 5-35 RemoteApp programs accessed through the RD Web Access portal

2. **Remote Desktop Web Connection**   The RD Web Access portal also provides a web-based RDP client. From within the web portal, when you click Connect To A Remote PC at the top of the page, you are presented with a standard set of RDP connection options. From here you can enter the name of an RDP-enabled resource on the network and connect using the security of your RDS infrastructure. Figure 5-36 gives you an example of the options made available through the Remote Desktop web client.

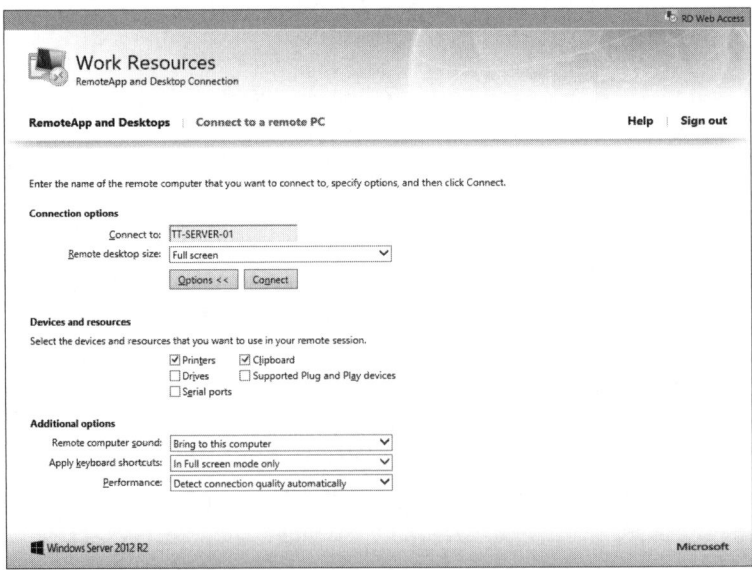

FIGURE 5-36 The RD Web Access portal provides a web-based RDP client

3. **RDWeb feed**   The RD Web Access role service hosts the RDWeb feed, which is essentially an XML file that contains your published RDS resources. When a client computer connects to the RDWeb feed the RDS resources automatically populate the user's Start menu, simplifying the navigation of remote services. The connection to the RDWeb feed is done on the client computer using the RemoteApp And Desktop Connection interface in Control Panel. Figure 5-37 shows an example of the Connection Wizard. In this example, you can see the connection URL, available program count, and available desktop count.

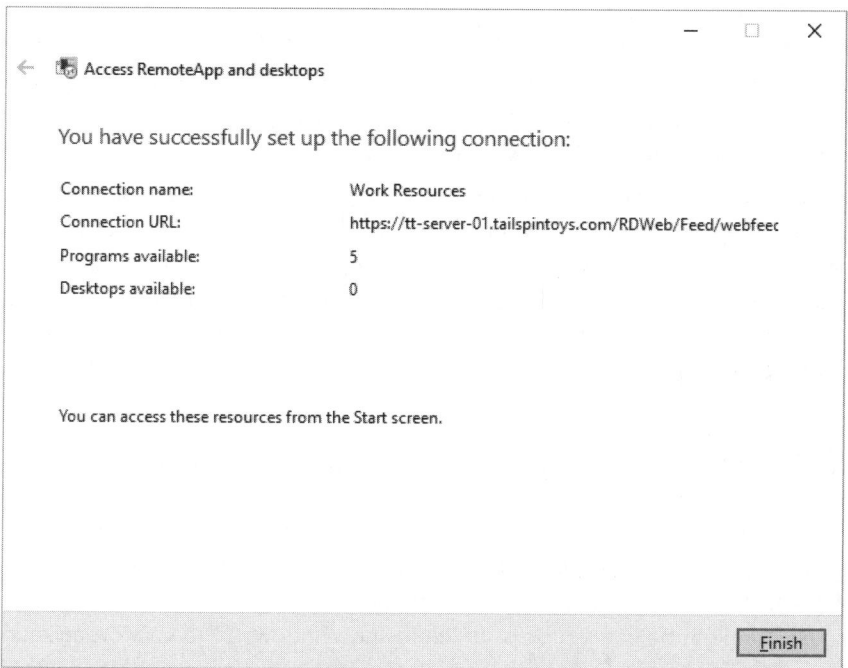

FIGURE 5-37 RemoteApp And Desktop Connections use a web feed to connect

In preparation for the exam, be sure to familiarize yourself with the functions that the RD Web Access role service provides. Understand why this role service is mandatory as part of your RDS infrastructure.

Summary

- There are three deployment scenarios for RemoteApp: on-premises, Azure RemoteApp in the cloud, and Azure RemoteApp hybrid.

- RemoteApps use RDP to stream applications and desktops to devices. RDP files can be signed or unsigned depending on the source. For optimal security, a GPO should be applied to control the implementation of RDP files.

- Azure RemoteApp enables you to create a custom template image that contains your organization's LOB applications and unique configurations. These template images can be imported into Azure RemoteApp for collection assignment and program publishing.

- Mobile devices are a big part of the RemoteApp solution. Microsoft has developed Remote Desktop app for both iOS and Android. The app is free to download, and the user experience is nearly identical between platforms.

- Programs that are published using RemoteApp can be accessed in multiple ways. Each method is dependent on the RDS infrastructure that the client is connecting to. Desktop clients can access RemoteApp through the RemoteApp And Desktop Connections UI in Windows, through the RD Web Access portal, or through the Azure RemoteApp client. Supported mobile devices can access RemoteApp through the Microsoft Remote Desktop mobile app.

Skill 5.2: Plan app support and compatibility

In today's enterprise market the lifecycle of an application isn't always in tune with that of the operating system. As a systems administrator it is your responsibility to support the business, and often times that involves addressing application support and compatibility requirements. For example, you might be in the process of deploying Windows 10 to your organization. Along the way you discover a finance application that isn't compatible with Windows 10. Rather than halting the upgrade, you leverage the Application Compatibility Toolkit (ACT) to create a fix that can be deployed until the application receives an upgrade. In another scenario you might just deploy Microsoft VDI. Your users would like to see their custom application and operating system settings synchronized between their local desktop and virtual desktop. To achieve this, you choose to implement User Experience Virtualization (UE-V), providing a streamlined user experience.

The topics discussed in this skill address solutions to problems that commonly appear when supporting Windows-based applications. Along the way you see both reactive and proactive formulas for identifying existing and potential compatibility issues. The goal is to use the tools at your disposal to prepare for product upgrades, browser changes, and even software updates.

Design for desktop app compatibility using Application Compatibility Toolkit

The first tool for managing application compatibility is the Microsoft Application Compatibility Toolkit (ACT). The ACT is a lifecycle management tool for your organization's application portfolio. Microsoft has made this tool available to all customers through the Windows Assessment and Deployment Kit (Windows ADK). Figure 5-38 is the visual illustration shown in the application, which briefly illustrates the workflow and features that ACT includes. Let's take a closer look at each of these.

FIGURE 5-38 The Application Compatibility Toolkit is a multi-step solution

- **Collect Inventory** The ACT collects valuable application and hardware data from the computers in your organization. That data is uploaded to a share and written to a SQL Server database for use in addressing potential compatibility issues.
- **Plan Testing** The ACT helps you identify the applications in your environment. From this data you need to organize and prioritize your test plan.
- **Test** Create and deploy runtime packages for assessing compatibility issues.
- **Analyze Results** Review the results of the runtime packages to assist in addressing compatibility issues or detecting problems before a major upgrade. Access a series of filtered reports to prioritize compatibility efforts.
- **Mitigate** The ACT offers mitigation strategies for identified compatibility issues, including custom compatibility fixes, previously referred to as shims. A compatibility fix is a small piece of code that intercepts communication between the application and the operating system, allowing you to redirect the malfunctioning application calls.

From a design standpoint, the ACT has a number of components. Depending on the size of your organization and your application compatibility strategy, ACT can be setup to run on a single consolidated server, or span multiple servers. Table 5-4 breaks down each of the components and how they interoperate.

TABLE 5-4 Application Compatibility Toolkit (ACT) components

ACT Component	Component details
Application Compatibility Manager (ACM)	ACM is the core of ACT. This tool enables you to configure, collect, and analyze application and device data in your organization so you can address any issues prior to deploying a new operating system or operating system update.
Data Collection Package (DCP)	DCPs are created using the ACM. The output is a Microsoft Setup Installation (MSI) file that you deploy to your client computers for evaluating compatibility issues. The DCPs are responsible for collecting applications, hardware, device information, and determining any associated compatibility issues based on new operating system or update planning.
ACT Log Processing Service	The ACT log processing service is responsible for processing the ACT log files uploaded from the computers running a DCP. The service gathers the logs and adds them to your ACT database.
ACT Log Processing Share	The ACT log processing share is the location where logs are written for processing. The ACT log processing service and all client computers require access to this share.
ACT Database	The ACT database requires Microsoft SQL Server. The inventory data collected by the DCPs and processed by the ACT log service is stored in the ACT database. This information can be viewed using a series of filtered reports through the ACM.
Microsoft Compatibility Exchange	The Microsoft Compatibility Exchange is a web service that Microsoft manages. Clients that participate send recorded compatibility issues to Microsoft. Clients also receive for updated compatibility information through this service. The process requires client computers to connect with Microsoft.

In this table you are introduced to the Data Collection Package (DCP), a client-side resource used to gather data and report it to your ACT database. In addition to collecting inventory, the DCP includes a number of compatibility evaluators. A compatibility evaluator is a run-time detection tool, designed to track behavior and identify potential compatibility issues as they occur. It is important to understand how these evaluators work in preparation for deploying a DCP because some evaluators collect large amounts of data. The following evaluators are part of the DCP:

- **User Account Control Compatibility Evaluator (UACCE)**   The DCP UACCE is used to identify compatibility issues between the user's application and the UAC framework. Application behavior is tracked through Protected Administrator (PA) and Standard User (SU) accounts.

- **Update Compatibility Evaluator (UCE)**   The DCP UCE is used to identify potential compatibility issues with upcoming Windows Updates. The data gathered by this evaluator can be used to prioritize the testing of Windows Updates in your organization.

- **Windows Compatibility Evaluator (WCE)** The DCP WCE is used to identify potential compatibility issues related to the release of a new operating system. The following components are tracked:

  - **Deprecated objects** This includes applications that use deprecated DLLs, EXEs, application-programming interfaces (APIs), COM objects, and Registry keys.

  - **Graphical Identification and Authentication (GINA) DLL** This includes applications that leverage the GINA DLL to handle identification and authentication, a feature that was deprecated with the introduction of Windows Vista.

  - **Session 0** This includes applications that still interface with Session 0, which handles all system services. With the introduction of Windows Vista applications are now required to interact with Session 1.

As you deploy ACT into your environment, you can identify some potential compatibility issues. In cases where a critical application is not operating correctly, the ACT includes a library of available fixes based on data received through the Microsoft Compatibility Exchange. Alternatively, the ACT enables you to generate a custom compatibility fix. A custom compatibility fixes can be created using the Compatibility Administrator. Before creating a custom fix, note that it is often times better to fix the application directly with the help of the developer. Deploying a custom fix typically results in long-term maintenance.

To create a custom compatibility fix using the ACT, follow these steps:

1. Open the Compatibility Administrator. The ACT includes both a 32-bit and 64-bit version of the Compatibility Administrator. Make sure you run the version matching the application architecture that you are working with.

2. On the tool bar, click the New icon to create a new custom compatibility database.

3. In the left pane of the Compatibility Administrator, under the Custom Databases heading, right-click the new database, click Create New, and click Application Fix.

4. On the Program information page of the Create New Application Fix Wizard, enter the name of the program, the vendor, and locate the .EXE associated with the program. Click Next.

5. On the Compatibility Modes page, you are given the option to set the compatibility mode of the application by operating system version, along with a variety of other pre-defined modes, as shown in Figure 5-39. Select your preferred mode settings and click Test Run to validate their behavior. Click Next to continue.

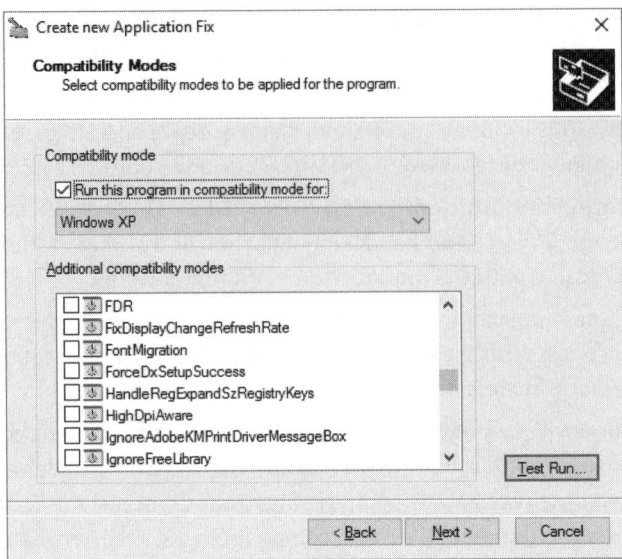

FIGURE 5-39 A compatibility fix supports different compatibility modes

6. On the Compatibility Fixes page, you are given the option to include a number of pre-defined fixes. Each fix can also include a custom command-line parameter, as shown in Figure 5-40. Select the check boxes for the fixes that apply to your application and click Test Run to validate the behavior. Click Next to continue.

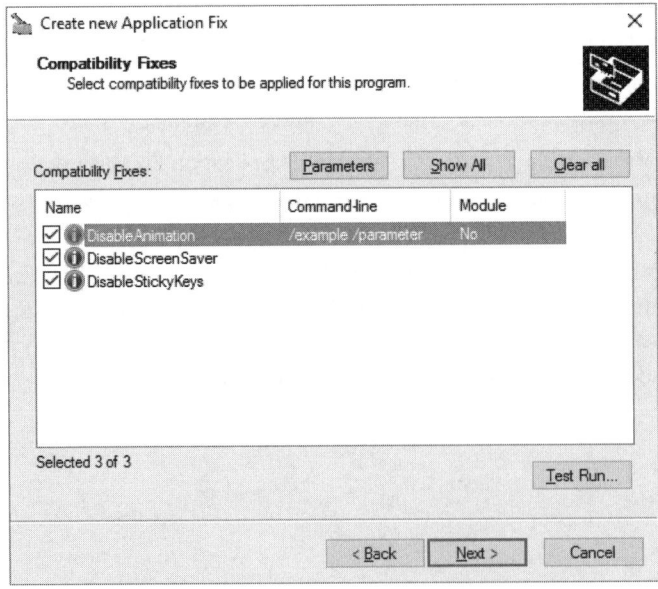

FIGURE 5-40 A compatibility fix supports different compatibility fixes

7. On the Matching Information page, you are given the option to associate additional criteria with your program for identification purposes. The tool gathers information from the program to populate what it can. Make any necessary changes and click Finish.

After creating a custom compatibility fix, you can save the .sdb database file and prepare it for deployment. All of the compatibility changes saved in this database file do not take effect until the file is applied to the client computers running the incompatible application. The compatibility fix database is applied using the sdbinst.exe command-line tool included with the Windows operating system.

Design desktop application co-existence

As covered in the beginning of this chapter the market for virtualization has directly impacted the old application support model. The idea of installing applications directly onto the local hard drive of a client computer is fading. Instead, you can take those vary same applications and virtualize them. This provides a number of benefits, including:

- **Access anywhere** Hosted solutions, like Azure RemoteApp, enable your users to connect to corporate applications and resources from anywhere.
- **Access from any device** Mobile apps, like Microsoft Remote Desktop, enable users to connect to corporate applications and resources using any of their devices, including personally-owned smartphones and tablets.
- **Simplified management** As an administrator, you no longer have to support or patch the same application across hundreds of machines. Instead, you maintain the application on the servers that host it, providing a more controlled environment.
- **Improved security** Your corporate applications and resources now remain in a secured datacenter, whether that is a Microsoft or company datacenter.. The services are hosted and encrypted.
- **Better user experience** In some cases users have the flexibility to use whichever device meets their needs. The constant changes that IT is required to push out for compliance drops noticeably. Patching and reboot interference is reduced. Gaining access to a new application or service is a smoother experience.

In this section, you review three tools to address application compatibility needs. Each of these tools use virtualization to accomplish the goal. You have the flexibility to deploy any of these solutions in whichever format best suites your environment. Some administrators might only leverage Azure RemoteApp, while others might incorporate all three.

Client Hyper-V

Hyper-V is a core component in many data centers around the world. Microsoft has taken this technology and included it in their client operating system. This enables Windows users to leverage a type-1 hypervisor on their daily computer. In the case of this skill, you are going to be focusing on Client Hyper-V for application co-existence. Client Hyper-V was first introduced with the 64-bit SKUs of Windows 8 Professional and Windows 8 Enterprise, and it continues to be an available feature with the release of Windows 10. The following requirements must be met in order to activate and use Client Hyper-V under Windows 10:

- The computer must be running Windows Professional 64-bit or Windows Enterprise 64-bit.

- The computer must be equipped with a 64-bit processor that supports Second Level Address Translation (SLAT).

- The computer must be equipped with at least four GB of RAM.

- The computer must support BIOS-level Hardware Virtualization.

After confirming that your computer meets these requirements, you can enable Client Hyper-V by following these steps:

1. Open Control Panel and navigate to Programs And Features.

2. In the left pane, click Turn Windows Features On Or Off.

3. On the Turn Windows Features On Or Off dialog box, scroll down the Hyper-V option and select the check box, as shown in Figure 5-41. This installs the Hyper-V hypervisor, services, management tools, and PowerShell module. Click OK to start the installation. When prompted, reboot the computer.

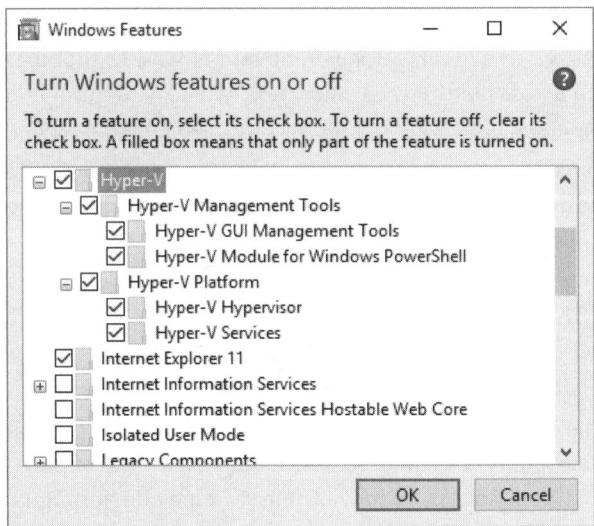

FIGURE 5-41 The Client Hyper-V feature is enabled through Programs and Features

Now that Client Hyper-V is installed, you can open Hyper-V Manager on your computer to begin creating and managing virtual machines. From an application standpoint, a guest operating system running on a user's existing computer enables them to run different versions and configurations of the same applications side-by-side. Additionally, developers and testers can leverage the checkpoint feature to apply changes and revert them. This experience can be very helpful for users that are comfortable using virtualization. It would not be a recommended solution for those that are less technical. Figure 5-42 demonstrates the co-existence of apps using Client Hyper-V. In this example, the host operating system is running Windows 10 Enterprise 64-bit. The guest operating system is running a copy of Windows 8.1 Enterprise.

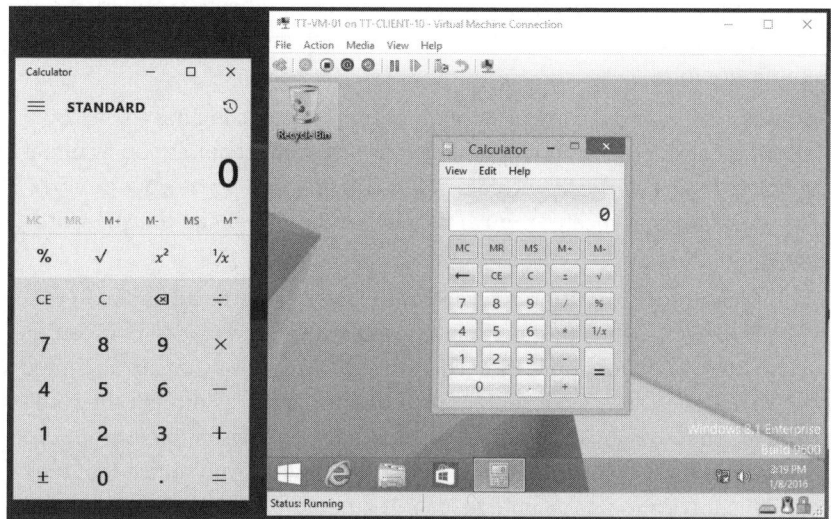

FIGURE 5-42 Client Hyper-V runs side-by-side with other programs

For large scale deployments of Client Hyper-V you have multiple tools at your disposal. Deployment automation can be achieved by installing the Client Hyper-V feature using PowerShell. To accomplish this run the following cmdlet from an elevated PowerShell window: Enable-WindowsOptionalFeature –Online –FeatureName Microsoft Hyper-V -All. Enabling the full Client Hyper-V feature also includes the Hyper-V PowerShell module, which can simplify the setup and configuration of VMs. Finally, a custom image template can be created from any Hyper-V instance. When you are ready to deploy it be sure to run sysprep, and then distribute the VHD to your target audience.

Azure RemoteApp

Azure RemoteApp extends application virtualization into the cloud, providing your users with access to corporate applications from any Internet-enabled devices that Azure supports. If you plan to deploy this solution in your environment, take note of the following design considerations:

- **Subscription**   Azure RemoteApp is a cloud-based service that requires an active subscription. As you design your deployment, review the subscription plans and determine the number of programs and users that you are supporting.

- **Deployment type**   Azure RemoteApp can be deployed in one of two configurations: cloud-only or hybrid. Before you deploy Azure RemoteApp, review the needs of your organization. If the application list is limited to items like Office 365, a cloud-only configuration is a good solution. If you need to provide Active Directory authentication with your applications, a hybrid deployment would be a better match.

■ **Template images**   Azure RemoteApp offers two template images that can be used for production. If you have a series of LOB applications that you need to publish, be prepared to create and manage a template image that contains your list of programs and custom configurations.

■ **Program updates**   The template image configuration uses static images to deploy programs. Take into consideration that updating your programs requires you to update an existing template image or possibly create a new one, depending on your deployment type. For programs that receive regular product updates, this can translate into additional management.

The process for publishing programs using Azure RemoteApp was reviewed earlier in this chapter. Be sure to familiarize yourself with the publishing workflow and the idea of creating and managing template images. As for application co-existence, programs that are streamed using Azure RemoteApp appear side-by-side with normal desktop apps, as though they are both running locally. In Figure 5-43 you are running Notepad through Azure RemoteApp (left), and on the local computer (right). If you take a look at the taskbar you notice that the RemoteApp version of the application has a unique icon. The RemoteApp version is also using a different color scheme. This is because the application is streamed from a Windows Server 2012 R2 template image.

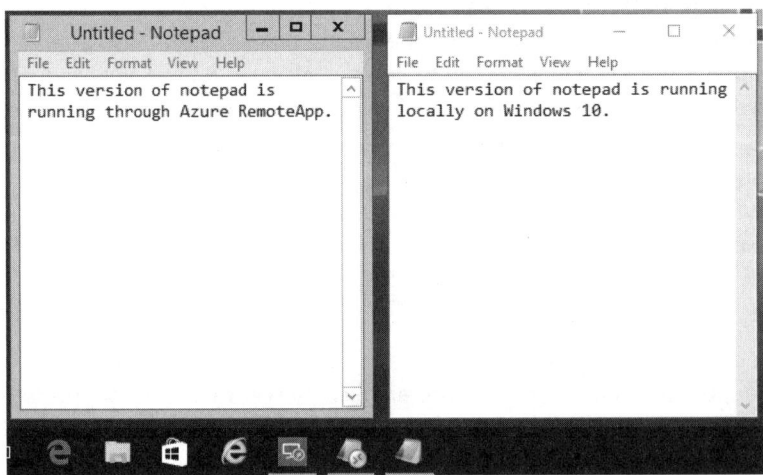

FIGURE 5-43 RemoteApp programs run side-by-side with desktop apps

Microsoft Application Virtualization (App-V)

Similar to RemoteApp, Microsoft Application Virtualization (App-V) is a solution targeted at providing a seamless application experience for the user. The product is included as part of the Microsoft Desktop Optimization Pack (MDOP). At its foundation, App-V shares many similarities with the Client Hyper-V and Azure RemoteApp solutions that were discussed previously. Before an application can be made available through App-V, it must first go through a sequencing process. This process records the installation parameters, configuration settings, and various other aspects about the install that you record. The application is them written to a self-contained virtual environment, or an APPV file. Computers running a compatible version of the App-V client can import the APPV file and run the contained program side-by-side with other App-V programs or local desktop programs. This virtualized environment operates as though the enclosed program were running on a dedicated guest operating system, similar to the Client Hyper-V scenario.

App-V also offers a streaming solution. With the proper App-V infrastructure, users can review a list of published programs using the App-V client. From there they can run the virtualized programs without having to wait, similar to the Azure RemoteApp scenario. In situations where streaming performance is not ideal, however, you can require the App-V package to be fully downloaded to the client computer before it runs. This gives them optimal performance and continues to provide access to the program even when the user goes offline.

In planning an App-V deployment, take the following design considerations into mind:

- **App-V Sequencer**   Building a library of App-V packages requires a dedicated virtual machine for sequencing. After each sequencing job the virtual machine must be reverted to a clean slate. This is achieved by using the checkpoint feature in Hyper-V.
- **App-V Client**   Client computers must have the App-V client installed in order to stream and run App-V content.
- **App-V Connection Groups**   App-V connection groups are created in scenarios where you have two or more virtualized programs that need to communicate with one another.
- **App-V full infrastructure model**   This deployment model leverages the App-V management server capabilities that come included with the product. A full infrastructure model includes the following roles: App-V management server, management web service, and streaming server.
- **App-V integrated model**   This deployment model leverages your existing System Center Configuration Manager 2012 server infrastructure. App-V packages can be imported and distributed using Configuration Manager, and streaming servers can be replaced by your existing Configuration Manager distribution points.
- **App-V standalone model**   This deployment model requires no server infrastructure. You can use any existing distribution tools to deploy the App-V client and packages, but streaming is not available.

In Figure 5-44, you are running two different versions of the 7-Zip archive utility on a single Windows 10 client computer. The version on the left, 9.20, has been sequenced and is running using the App-V client. The version on the right, 15.14, was installed directly on the client computer. Both applications run side-by-side in a seamless fashion, and appear in the Start menu as though they are locally installed on the client computer.

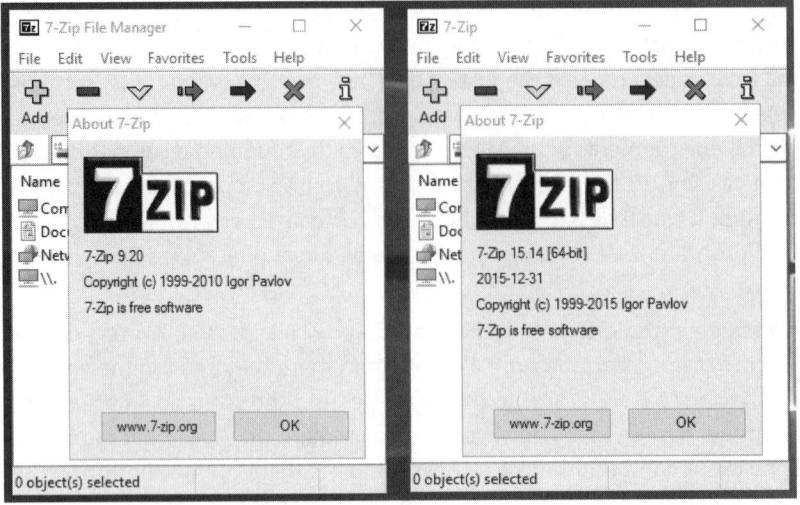

FIGURE 5-44 App-V programs run side-by-side

Install and configure User Experience Virtualization (UE-V)

Microsoft User Experience Virtualization (UE-V) is designed to provide users a consistent experience when working between multiple devices. For example, imagine you support a team of sales engineers at your organization. These engineers move between a Surface Pro, a hosted virtual desktop, and a stand-alone desktop at their home office. In the past these users simply repeated the same steps multiple times when setting up a new application or changing a setting between their environments. With UE-V, when a user changes the desktop background on one computer and locks the screen, that same background is applied to the next computer when he logs in. UE-V offers several default templates for managing which settings are synchronized, along with the ability to customize your own application templates.

UE-V is another tool included with MDOP. Refer to the following list of components and planning considerations when deploying UE-V in your environment:

- **UE-V Agent** The UE-V agent must be installed on each computer that plans to use UE-V for roaming application and Window settings. The agent setup can be distributed using a deployment tool like Configuration Manager. The agent also supports a series of command-line parameters for custom configurations.

- **Custom applications** Are you planning to support custom applications? This includes third-party or LOB applications. If the answer is yes, you need to install the UE-V generator.

- **UE-V Generator** The UE-V generator is a tool used to create a custom XML settings location template for your custom applications.

- **Settings Location Templates** UE-V uses an XML file, also referred to as a settings location template, to define the application and Windows settings that the UE-V agent can track and apply.

- **Settings Packages**   Settings packages contain the files identified for synchronization based on the settings defined in your settings location templates.

- **Settings Storage Location**   The settings storage location is used by UE-V to store application and Windows settings associated with your UE-V users.

- **Settings Template Catalog**   The settings template catalog is a folder path on computers running the UE-V agent, or an SMB network share centrally stores the custom settings location templates that you create.

- **Offline files**   The offline files feature in Windows needs to be enabled for client computers that disconnect from the corporate network. Virtual desktops that are connected to the network all the time do not require offline files to be enabled.

Now that you have a basic understanding of the components associated with UE-V, let's take a look at installing and configuring UE-V. Before you begin, download a copy of MDOP and extract the UE-V folder to your desktop. In the following example, you create an Active Directory security group and setup your settings storage location:

1. Open Active Directory Users and Computers.

2. Create a new security group for UE-V. For this example, you call your group UEV-USERS. Add yourself to this group so you can test and validate the sync operation. This group is used to regulate access to the Settings storage location.

3. Identify a server to host your Settings Storage Location. You specify a settings storage location in this example, but if one is not specified, UE-V uses Active Directory to store settings packages. Create a new shared folder with the following permissions:

```
Share permissions = Full Control for the UEV-USERS group.
ACL permissions = List folder / read data and create folders / append data for the
UEV-USERS group.
```

Next, you install the UE-V agent on a few test computers and start the synchronization process.

1. Copy the AgentSetup.exe to two of your computers. Run the AgentSetup.exe to start the installation.

2. On the Welcome page of the UE-V Agent Install Wizard, click Next.

3. On the End-User Licensee Agreement page, select the check box to accept the license. Click next.

4. On the Microsoft Update page, select Do Not Use Microsoft Update. Click Next.

5. On the Customer Experience Improvement Program page, select Do Not Join The Program At This Time. Click Next.

6. On the Options page, under Synchronization Method, select SyncProvider. Under the Settings Storage Location Path, type the UNC path to the file share you created earlier, as shown in Figure 5-45. Click Next.

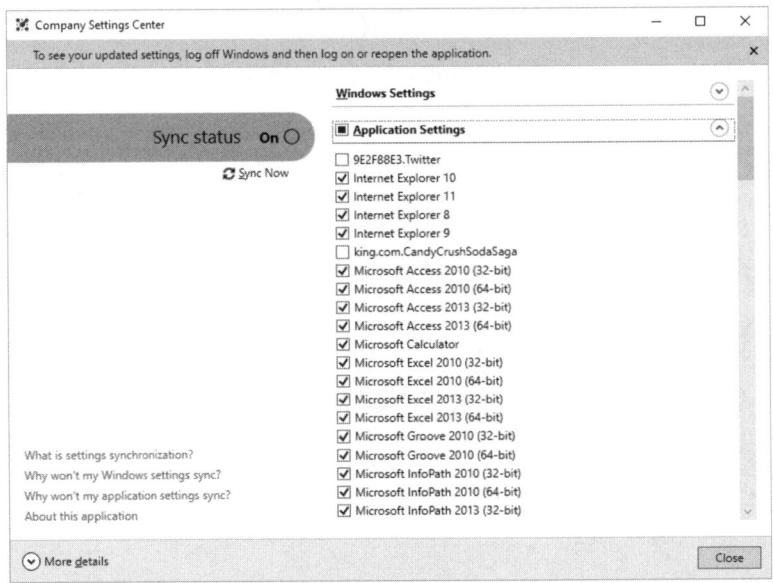

FIGURE 5-45 The UE-V agent requires a settings storage location path

7. On the Begin Installation page, click Install. You are required to restart your workstation following the install.

After restarting both computers and logging in for the first time, you should see a new icon in the system tray called Company Settings Center. This is the UE-V agent. Users can interact with the agent and customize the applications and windows settings that get synchronized, as shown in Figure 5-46.

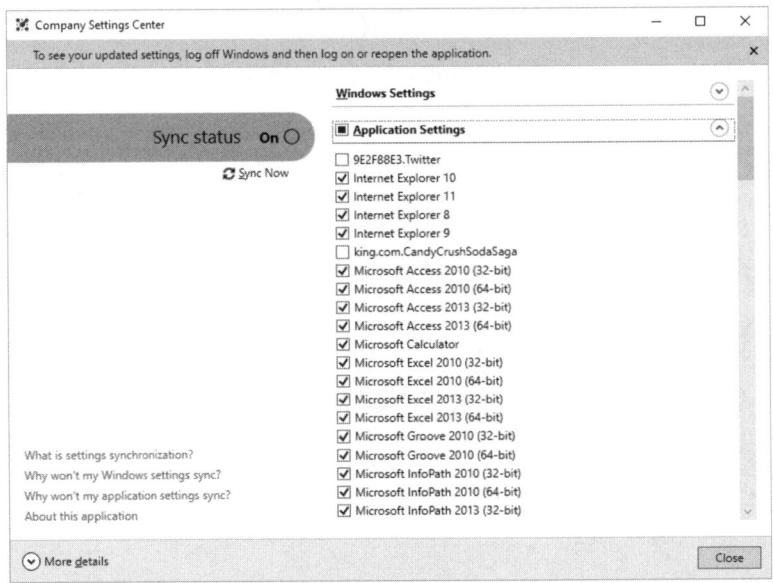

FIGURE 5-46 The UE-V desktop agent includes a few client-facing options

A sync of user settings can be manually initiated using the agent, as shown in Figure 5-45. The agent itself operates on its own using a feature called sync trigger events. These triggers tell the agent when to check for setting changes. In Table 5-5 you can see a list of the supported sync triggers when using the SyncProvider sync method.

TABLE 5-5 UE-V Sync Triggers

UE-V Sync Trigger Event	Action
Windows Logon	Application and Windows settings are imported to the local cache from the settings storage location. Asynchronous Windows settings are applied. Synchronous Windows settings are applied during the next Windows logon. Application settings are applied when the application starts.
Windows Logoff	Store changes locally and cache and copy asynchronous and synchronous Windows settings to the settings storage location server, if available.
Windows Connect (RDP) or Unlock	Synchronize any asynchronous Windows settings from settings storage location to local cache, if available. Apply cached Windows settings.
Windows Disconnect (RDP) or Lock	Store asynchronous Windows settings changes to the local cache. Synchronize any asynchronous Windows settings from the local cache to settings storage location, if available.
Application Start	Apply application settings from local cache as the application starts.
Application Exit	Store any application settings changes to the local cache and copy settings to settings storage location, if available.

Plan for desktop apps using Microsoft Intune

As a Microsoft Intune subscriber, you have the ability to package, upload, and distribute desktop applications to enrolled Intune devices. These tasks are accomplished through the Intune admin console and share many similarities with application management in Configuration Manager. These tasks include packaging the application, assigning requirements, using detection methods, and deploying to a collection of devices. Before you begin managing apps through Intune, here are a few planning considerations to be aware of:

- **Intune Software Publisher** The process of publishing an application within Microsoft Intune revolves around using the web-based Intune admin console. The first time you click to add an application, a dialog box appears prompting you to install the Intune Software Publisher, as shown in Figure 5-47. In addition to accepting the install, you need to be using Internet Explorer 11 or later, and have the full version of Microsoft .NET Framework 4.0 or later installed.

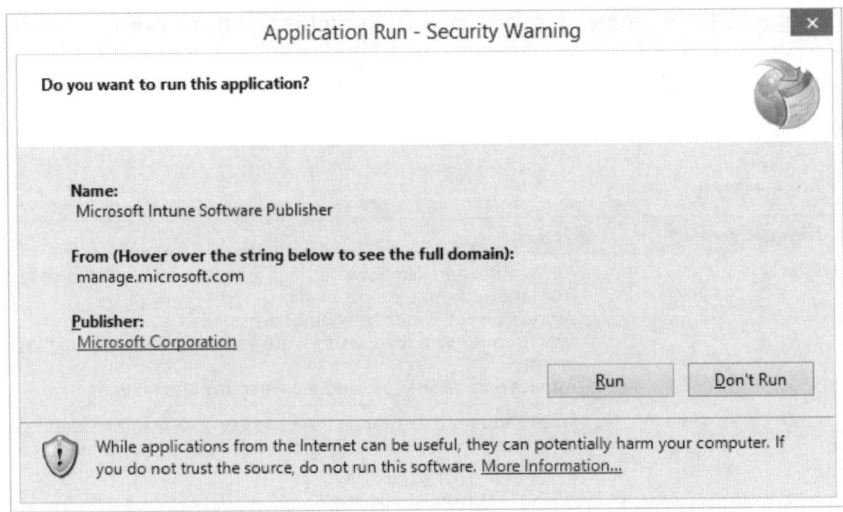

FIGURE 5-47 The Microsoft Intune Software Publisher raises a security warning

- **Cloud storage space** Microsoft Intune subscribers are allocated 20 gigabytes (GB) of cloud-based storage, with the option to purchase more space as needed. Any applications that you publish using the software installer option must be uploaded to your Intune cloud storage. You can manage your cloud storage usage through the Intune admin console by clicking the admin workspace, and clicking the storage use task, as shown in Figure 5-48. From this page, you can view the properties for any of the applications, delete an application, and purchase additional storage.

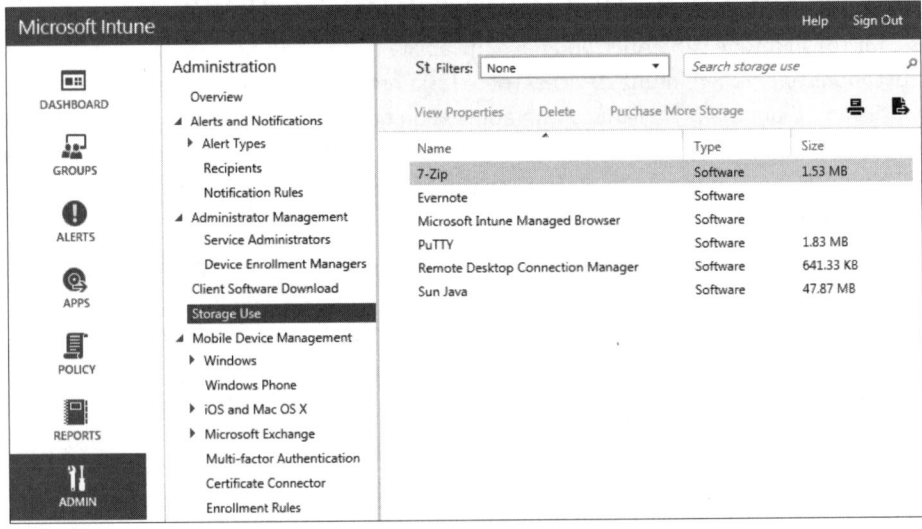

FIGURE 5-48 Intune cloud storage is used to store your applications

- **Source files** Application installers come in different sizes and flavors. Some installers are a single .msi file, while others include a directory structure that is required for the install to complete successfully. For each application that you publish to Intune using the software installer option, the source files must meet the following requirements:

  - All of the files associated with the application installer must be included in the up-load. This is achieved by having the files in the same directory and checking the box to include additional files and subfolders, which you can see shortly in an upcoming example.

  - Files that are uploaded cannot exceed 2 GB in size.

  - The uploader must have a minimum Internet speed of 768 kbps.

With these planning considerations in mind, take a closer look at the available file types that Intune supports for desktop applications. The supported desktop application file types and their associated requirements include the following:

- **Windows Installer (*.exe, *.msi)** The Windows Installer file type requires a silent in-stallation, typically achieved using a /q or /s switch as part of the installation command. Refer to the developer's documentation for the supported options. Windows Installers are uploaded to your Intune cloud storage for distribution. This file type is used for the distribution of most desktop applications.

- **Windows app package (*.appx, *.appxbundle)** The Windows app package file type requires an enterprise mobile code-signing certificate. For Windows 8 and Windows 8.1 devices, this must be a Symantec code signing certificate. For Windows 10 devices you can use a non-Symantec code signing certificate. Windows app packages are uploaded to your Intune cloud storage for distribution. This file type is utsed for the distribution of the Universal Windows Platform (UWP) line of business applications.

- **Windows Installer through MDM (*.msi)** The Windows Installer through MDM file type includes the following considerations:

  - You can only upload a single file using the .msi extension.

  - The file's product code and product version are used for app detection.

  - The default restart behavior of the app is used. Intune does not control this.

  - Per user, MSI packages are installed for a single user.

  - Per machine, MSI packages are installed for all users on the device.

  - Dual mode MSI packages currently only install for all users on the device.

  - App updates are supported when the MSI product code of each version is the same.

The Windows Installer through MDM file type is uploaded to your Intune cloud stor-age for distribution. This file type is used for the distribution of desktop applications to Windows 10 devices that were enrolled to Intune using auto-MDM.

Now let's dive into a real-world scenario using the Intune Software Publisher. In the fol-lowing example, you are going to take a copy of the Microsoft Remote Desktop Connection Manager and publish it using Microsoft Intune.

1. Sign-in to the Microsoft Intune admin console at *https://manage.microsoft.com/.*

2. Click the **Apps** workspace.

3. On the Apps page, click Add Apps. This launches the Intune Software Publisher. If this is your first time using the Intune Software Publisher, you are prompted to accept a security warning and install the web app.

4. On the Before You Begin page, click Next.

5. On the Software Setup page, under Select How This Software Is Made Available To Devices, select Software Installer as shown in Figure 5-49.

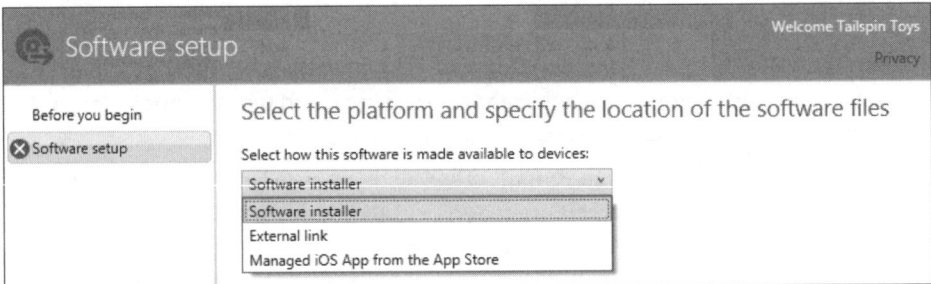

FIGURE 5-49 Software installers are used to deploy and install applications

6. By selecting Software Installer, you are presented with a new drop-down list that reveals the supported file types, as shown in Figure 5-50. The other options in the list point to outside sources, such as hosted web apps or app store URLs. Under Select The Software Installer File Type, select Windows Installer (*.exe, *.msi).

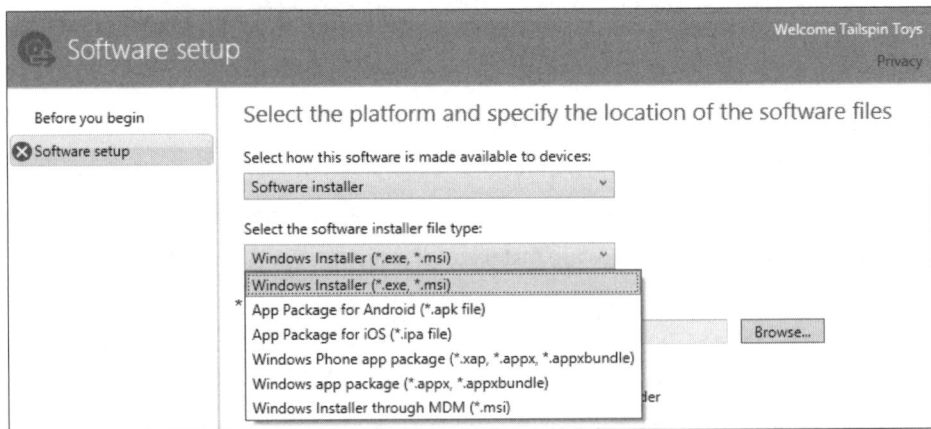

FIGURE 5-50 Intune supports various file types

7. Under Specify The Location Of The Software Setup Files, click the Browse button, locate the installer, and click Open. After the file path is validated, a list of remaining steps appears on the left, along with the option to click Next, as shown in Figure 5-51.

If your installer requires any additional source files or subdirectories, you need to select the Include Additional Files And Subfolders From The Same Folder check box. In the case of this example, you are using a standalone installer. Click Next.

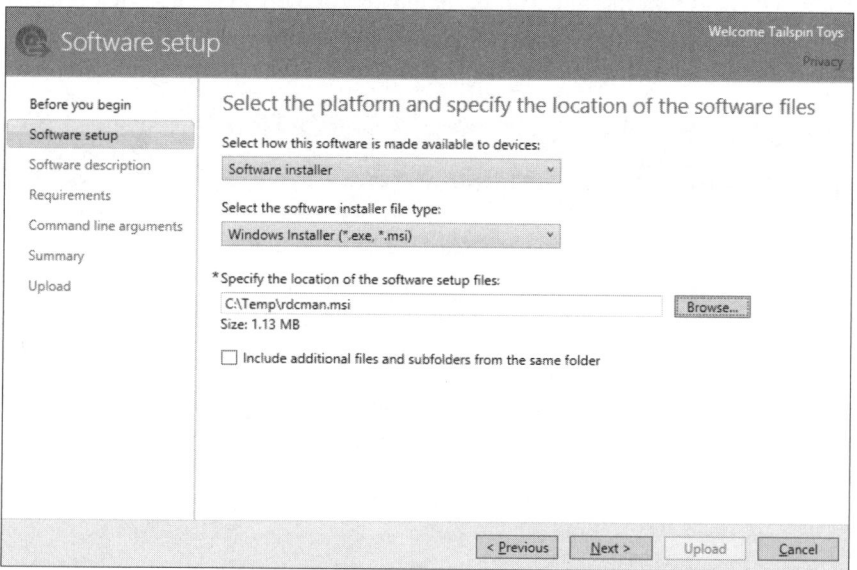

FIGURE 5-51 Intune requires all files associated with the install in the same folder structure

8. On the Software Description page, fill in all of the required fields (*), as shown in Figure 5-52. Click Next to continue.

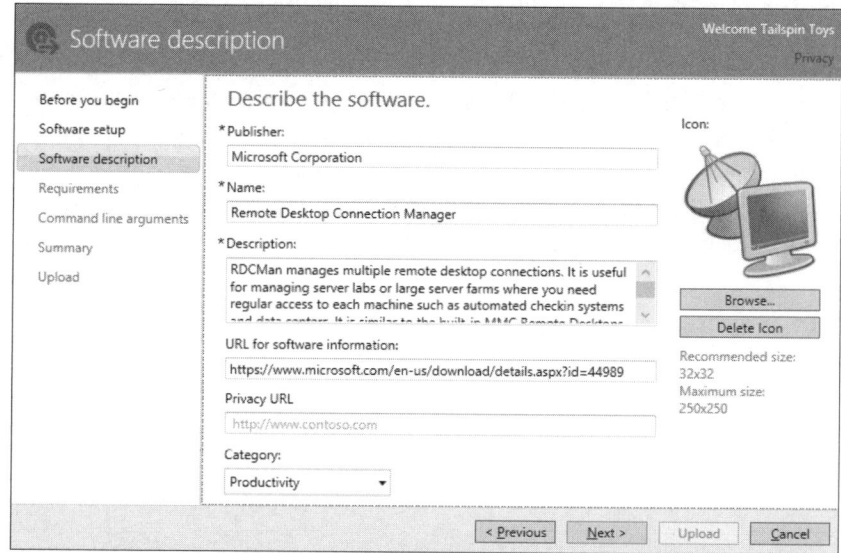

FIGURE 5-52 Intune Software Publisher gives you options to customize the icon and product information

9. On the Requirements page, select the required architecture and operating system for your application. In this example, your application is technically compatible with any of the configurations, but you must use the requirements page to narrow the scope of devices that this application can be installed on. In Figure 5-53, you have chosen the 64-bit architecture requirement, as well as narrowed the supported operating system from Windows 7 to all newer versions. Click Next to continue.

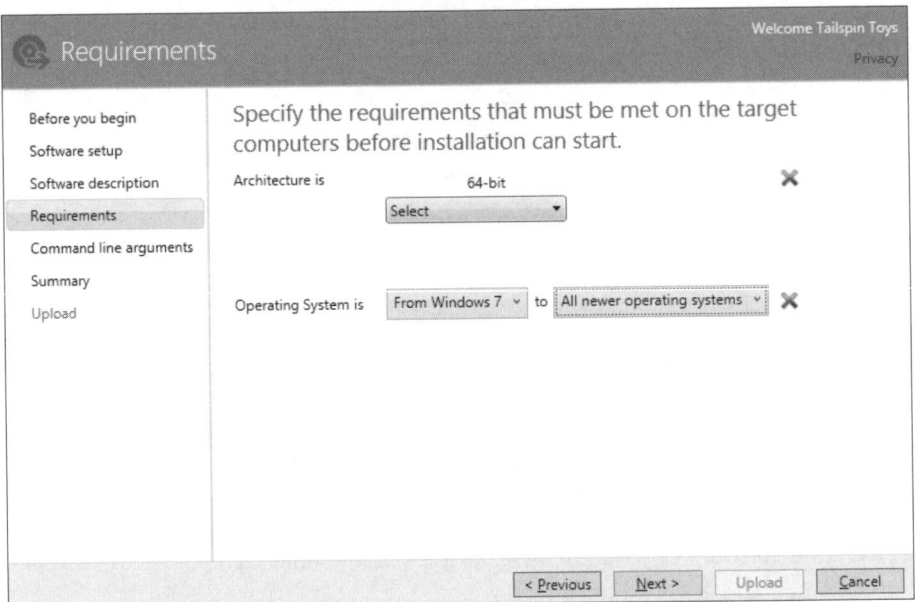

FIGURE 5-53 Application requirements are added in the Software Publisher

10. On the Command Line Arguments page, select Yes and type **/qn** in the Command Line Arguments box, as shown in Figure 5-54. Depending on the installer, there is typically a command line argument for initiating a silent install. Remember that this is a requirement for the Windows Installer file type, and it is on this page where those arguments are declared. In the case of this example, you are using a .msi, which recognizes /qn as quiet mode with no UI. Click Next.

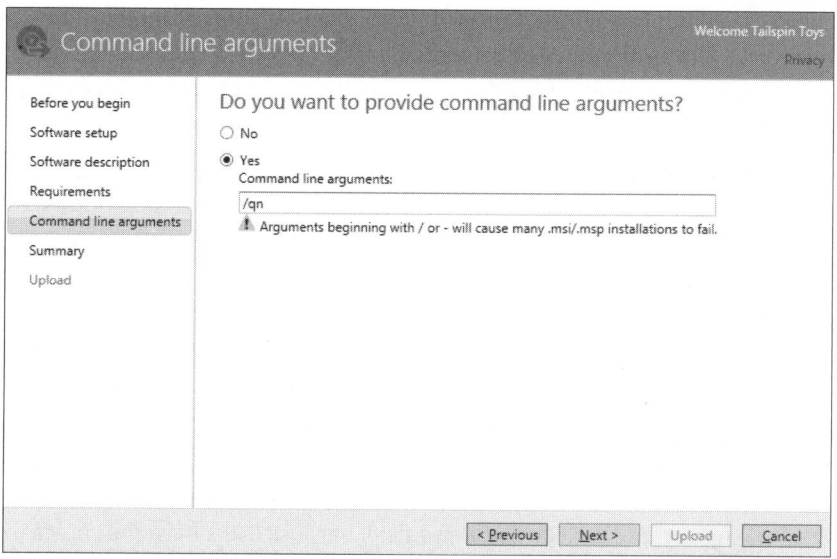

FIGURE 5-54 Applications accept command-line arguments

11. On the Summary page, confirm that all options and details have been entered correctly. Click Upload to proceed with uploading the source files to the Intune cloud storage. Note that this reinforces the requirement to upload content because this is the only available option to complete the application publishing process.

12. On the Upload page, confirm the upload completed successfully and click Close.

   Once uploaded, the application appears in the Apps workspace, as shown in Figure 5-55. From this location, you can review the current status of the application and the associated properties, as well as manage the deployments.

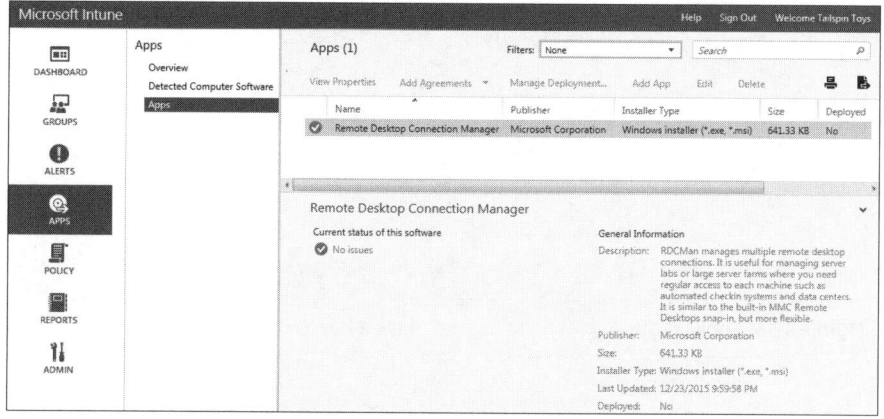

FIGURE 5-55 Applications created in Intune appear in the admin console

13. On the Apps page, select the desired app from the list of available applications, and click Manage Deployment from the menu bar. This opens the Deployment Wizard.

14. On the Select Groups page of the Deployment Wizard, identify the desired deployment groups and add them to the selected groups list. Click Next.

15. On the Deployment Action page, under the Deployment column, click the drop-down list and choose an option appropriate for your deployment. For the Windows Installer file type your options vary based on the group type that you selected on the previous page. The options are broken down in the following order:

   - **Device deployments** Deployments that target a group of devices are limited to the following options under the Deployment column: Required Install and Uninstall. Device deployments also have a scheduling component, which is coordinated by the option selected from the Deadline column.

   - **User deployments** Deployments that target a group of users are limited to the following option under the Deployment column: Available Install. Making an install available by design does not use a deadline, because users have the choice to install the application at their discretion.

16. For this example, you deploy the application to a group of users, and make that deployment available so that they can install the application at their leisure. These settings are shown in Figure 5-56. Click Finish to enable the deployment.

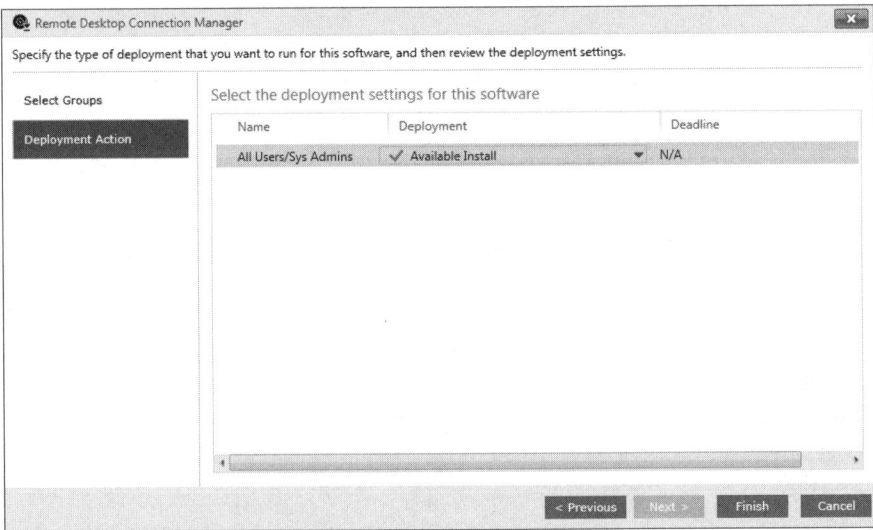

FIGURE 5-56 Applications deployed to users can be set as available

At this stage you should have a good understanding of the planning considerations for managing desktop applications with Microsoft Intune. For the exam, you should be familiar with the publishing process, the supported file types, and the requirements that each file type needs.

> **NEED MORE REVIEW?** **PLAN FOR APP DEPLOYMENT IN MICROSOFT INTUNE**
>
> To review more about the application management considerations for Microsoft Intune, visit *https://technet.microsoft.com/library/dn646955.aspx*.

Summary

- The ACT can be deployed in a single-server configuration, or expanded to accommodate larger environments.
- The ACT requires a SQL Server database for storing inventory.
- Client Hyper-V, Azure RemoteApp, and App-V are all tools that provide virtualized applications.
- RemoteApp and App-V both offer a steaming solution. App-V and Client Hyper-V both offer an offline solution.
- Client Hyper-V is not a feature that is enabled by default. Client computers must meet the necessary hardware requirements. The feature can be enabled through the GUI or through PowerShell.
- UE-V is an agent-based solution that addresses user roaming scenarios where application and Windows settings need to be synchronized. The agent supports several command-line parameters for easier distribution and configuration.

Thought experiment

In this thought experiment, demonstrate your skills and knowledge of the topics covered in this chapter. You can find the answer to this thought experiment in the next section.

You are a systems administrator for Contoso, Ltd, a marketing corporation in Los Angeles that provides web-based services and solutions for startup companies. Over the last six months the company has doubled in size, totaling 400 employees. The new employees were brought onboard to support customers on the East Coast. Prior to this unexpected growth, you and your team used System Center Configuration Manager 2012 to package, deploy, and support the company's application portfolio. This solution has been sufficient up until this point, but the added growth has put a lot of strain on your small team. You have identified the following areas that need improvement:

- Contoso's single data center in the Los Angeles area suffers from high latency due to the remote offices on the East Coast. This slows down data transfers with Configuration Manager, and makes application deployments take a long time to complete.

- Employees are asking to use their smartphones and tablets to access corporate resources.

- The help desk team is spending 10 hours every week uninstalling and reinstalling applications to resolve compatibility issues.

- There are two business-critical finance applications that need to authenticate with Active Directory.

You need to identify an application management solution that can address these issues, while reducing costs and limiting the reliance on a west coast datacenter. Choose the best solution that can address these points.

    A. An on-premises installation of RemoteApp

    B. A cloud-only configuration of Azure RemoteApp

    C. A hybrid configuration of Azure RemoteApp

    D. Client Hyper-V

    E. Microsoft Application Virtualization (App-V)

    F. Microsoft Intune

    G. Microsoft User Experience Virtualization (U-EV)

Thought experiment answer

This section contains the solution to the thought experiment.

The correct answer is C, a hybrid deployment of Azure RemoteApp. This solution provides Contoso with the flexibility of regional datacenters through the Microsoft cloud, which will help reduce existing latency issues. It also enables employees to connect to RemoteApp from mobile devices, offers a simplified application compatibility model, and provides Active Directory authentication to LOB applications.

CHAPTER 6

Plan updates and recovery

This chapter reviews the methods around planning for Windows Updates on computers and devices, including the recovering of computers in the event that it becomes unusable. File recovery is also an important aspect of managing clients and devices in an enterprise environment. Users may store files in a variety of places, some of which are critical to an organization. An appropriate file recovery strategy is key to retrain these important files. Finally, devices updates are increasingly hard to manage. With most organizations implementing a BYOD strategy, devices and mobile application updates are critical to maintaining functional and secure mobile. You should expect the certification exam to highlight all of these scenarios.

Skills covered in this chapter:

- Plan for system recovery
- Plan file recovery
- Plan device updates

Skill 6.1: Plan for system recovery

Planning for system recovery has many options, including USB drivers, restore points, and resetting the computer. Having a solid system recovery plan in place is important, because eventually any device can experience a hardware problem. This could be as simple as a fan that stops working, or something more critical such as a hard drive failure. Having the ability to recover from most any scenario will give you more agility and create less downtime in any environment. The best option to use depends on the scenario that is being presented, because each has pros and cons for use. This skill focuses on planning for different system recovery options and also on designing options for Windows 10 clients.

> **This section covers how to design for:**
> - Recovery drive
> - System restore
> - Refresh or recycle
> - Driver rollback
> - Restore points

Design for recovery drive

The use of a recovery drive is one aspect of planning for system recovery. Administrators routinely have help desk tickets or reports of a computer that does not boot, or has some type of corruption. Using enterprise tools to redeploy a computer with a corporate image is a good remediation step, depending upon the environment. Sometimes, using a recovery drive is helpful to minimize the downtime for the user. Recovery drives are useful when a user has several applications installed that are not deployed as part of a corporate image. Therefore, redeploying the image would cause additional overhead and recovery time. A recovery drive is a USB device that contains the settings and system files for a particular Windows 10 installation. Some OEM vendors include a dedicated recovery partition when using out of the box solutions. Figure 6-1 shows a computer with a Windows 10 operating system and an OEM recovery drive.

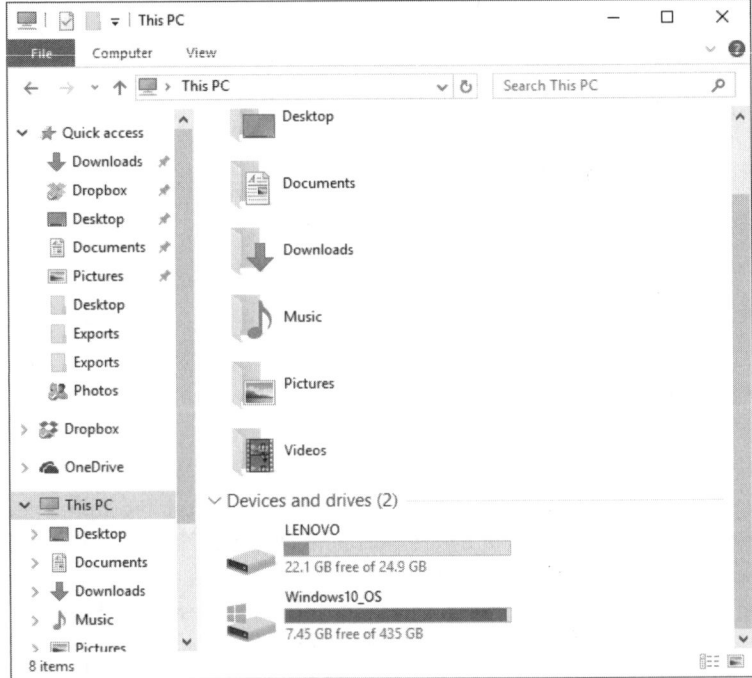

FIGURE 6-1 An OEM recovery drive

You can reclaim the disk space reserved on this partition by creating a USB recovery drive, which can be used to quickly restore a computer that won't start. In a situation where the computer won't start at all, often due to file corruption, a recovery drive helps you reinstall the Windows operating system without losing your settings or data. Designing a scenario for a recovery drive is not a complex matter. Because of this, operational and configuration scenarios might be present on the exam.

1. To create a recovery drive, launch the Create A Recovery Drive Wizard in Control Panel. Figure 6-2 shows the first page of the Recovery Drive Wizard.

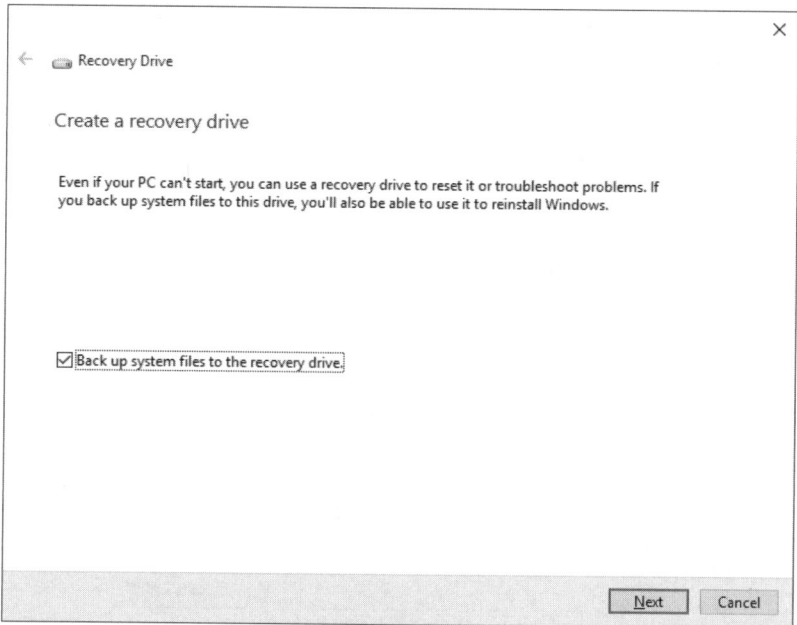

FIGURE 6-2 The Create A Recovery Drive Wizard

EXAM TIP

The Recovery Drive Wizard can also be started from the command line by using RecoveryDrive.exe.

Typically, it is a good idea to select the Back Up System Files To The Recovery Drive check box in the wizard. This ensures that the recovery drive is able to restore the system, even if some system files have been corrupted.

2. The wizard scans the system for available drives to be used as a recovery drive. If there is more than one drive available, select the drive from the list, and then click Next.

Note, that the wizard deletes the contents of the drive during the process because the drive is formatted as FAT32. The necessary size of the drive can vary depending on the size of the system, and can range from 4 gigabytes (GB) to 32 GB. Figure 6-3 shows the page with a list of available drives. It also tells you how much space is required on the drive.

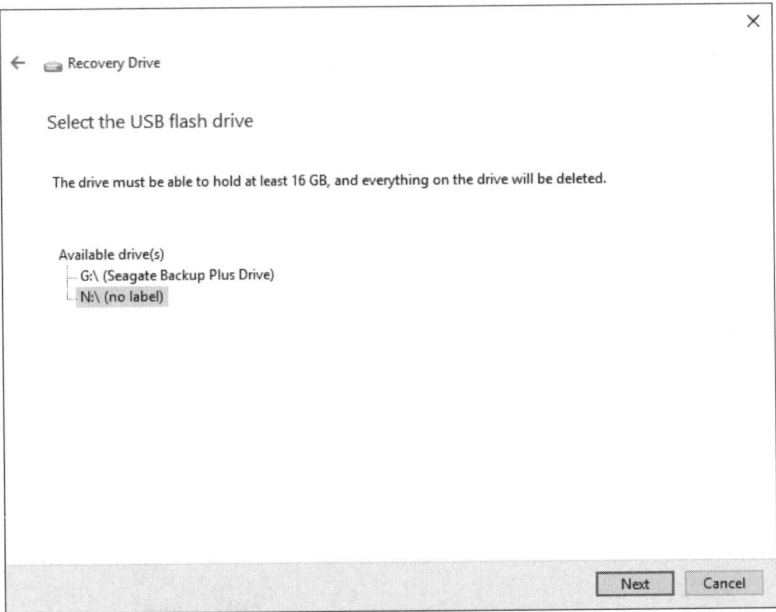

FIGURE 6-3 Select the USB flash drive

3. After selecting the drive, a warning message, shown in Figure 6-4, tells you that the contents of the drive will be deleted. You are reminded that you should back up any files you might need before proceeding. If you have backed up your files and you are ready to proceed, click Create.

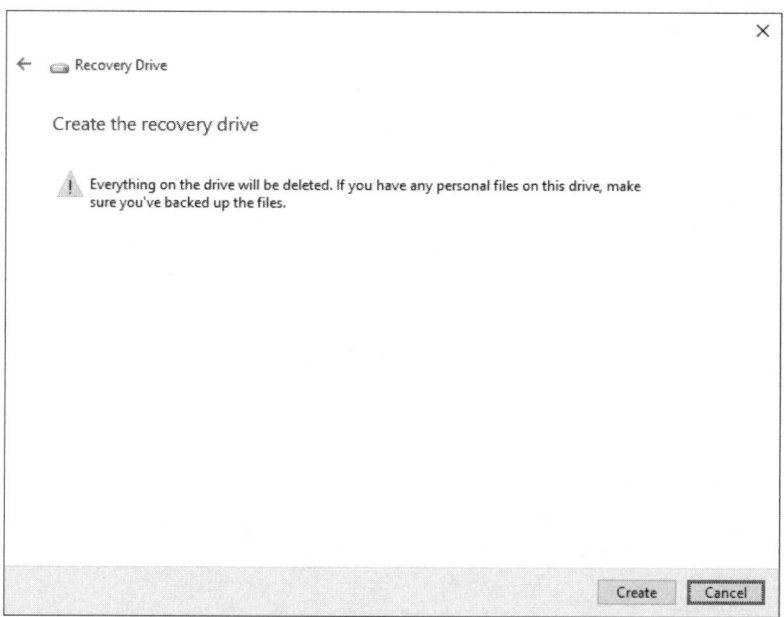

FIGURE 6-4 A warning that everything on the drive will be deleted

Design for System Restore

Another method of planning a system recovery is to use a system restore point. A restore point is typically useful when a computer starts but isn't working as expected. This could be from an application install or update, a change in device drivers, or a Windows Update.

System Restore is configured locally on systems from the System Protection menu of the System Properties. The system restore can be set for disk drives that the system is aware of. This includes before any application and Windows Update installations. Figure 6-5 shows the system protection configuration with multiple drives available.

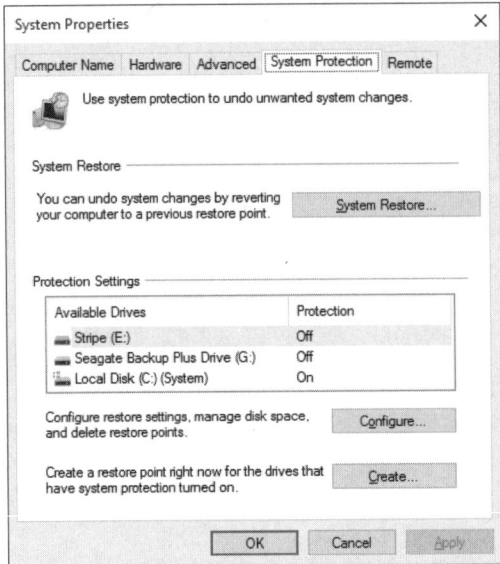

FIGURE 6-5 System protection settings

System Restore can be enabled and configured for each of the drives that are listed. In the example, protection is enabled on the local disk but not for the additional disks. For each drive that is enabled, the maximum amount of space that is to be used by System Restore can be configured. Figure 6-6 shows the slider bar that controls the percentage of the drive that is used.

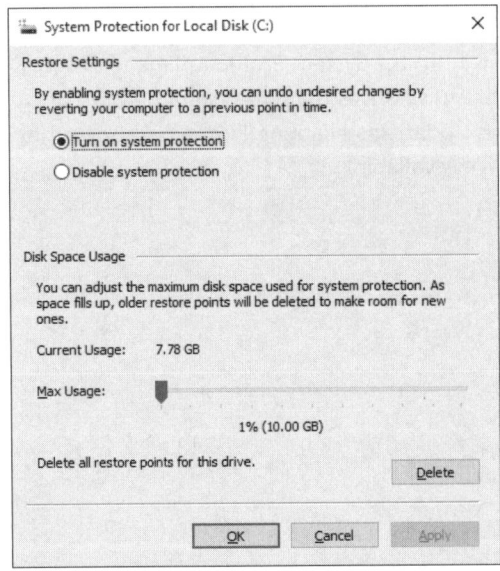

FIGURE 6-6 Configured System Restore settings

When enabled, System Restore automatically creates restore points at critical moments. These critical restore points always occur before new application installations, and before installing Windows Updates. If you plan on making a system change, it is recommended that you manually create a restore point before implementing the change. This can be accomplished by using the same System Protection menu of the System Properties. Creating a restore point manually is very simple, and only requires a description, as shown in Figure 6-7.

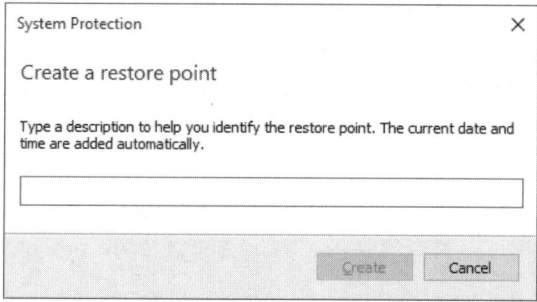

FIGURE 6-7 Manually creating a restore point

Finally, System Restore can be disabled by using a Group Policy Object (GPO). This is typically used to ensure that any problems that occur are handled appropriately by the IT department, instead of by the user. This could either be by standardizing corporate images, or providing a temporary computer while the other is being repaired.

The System Restore menu is under the System node of Administrative Templates. Figure 6-8 displays two available options in a GPO.

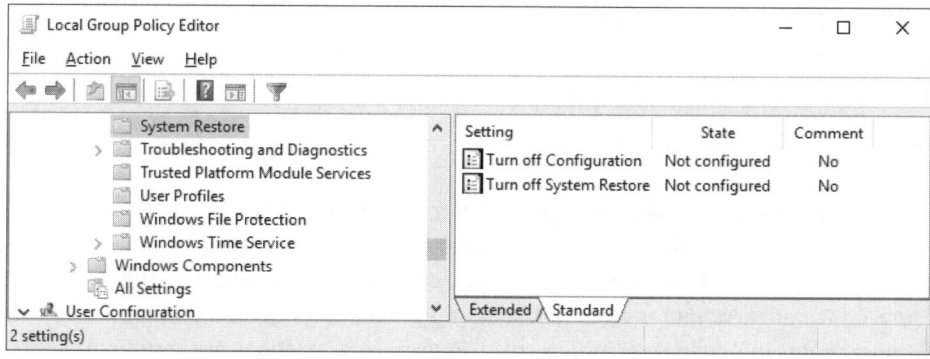

FIGURE 6-8 Available GPO option

The two available options are:

- **Turn Off Configuration**   When enabled, this setting controls whether or not the configuration can be edited by users. Note that a restore point can still be created manually. Users can also select a restore point to restore to.

- **Turn Off System Restore**   When enabled, this setting ceases all System Restore operations. Users will not be able to create, modify, or restore from a restore point. This setting overrides the Turn Off Configuration setting, even if it is configured.

Figure 6-9 displays the available options if the Turn Off Configuration GPO setting has been enabled. Note that the Configure button to modify the system protection settings is not available.

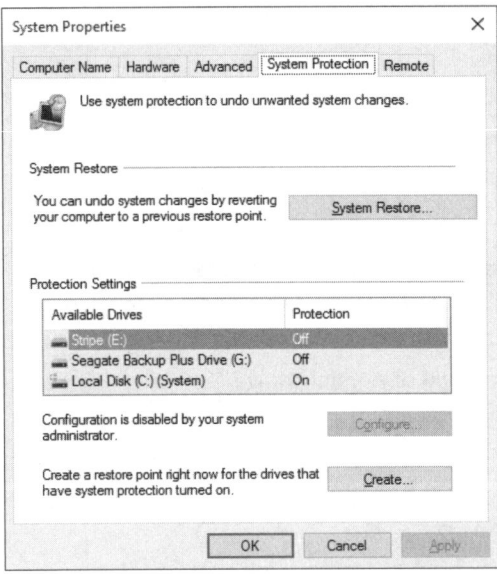

FIGURE 6-9 System protection options with a GPO enabled

There are many other options available to configure recovery and restore settings throughout the Computer Configuration Administrative Templates section of a GPO. The Computer Configuration portion is an important aspect, because the settings in the GPO must be applied towards computer objects. Configuring these settings, and then applying them to user accounts in Active Directory Domain Services (AD DS), does not result in the settings being configured appropriately. Table 6-1 describes the available GPO settings and their location.

TABLE 6-1 System recovery and restore GPO settings

Location	Setting	Description
System\Device Installation	Prevent Creation Of A System Restore Point During Device Activity That Would Normally Prompt Creation Of a Restore Point	This setting enables you to stop the automatic creation of a restore point when a driver change or installation is performed.
System\Recovery	Allow Restore Of System To Default State	This setting enables users to restore their computer back to a factory-installed image, or a created image. This function is available from the Advanced settings of the client recovery options.
Windows Components\Windows Installer	Turn Off Creation Of System Restore Checkpoints	This setting disables the use of restore points when installing an application. By default, a restore point is created before an application is installed. If enabled, a restore point would not be created.

✔ **Quick check**

- You just configured a GPO that is linked to all client computers in the domain. You enable the Turn Off System Restore setting. Which system protection settings can a user configure on the client with this enabled?

Quick check answer

- With the Turn Off System Restore setting enabled, none of the system protection settings are available to the client.

Design for refresh or recycle

This skill can be interpreted in one of the following two ways:

- **The built-in reset functionality of Windows 10**   This is used to recover a system that is having trouble and gives you the option to keep or remove all files. Because the skill title is "Plan for system recovery," you should focus primarily on this task.

- **The lifecycle management of devices in an enterprise**   This is the concept of refreshing and recycling PCs that are older, but still meet the minimum requirements of running Windows 10. You should be aware of this concept in case it appears in the exam.

The terms *reset* and *refresh* for recovering a computer were introduced with Windows 8. In Windows 10, this has been simplified to strictly use the term reset. Figure 6-10 shows the available options when resetting a computer with Windows 10.

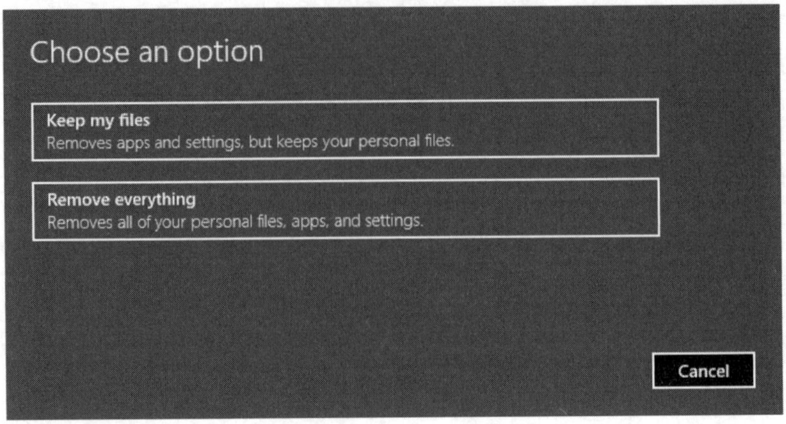

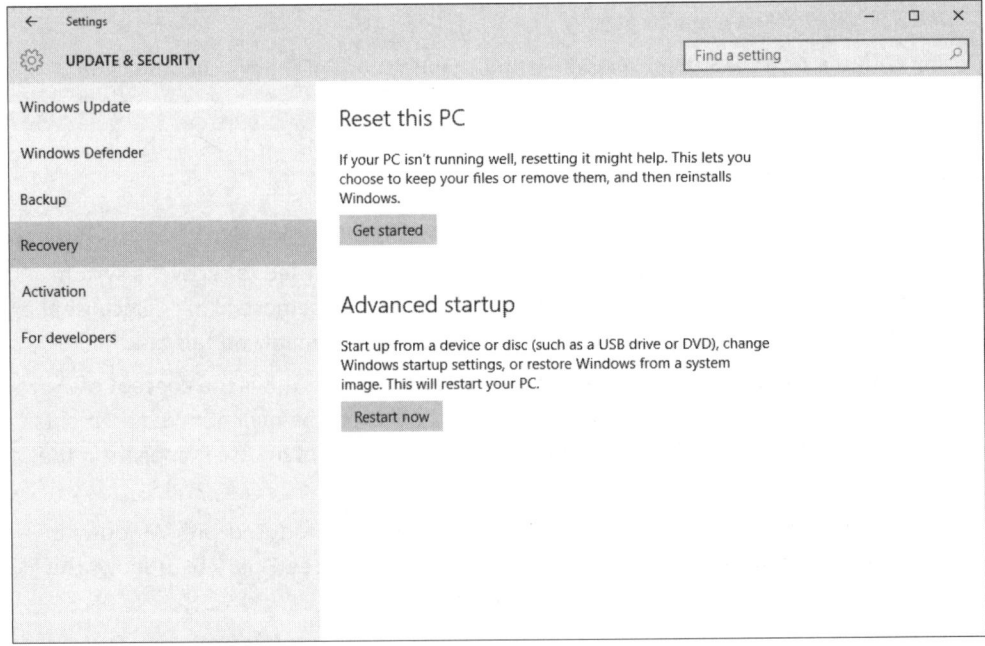

FIGURE 6-10 The computer reset options

When resetting a computer, you can either keep files while removing applications and settings, or you can remove everything. It is recommended that you back up any critical data before using either option for resetting a computer. With the reset option, a disc or operating system image is not required.

The Recovery option in the Settings menu also provides alternative methods of recovering a computer. Figure 6-11 shows the options to reset or perform advanced recovery options on a computer.

FIGURE 6-11 Recovery options

The advanced startup options include:

- **Booting from a USB device or disc**   This option enables you to boot from a CD/DVD, or a USB device. For example, you can launch a Windows To Go session from a USB drive.

- **Modifying startup settings**   This is the order of startup if multiple operating systems are installed.

- **Restoring from a system image**   This is necessary to recover from a system image, as discussed in the "Design for recovery drive" section of this chapter. You boot from the recovery drive that you created.

The other interpretation for refreshing or recycling computers is the lifecycle management of a computer. With Windows 10, the system requirements are generally identical to older operating systems. Therefore, you can reliably run Windows 10, even on most older hardware. To refresh, you can install Windows 10 over an older operating system as long as the computer supports it. Otherwise, you might need to recycle the computer and replace it with a computer that meets the recommended system requirements. Either method of refreshing or recycling can be daunting depending on the number of computers in the environment, and the administrative effort that is required to replace them.

EXAM TIP

Be aware of wording like "least administrative effort." This typically means an effort that requires the least amount of work for the administrative staff, even if it is not a best practice, or is not always cost-effective.

Design for driver rollback

Rolling back a driver might be necessary if a Windows Update, or manual driver update, causes a peripheral or component to act unexpectedly. For example, an administrator can install a network card driver update, which then causes a glitch with the network speed negotiation.

In this scenario, one available option for troubleshooting is to roll back the network card driver to a previous version. The current version and rollback options for device drivers are located in the Properties menu of the Device Management console. Figure 6-12 displays the driver details for a network card installed in a Windows 10 computer.

FIGURE 6-12 Driver details of a network card

This page provides you with the ability to see the current version number, and update the driver if you have manually downloaded a different version. To roll back a driver, click the Roll Back Driver button in the driver properties of the selected device. Figure 6-13 displays the Roll Back Driver confirmation.

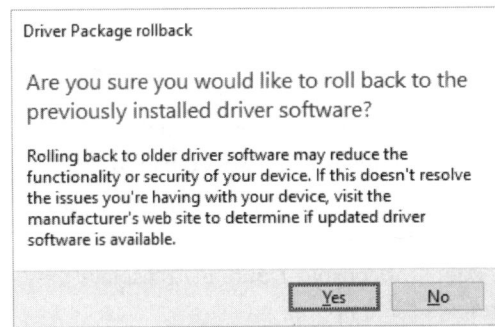

FIGURE 6-13 Roll back confirmation for a specific device driver

After a driver has been rolled back, the interface removes the option for clicking the Roll Back Driver button, and the date is modified to reflect the date of the older driver. The layout of this screen, with the Roll Back Driver button unavailable, is how a system acts when there is no previous version of a driver available to roll back to. Figure 6-14 displays the driver details after rolling back a driver.

FIGURE 6-14 Driver details after rollback

Design for restore points

Restore points are associated with the System Restore function that is discussed in the "Design for System Restore" section in this chapter. In addition to enabling and configuring the system protection settings, you can also delete restore points. This can be accomplished through the same menu that sets the storage limits.

Windows PowerShell can also manage System Restore and restore points, using the following PowerShell cmdlets:

- **Disable-ComputerRestore**   This cmdlet disables System Restore for a drive or drives that are specified.

- **Enable-ComputerRestore**   This cmdlet enables System Restore for a drive or drives that are specified.

- **Get-ComputerRestorePoint**   This cmdlet retrieves the available restore points on the system, displaying the reason why the restore point was created in the Description property. An example for a restore point might be: "before an application installation, or before Windows Updates were installed.

- **Restore-Computer**   This cmdlet restores the computer back to the specified restore point.

There are several recovery scenarios, and Figure 6-15 identifies these topics discussed in this section.

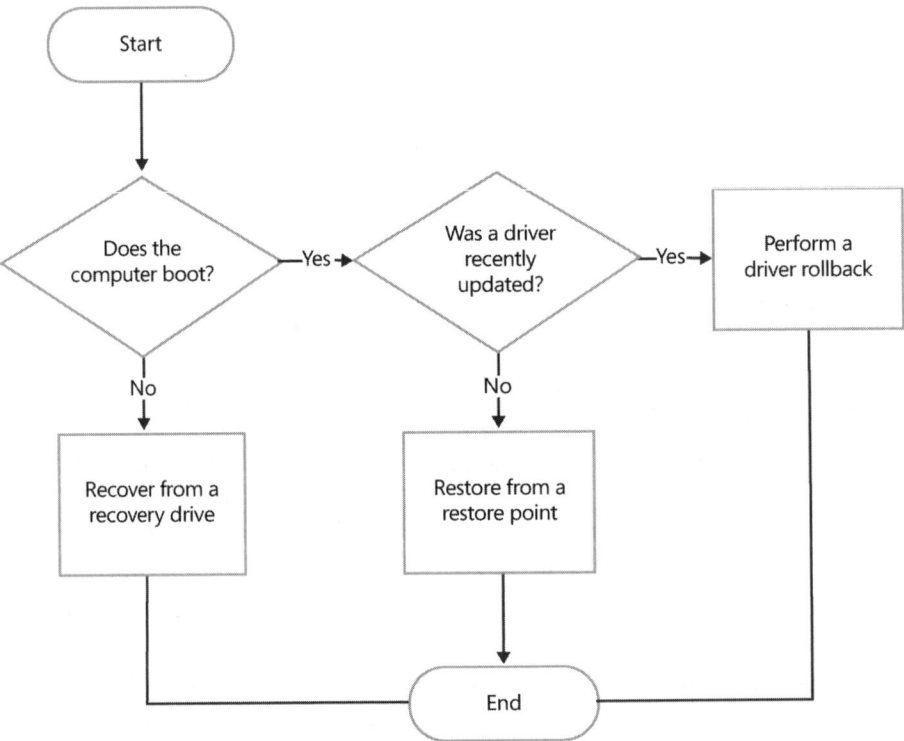

FIGURE 6-15 A decision tree that identifies the basic recovery scenarios

In summary, a recovery drive should be used only if the computer does not boot. Using a recovery drive when simply rolling back a driver or using a restore point increases administrative overhead and is not recommended. Likewise, using a restore point to attempt to resolve a driver issue is not recommended. Use the appropriate action to troubleshoot or resolve a suspected issue.

Summary

- Recovery drives are FAT32 formatted drives that range from 4 GB to 32 GB to restore a system that cannot boot.
- Restore points are created automatically before an application is installed or Windows Updates are applied.
- System protection can be controlled by using a GPO.
- Resetting a Windows 10 computer can be accomplished without a disc.
- If a device update causes stability or usability issues, rolling back a driver to a previous version can be used for troubleshooting or as a workaround.
- For the purposes of the exam, restore points and system restore are generally the same topic.
- PowerShell cmdlets for System Restore require administrative privileges.

Skill 6.2: Plan file recovery

There are several methods to enable file recovery for users. With client features like restore points, previous versions, File History, and OneDrive, users can recover old, deleted, or corrupted files without contacting the IT department of their organization. This reduces the administrative overhead for an organization, and doesn't require the use of a centralized storage system with regular backups. While centralized storage might be a better solution in large organizations, client-side recovery methods are reliable for mobile users and users that do not access file shares. Cloud storage also provides an easy method of enabling users to recover deleted documents, or in the case of Office 365, previous versions of documents. Both OneDrive and OneDrive for Business have a dedicated Recycle Bin that users can retrieve documents from.

> **This section covers how to:**
> - Design for previous versions of files and folders
> - Design File History
> - Recover files from OneDrive
> - Design for previous versions of files and folders

Previous versions provide an easy and effective method of restoring old versions of files and folders on a client operating system. Even in enterprise environments that have several file shares and centralized storage, users must still manage to save files locally on their computer. While previous versions are not managed or centralized, it does provide a way for users to access an older copy of a file or folder in case of corruption or accidental deletion.

Previous versions of files or folders are used in conjunction with restore points, or with File History. For a previous version of a file or folder to be available, a restore point must be available, or File History must be enabled. Restore points were discussed earlier in this chapter, and File History is discussed in the next section. It is important to note that if previous versions are used in conjunction with restore points, the previous version of a file is only the restore point copy. For example, if a Microsoft Word document is included in a restore point, and it has been modified several times since the restore point was taken, only the version at the time of the restore point is available. Restore points do not keep version by version histories of a file or folder.

If no previous version is available, due to either no restore point or File History has not been enabled, the Previous Versions tab of a file in the folder Properties dialog displays that there are no versions available, as shown in Figure 6-16.

FIGURE 6-16 No available previous versions

If a previous version of a folder is available, the date of the previous version is displayed, and the Open button is enabled. Figure 6-17 displays the Previous Versions tab with an older version available. In this example, the previous version is available from a restore point. When clicking the Open button, a browsing window appears that allows you to select a file or folder. The browser is a typical File Explorer window, which enables you to copy any existing file to restore back to the original location or to a new location.

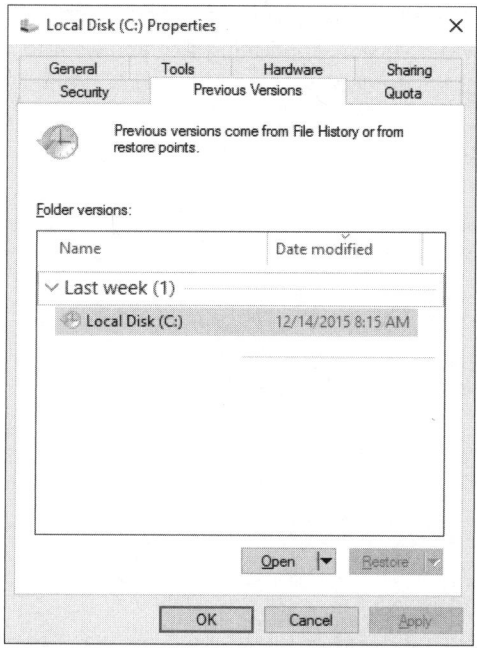

FIGURE 6-17 An available older version

When using previous versions with an individual file, the Restore button is also available. The Restore button automatically replaces the current file with the previous version that is available. The previous version of an individual file is displayed in Figure 6-18.

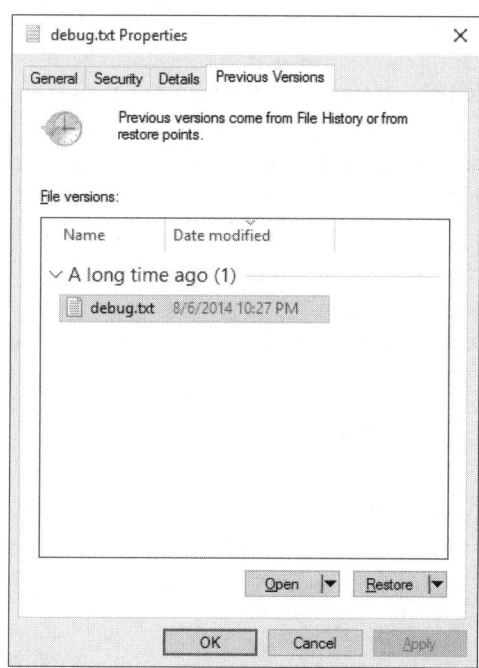

FIGURE 6-18 A previous version that is available for restore

When clicking the Restore button, the existing file is overwritten with the previous version. You are prompted whether to restore this previous version, confirming that you want to overwrite the version that is currently available. Figure 6-19 displays the confirmation prompt before restoring an individual file.

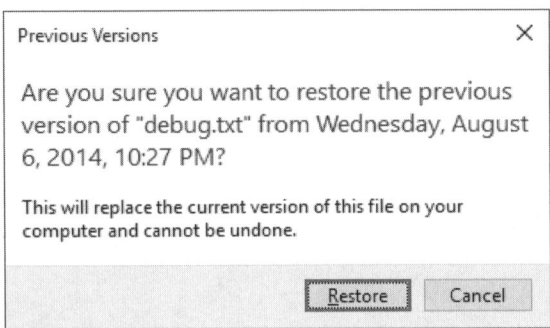

FIGURE 6-19 Confirming a previous version restore

EXAM TIP

The underlying metatable for previous versions is based on the filename of the file that you would like to recover. Therefore, if you rename a file, the previous versions of that file are unavailable from the Previous Versions tab. However, if you return the filename to the original name, the previous versions of the file are available.

Design File History

File History is a client-side utility that can be used to backup files and folders to either a local drive, or a network location. This is a useful tool, even in enterprise environments, to allow users to store files locally, and still have a centralized backup on a network share. This also enables the users to perform file recoveries on their own, without having to contact the IT department.

When enabled, File History by default backs up the libraries, desktop, contacts, and favorites of the user profile to the specified location. File History can also be used with local disks, including flash drives. Figure 6-20 displays the available option to configure File History with a flash drive.

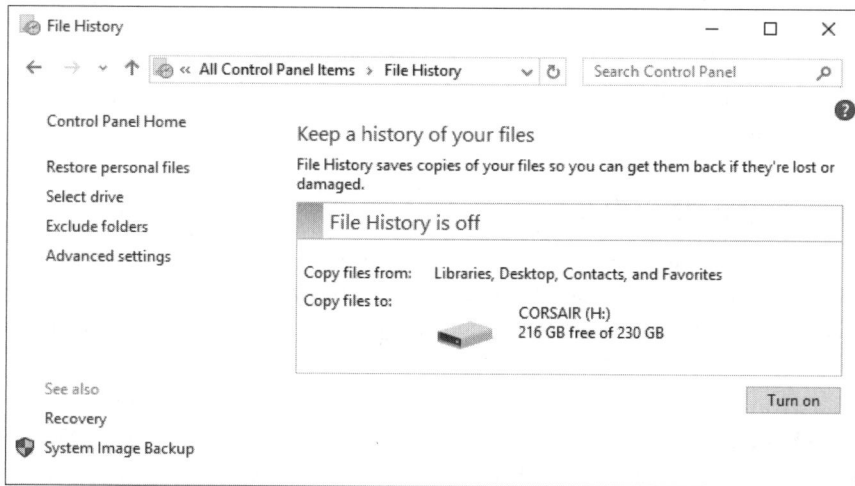

FIGURE 6-20 File History configuration

Enabling File History is as simple as clicking the Turn On button in the File History window. The default settings for File History is to save copies of files every hour, and to keep these multiple versions for forever. Figure 6-21 displays the Advanced Settings of File History, where this can be configured.

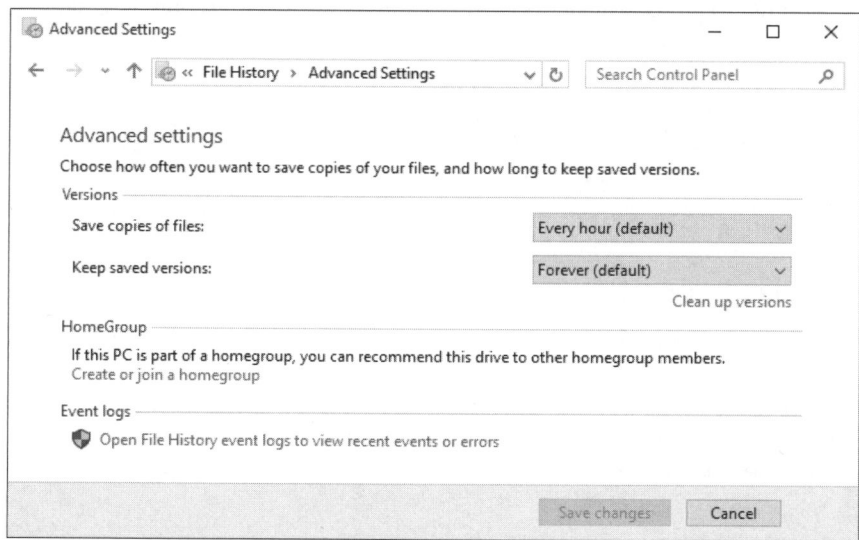

FIGURE 6-21 File History Advanced Settings

File History can be disabled by using a GPO. This is necessary if company policy requires users to communicate with the IT department to retrieve a backup of a file. However, this also means that all storage must be centralized to ensure that users can request backup copies. File History can be disabled in the Computer Configuration\Policies\Administrative Templates\

Windows Components\File History node of a GPO. Note that this is computer configuration, so the GPO that defines the settings must be applied to computer objects within AD DS.

Like other recovery options, the File History interface can also be launched by running **FileHistory.exe** from the command line.

Recover files from OneDrive

Microsoft OneDrive is a cloud storage solution with a built-in client for Windows devices, and mobile applications for third-party devices. OneDrive enables you to synchronize local files to the OneDrive cloud service. These files are then accessible from other devices, such as Windows Mobile devices or tablets.

There are two different versions of OneDrive: A consumer version available with all modern versions of the Windows client, and OneDrive for Business, which is part of an Office 365 subscription. The primary differences between the two versions are the capabilities of the cloud storage. From a recovery perspective, OneDrive for Business is required to restore previous versions of an individual file. Both versions enable you to restore deleted documents from the OneDrive portal.

To use OneDrive, you must have a Microsoft account. Microsoft account, compared to a Work or School account, is a free account to access Microsoft online services. A Work or School account is an organizational account that is typically associated with a paid subscription of Office 365 or other Microsoft online services. Creating a Microsoft account is as simple as creating a new email address, as shown in Figure 6-22. You can also use an existing email address or phone number to create a Microsoft account.

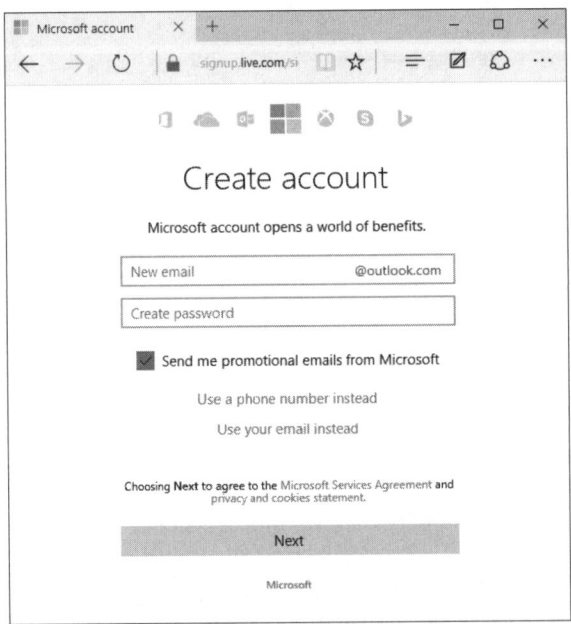

FIGURE 6-22 The Create Account page for a Microsoft Account

After creating a Microsoft account, you can configure the desired client to access the cloud storage. In Windows 10, you can click the OneDrive option in the quick access menu of the File Explorer. Figure 6-23 displays Welcome page for setting up OneDrive with Windows 10.

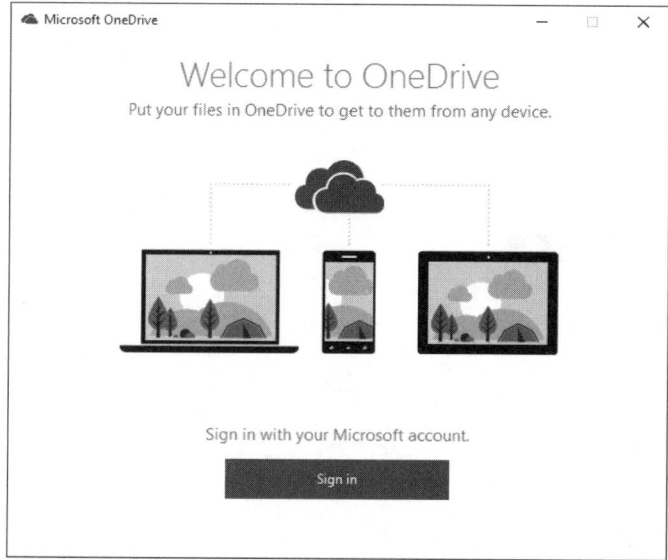

FIGURE 6-23 Welcome to OneDrive

Click the Sign In button to be prompted for credentials required for the Microsoft account. Figure 6-24 displays the Sign In page of the OneDrive client.

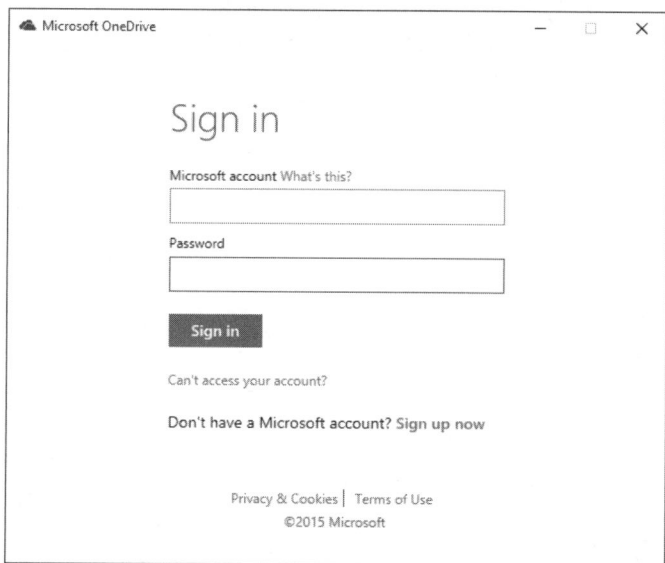

FIGURE 6-24 Signing in to OneDrive

After providing the credentials, OneDrive confirms the creation of the OneDrive folder. By default, the OneDrive folder is placed in the user's directory. This means, for User1, it is placed in C:\Users\User1\OneDrive. Alternatively, the OneDrive folder location can be configured to anywhere on the local computer, including an existing directory. Figure 6-25 displays the OneDrive folder configuration.

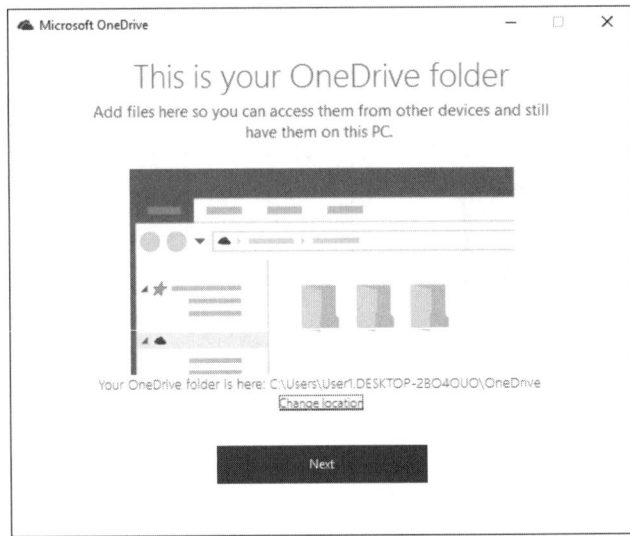

FIGURE 6-25 The default OneDrive folder creation

After specifying the location, you are prompted to confirm which directories and files you would like to upload to the cloud service. By default, the Documents and Pictures libraries are synchronized. Figure 6-26 displays the default OneDrive folder synchronization settings.

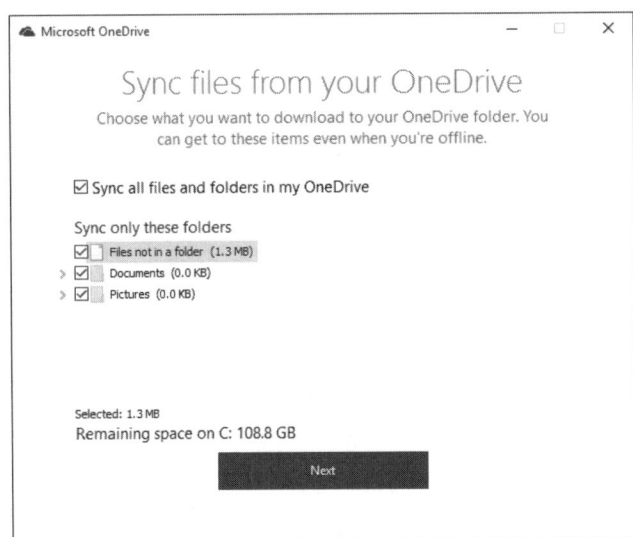

FIGURE 6-26 The default OneDrive folder synchronization settings

After selecting the folders that you would like to synchronize, the OneDrive client is ready for use. You can open the OneDrive folder and view the contents that are being synchronized. Figure 6-27 displays the final page of the OneDrive Configuration Wizard.

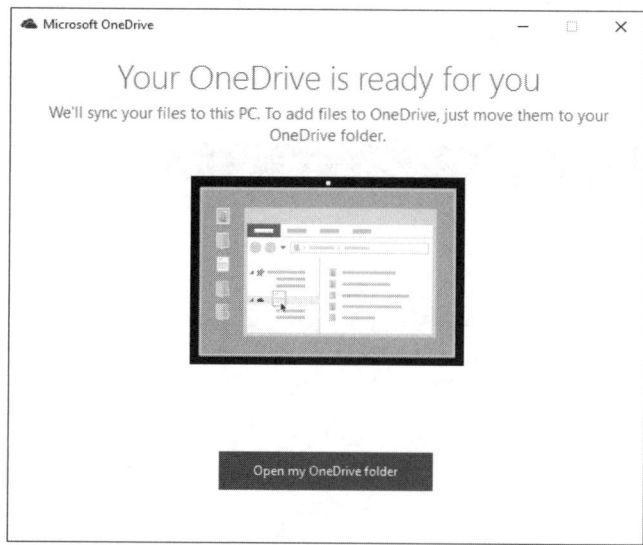

FIGURE 6-27 The completed OneDrive Configuration Wizard

After completing the setup of OneDrive, the files stored in any of the synchronized folders are sent to the OneDrive cloud. After a user has begun to use the OneDrive cloud storage, they can recover deleted documents from the OneDrive portal. The OneDrive portal is available at *https://onedrive.live.com*. Log on to the Portal, and then from the menu click Recycle Bin, as displayed in Figure 6-28.

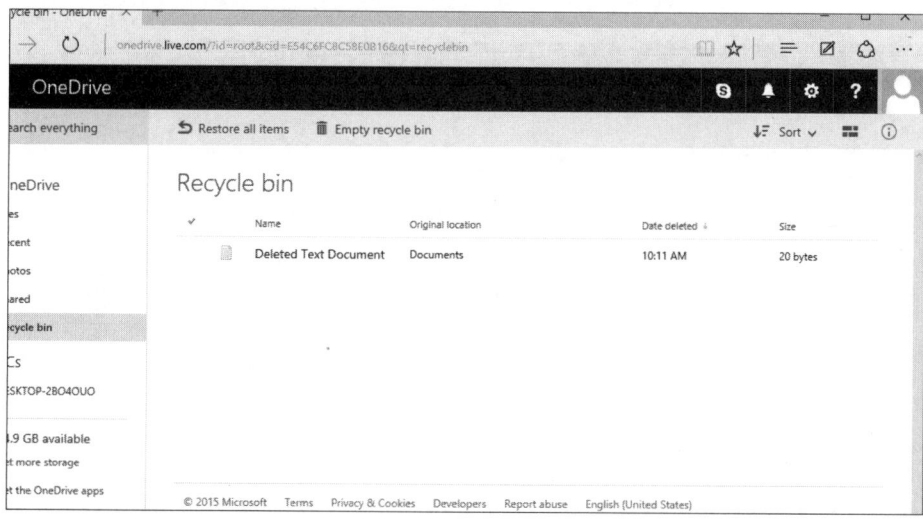

FIGURE 6-28 OneDrive portal

In this example, there is a text document that has been deleted from the client. The deleted document is available in the Recycle Bin. Note that if the document is deleted from a Windows 10 client, the document is also available in the local Recycle Bin on the desktop. To restore the deleted document from the OneDrive portal, select the document, and then click Restore. Figure 6-29 displays the selected document and the Restore button.

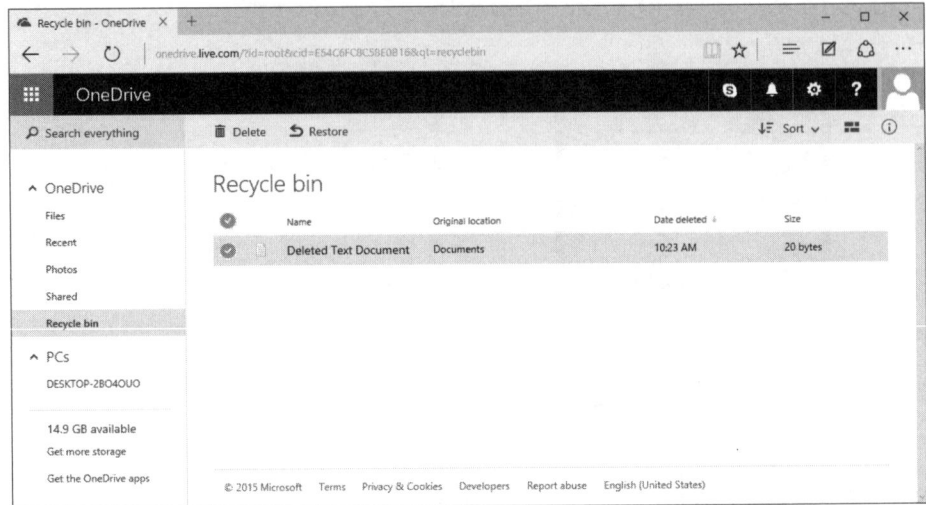

FIGURE 6-29 OneDrive Recycle Bin with a document selected

When restoring a document from the portal, it is restored to the original location on the OneDrive client. The OneDrive interface also displays the success of the restore in the browser, as shown in Figure 6-30.

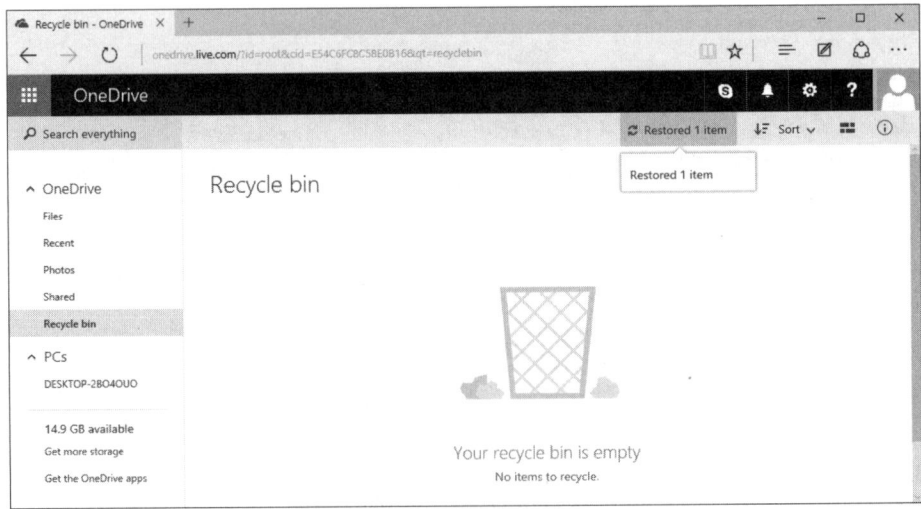

FIGURE 6-30 A successful document restore

If you are using OneDrive for Business with an Office 365 subscription, you can also restore previous versions of an individual file. The key difference in using OneDrive for Business, is that instead of using a Microsoft account as previously discussed, use the organizational account that has been assigned to you by the Office 365 administrator. Like restoring an individual file, this process is also accomplished by using the OneDrive portal. Simply select the document in the portal, click the Ellipses, and then click Version History. Figure 6-31 displays the version history for a document in the OneDrive portal.

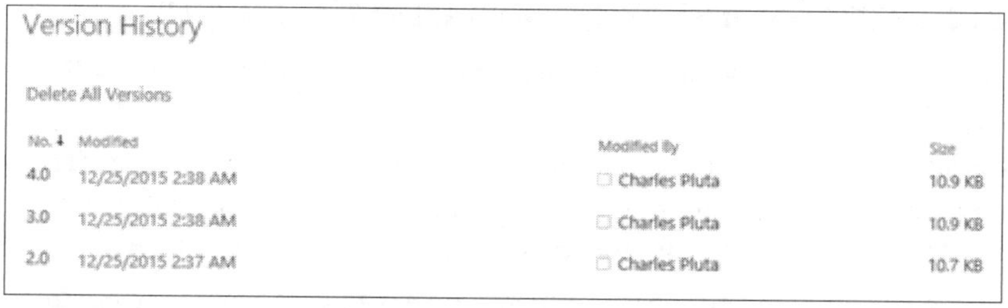

Version History

Delete All Versions

No. ↓	Modified	Modified By	Size
4.0	12/25/2015 2:38 AM	☐ Charles Pluta	10.9 KB
3.0	12/25/2015 2:38 AM	☐ Charles Pluta	10.9 KB
2.0	12/25/2015 2:37 AM	☐ Charles Pluta	10.7 KB

FIGURE 6-31 The version history of a document

Similar to File History, OneDrive can also be disabled by using a GPO. The GPO setting is figurelocated in Computer Configuration\Policies\Administrative Templates\Windows Components\OneDrive. Remember, because the policy is a computer configuration item, the GPO that contains the specified setting must be linked to computer objects. Linking the GPO to user objects results in the policy not having the desired effect.

Summary

- Previous versions require filenames to remain exactly the same, or the file will be unavailable to restore.
- Restoring an individual file from a previous version overwrites the current file.
- File History saves files to an external drive or network location by default every hour.
- File History and OneDrive can be disabled by using a GPO.
- Both versions of OneDrive provide a dedicated Recycle Bin.
- OneDrive for Business enables version history for individual files.

Skill 6.3: Plan device updates

Updating devices can be managed in a number of ways, depending on the environment that the device is being used in. For example, in a very small office, you can configure and trust each individual computer to receive updates directly from Microsoft. In a larger enterprise environment, you would use either Windows Server Update Services (WSUS) or System Center Configuration Manager to deploy and manage the updates. Finally, for mobile devices and other clients, a Windows Intune subscription is very useful. However, this subscription does not manage servers.

Design update settings and Windows Update policies

There are several methods of controlling update settings for Windows clients and devices. These methods include:

- **Local configuration**   These are the local Windows Update settings that can be configured on each individual computer or device. This is typically used in a workgroup environment, or for computers that are not joined to a domain for a specific reason.

- **Group Policy**   A GPO can be applied to computer objects to determine the update policy to apply to the computer. GPOs can be used with the regular update service, or with a Windows Server Update Services (WSUS) server to manage the update process for the computers.

- **Configuration Manager**   This is part of the System Center Suite that is available from Microsoft to manage several aspects of clients and servers, including updates. Windows Updates can be managed and deployed by using the Configuration Manager, as well as third-party updates with the System Center Updates Publisher (SCUP).

- **WSUS**   A WSUS server manages the update process for the computers that are configured to use the server in an environment. WSUS manages the download and installation process for all computers that report to it.

- **Microsoft Intune**   Intune can manage the update process for clients and devices, similar to WSUS. However, Intune is a subscription-based cloud service that requires the devices to be enrolled through the service portal. A major difference with Intune is that is cannot manage updates for a Windows Server operating system.

In this section, we discuss each of the available options to update clients and devices.

Local configuration

For Windows computers that are in a workgroup, a common configuration for Windows Updates is to use the local configuration options. By default, updates are scheduled to install automatically. The Windows Update page also gives you the option to install these updates at a later time, or to restart now to finish the installation. Figure 6-32 displays the Windows Update page for a Windows 10 client that has automatic updates enabled.

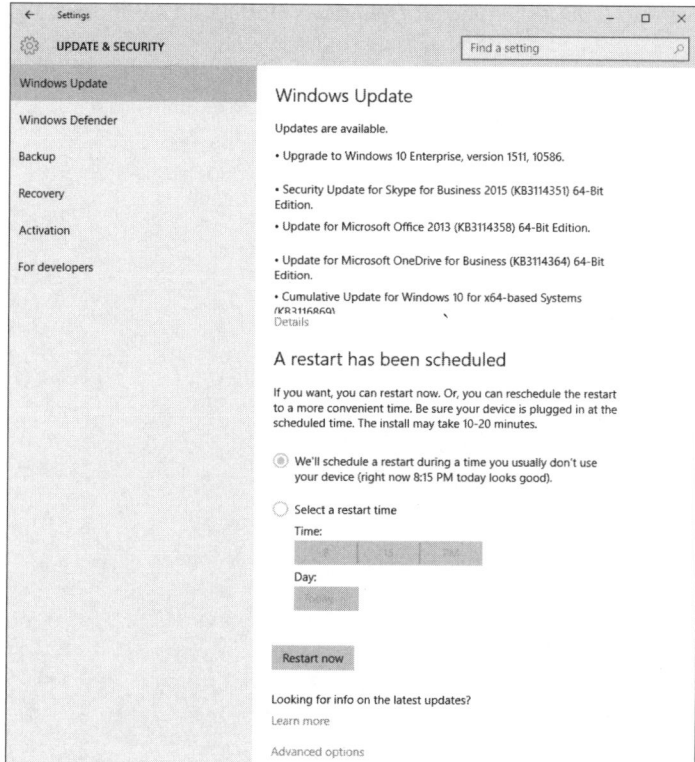

FIGURE 6-32 The Windows Update page

The Advanced Options of Windows Updates is how you can modify the update settings for the local client. This includes whether restarts are set automatically, or whether the user is notified to schedule a restart time for the updates to install. Beginning with Windows 10, the installation of updates are automatic. The only control the user has is the timing of the restart to finalize the installation.

Certain editions of Windows 10 can also defer upgrades. Upgrades are defined as new features that are added to Windows, but do not include security updates. The Advanced Options page is where upgrade deferment can be enabled on a Windows 10 client.

EXAM TIP

Windows 10 Professional, Enterprise, and Education editions can defer upgrades. Windows 10 Home, Mobile, and Internet of Things (IoT) editions cannot defer upgrades.

Insider builds are previews to future releases of the Windows client. This provides users with the latest versions of features, and the ability to provide feedback on those features. The Inside Builds are recommended only for enthusiasts, and are not recommended for production use.

Figure 6-33 displays the Advanced Options settings for Windows Update on a Windows 10 client.

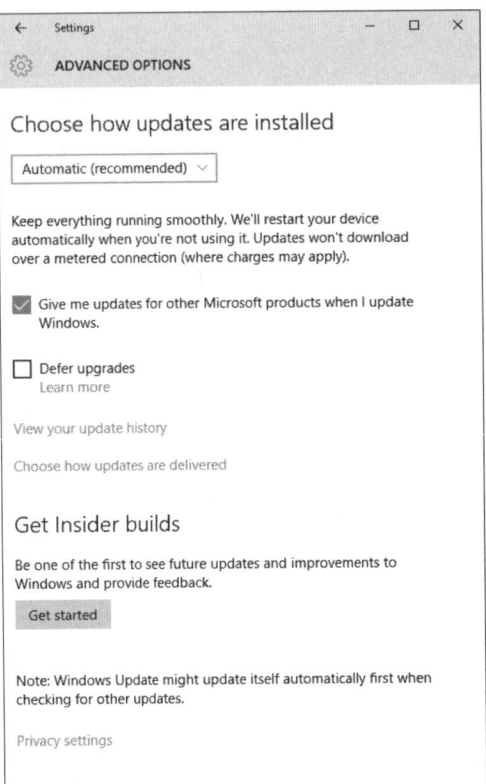

FIGURE 6-33 Windows Update Advanced Options

Group Policy

Group Policy is a common method of controlling the client settings for Windows Updates. There are many Group Policy settings that can be used to manage the client settings for updates. While there are several settings that are available to configure update settings through Group Policy, you'll look only at the settings that apply to modern operating systems. These are the settings included:

- **Always automatically restart at the scheduled time**   This setting is for Windows 8 or newer, and Windows Server 2012 or newer clients. If enabled, this setting forces a restart to finalize the Windows Updates for the client.
- **Configure Automatic Updates**   This setting is for any Windows modern client or server operating system, except Windows RT. This setting controls how the update settings are configured on the selected client. You can choose the update settings, as well as specify the scheduled installation time.

- **Specify intranet Microsoft update service location**   This setting is for any modern Windows client or server, except Windows RT. If enabled, this setting specifies the internal location of a WSUS server to receive updates from.

- **Defer Upgrade**   This setting is only for the Pro, Enterprise, or Education versions of Windows 10. If enabled, this setting enables you to delay upgrades from Microsoft. It is important to understand the difference between upgrades and updates, because a Windows 10 client automatically receives security updates, even if upgrades have been deferred. Upgrades are defined as new features that are added to the client functionality.

- **Automatic Updates detection frequency**   This setting is for any modern Windows client or server, except Windows RT. If enabled, this setting controls the interval number, in hours, that the client checks for updates. By default, this is configured to 22 hours, but can be configured anywhere from one to 22.

- **Do not connect to any Windows Update Internet locations**   This setting is for any modern Windows client or server. If enabled, this setting prevents the client or server from receiving information from the Microsoft Update services. This is possible even if an Intranet location has been specified. It is important to note that with this disabled, some public services such as the Windows Store might stop working.

- **Allow non-administrators to receive update notifications**   This setting is for any modern windows client or server, and is enabled by default. When enabled, this setting enables all users of the computer to receive automatic updates from the update source. If disabled, then only administrative users can receive updates.

- **Enable client-side targeting**   This setting applies to all Windows client and server operating systems, except Windows RT. This setting is used in addition to specifying an Intranet location for updates, and if enabled, specifies the WSUS target group the computer should receive updates from. This setting is only in effect if an Intranet location has been specified to receive updates from.

- **Allow signed updates from an intranet Microsoft update service location**   This setting is for all modern Windows client operating systems, except Windows RT. If enabled, this setting receives signed updates from an Intranet location through Windows Updates. The signed certificate must be placed in the Trusted Publishers certificate store of the local computer. If this is disabled or not configured, the updates must be signed by Microsoft.

WSUS

One of the most popular ways that updates can be deployed to clients and devices is by using the Windows Server Update Services server role. WSUS enables you to control the Windows Updates that are deployed to the devices that are being managed by WSUS. WSUS also provides additional features for managing these updates, such as reporting and compliance levels of the computers that are being managed.

Depending on the environment, multiple WSUS servers might be necessary. You can configure WSUS in one of two ways:

- **Obtain updates directly from Microsoft**   With this method, each individual WSUS server downloads the configured updates directly from the Microsoft Internet servers. This increases bandwidth usage on the networks that WSUS is deployed in. Figure 6-34 displays an organization's environment with a single WSUS server that provides updates to clients and servers.

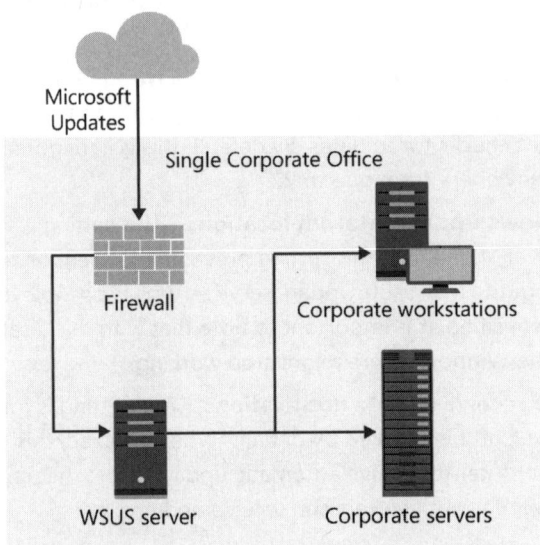

FIGURE 6-34 A single WSUS environment

- **Obtain updates from another WSUS server**   With this method, a single WSUS server must still obtain the updates directly from Microsoft. This server is typically named an upstream server. However, additional servers named downstream servers, download the updates directly from the internal location. This is useful for additional offices that might have a low bandwidth connection, or have a direct connection back to a primary datacenter. Figure 6-35 displays an organization's environment with multiple offices. The corporate office receives updates directly from Microsoft to an upstream WSUS server. A downstream WSUS server in the branch office then receives updates only from the upstream server. Both servers provide the approved updates to the clients and servers in their respective offices.

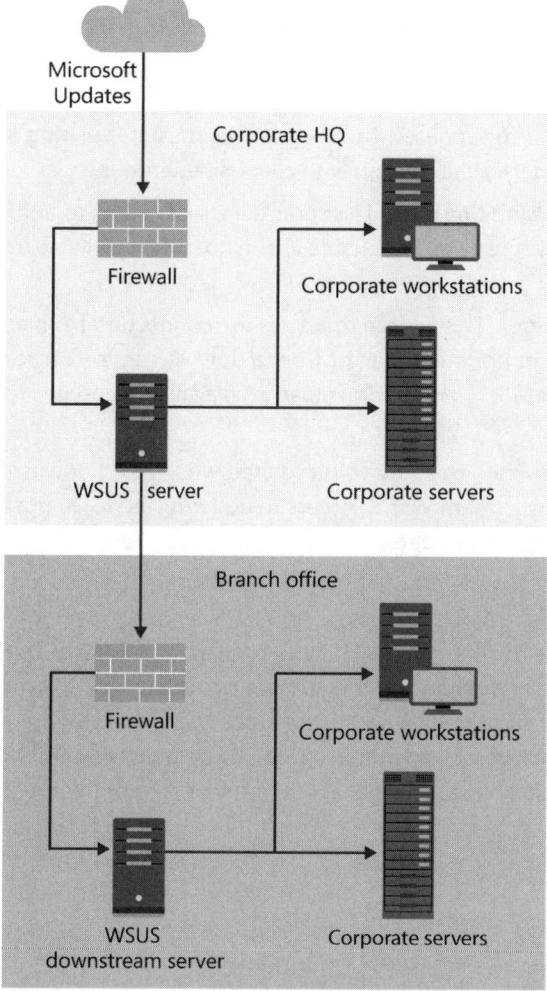

FIGURE 6-35 A WSUS environment with multiple servers

By default, the database that WSUS uses is a Windows Internal Database (WID). Alternatively, during the installation process, the WID can be deselected and support for a SQL server connection can be added. The WSUS services that are installed with the server role are:

- **Update Service** This is the primary service that is used to run WSUS and download the updates from Microsoft or an upstream server.

- **Reporting Web Service** This service powers the reporting capabilities of the WSUS server role, and must be running to obtain reports for the managed computers.

- **API Remoting Web Service**   For custom written reports, an API is available to connect to the WSUS server. This service must be running for the API to respond and for custom reports to run successfully.

- **Client Web Service**   This service responds to the client computers that have been configured to contact the server for updates. This web service must be running for clients to receive updates, or specify their group with client-side targeting.

- **Simple Web Authentication Web Service**   This service enables clients to authenticate with the WSUS server. This is required to enable clients to obtain updates from the WSUS server.

- **Server Synchronization Service**   This service manages the connection from an upstream server to the Microsoft update server, or from a downstream server to the upstream server. The synchronization service downloads the available updates to the WSUS server.

- **DSS Authentication Web Service**   This service is required when using downstream WSUS servers. This enables downstream WSUS servers to authenticate to an upstream WSUS server to receive the approved updates.

When installing WSUS, a design requirement is that the storage for the downloaded updates must be specified. This location can be locally on the WSUS server, or a remote UNC path. The destination storage must be formatted as NTFS, and have at least 6 GB of free disk space. The total required disk space depends on the number of updates and products that you plan to manage updates for. If the updates are stored locally, the updates are not downloaded until they have been approved by an administrator. By default, when updates are approved, they are downloaded for all languages. Figure 6-36 displays the configuration page for storing updates locally.

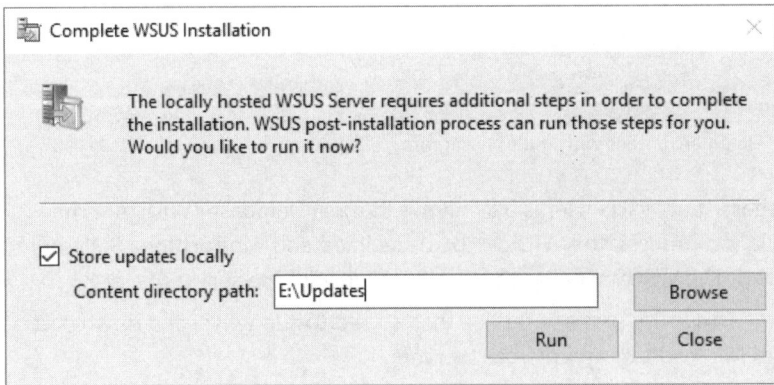

FIGURE 6-36 The Complete WSUS Installation dialog box

Once setup, a number of settings can be customized to fit the design requirements of an environment. Table 6-2 lists the options that are available for customization with WSUS.

TABLE 6-2 Configuration Manager components

Option	Description
Update Source and Proxy Server	This specifies where the WSUS server receives the updates from, and if a proxy server is required to access the source. Proxy servers are typically only required in highly secure environments, where the WSUS server would not have direct access to the Internet.
Products and Classifications	The products and classifications section specifies which updates are downloaded from the source. This could be by operating system, such as Windows 10 or Windows Server 2016, or other products such as Microsoft Office 2016.
Update Files and Languages	This setting enables you to specify the languages to download the updates in. By default, the language of the operating system is selected during setup. Additional languages can be downloaded automatically for each update.
Synchronization Schedule	The synchronization schedule can be set to either Manual or Automatic. If automatic is selected, you can specify the starting time and frequency (up to 24 times per day) that the WSUS server can contact the source for the latest updates.
Automatic Approvals	This option enables you to approve specific types of updates, such as security updates, for selected groups.
Computers	This option controls how computers are assigned to groups. By default, the Update Services console is used to assign computers. You can also configure the option to use Group Policy, or use the Windows registry.
Server Cleanup	This option removes old updates, computers, and update files from the repository on the server.
Reporting Rollup	This option configures how downstream servers are reported. By default, the downstream servers' status is rolled up.
EMail Notifications	This option enables you to configure WSUS to send an email when new updates or reports are available.
Personalization	This option configures how downstream server data is displayed. The To Do List items can also be customized to show only items you want to see.
WSUS Server Configuration Wizard	This is the initial wizard that can be launched again. The wizard enables you to modify most of these settings in a step-by-step manner.

There are also a number of reports that can be created and viewed from the WSUS server. These reports include update status, computer reports, and synchronization results. The list of default reports are displayed in Figure 6-37.

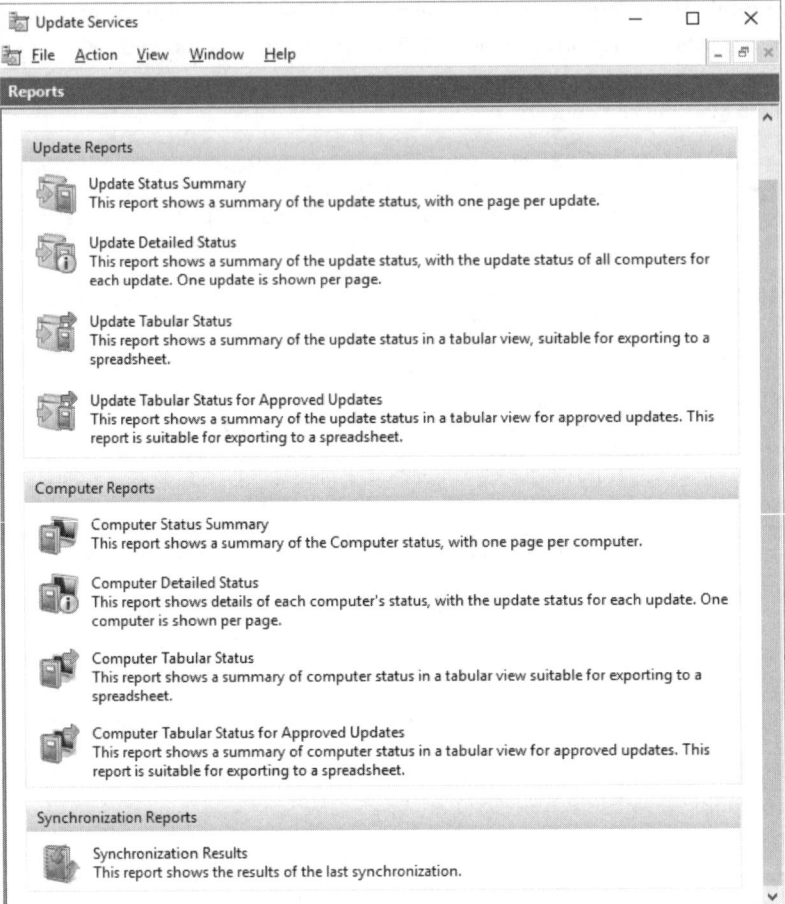

FIGURE 6-37 Available reports in WSUS

System Center Configuration Manager

Configuration Manager can also be used to manage the update process for Windows clients and servers. For Windows Updates, a WSUS server is still typically used. However, the WSUS server only provides metadata information, and the actual update deployment and installation process is managed by Configuration Manager. Table 6-3 displays the components and usage of each component when using Configuration Manager to manage the Windows Update process.

TABLE 6-3 Configuration Manager components

Component	Description
Software update point	A software update point is a WSUS server that clients can communicate with. However, with Configuration Manager, the clients do not download the updates from the WSUS server.
Deployment package	A deployment package contains the source files for the update. The managed client downloads the package and receive the included updates.
Software update groups	A software update group is used to deploy updates to devices, and verify the compliance status of the included devices.
Deployments	Deployments contain the information that is necessary to schedule and install the included updates. Deployments are created with the Automatic Deployment Rules, or can also be created manually.
Software update templates	Templates provide collections of pre-defined computers and methods of completing an update.
Maintenance window	The maintenance window defines when updates can be installed, and when the clients can be restarted.
Automatic Deployment Rule	Deployment rules define the time, updates, and other settings for completing a deployment. There are automatic rules that can be created, or you can create rules manually.

Understanding each of the component roles can be critical in understanding how to design a Windows Update deployment. Ask yourself, how many deployment packages do you need for your environment? Should each deployment package have its own deployment rule? How does the maintenance window impact how you plan to deploy updates? All of these questions, and more, impact how update deployments can be designed.

Configuration Manager can also be used with System Center Updates Publisher (SCUP) to deploy third-party updates. SCUP enables you to publish the required deployment information to a WSUS server, which in turn allows Configuration Manager to manage the update lifecycle.

To use SCUP with third-party updates, you must import the update catalogs for the vendor that you plan to perform updates for. There are two types of files that are included with catalogs:

- **CAB files**   These are the actual catalog of updates that are available for the application.
- **XML files**   This is a hash that can be compared to the CAB file to verify the integrity of the CAB file. The XML file is not required, and is not provided by all software vendors.

After you have obtained the files from the vendor, you can import them into SCUP. SCUP prompts you to verify that the catalog is valid, and present the products that are included in the catalog. Always follow the vendor's required method or order of installation. For example, some software vendors might require that updates be installed in a specific order for the software to operate properly. In this case, you must use the prerequisite rules to ensure that the updates are installed in the proper order. If an update does not have any prerequisite rules in place, it can be deployed as is. Additionally, if the update does not have an Installed Rule in place, it is not deployed to the client.

Microsoft Intune

Microsoft Intune provides an easy method of configuring updates to Windows clients and devices that are being managed by a Microsoft Intune subscription. However, Microsoft Intune cannot manage updates for Windows Server operating systems. Therefore, using Microsoft Intune to manage the update process is typically used in small environments, or together with System Center tools for enterprise environments.

To manage updates for Windows clients and devices with Microsoft Intune, you can customize one of two policy templates:

- General Configuration (Windows 10 Desktop And Mobile And Later)
- Microsoft Intune Agent Settings

The General Configuration template has options to configure automatic updates on clients that the policy is applied to. The two options that can be configured are:

- Allow Automatic Updates. When enabled, the available options are:
  - Notify download
  - Auto install at maintenance time
  - Auto install and reboot at maintenance time (default)
  - Auto install and reboot at scheduled time
- Allow Pre-Release Features. When enabled, the available options are:
  - Not Allowed (default)
  - Yes - Setting Only
  - Yes - Settings and Experimentations

Figure 6-38 displays the available settings in the Microsoft Intune console.

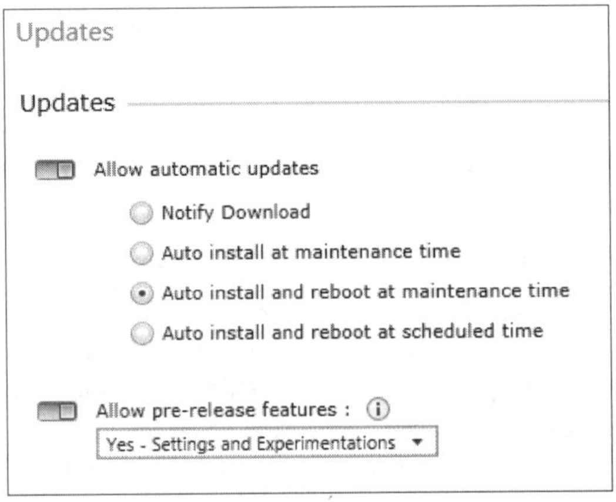

FIGURE 6-38 Update policies in Microsoft Intune

The Microsoft Intune Agent Settings template controls the settings that you can configure locally on the client. These settings include:

- Detection frequency
- Restart schedule
- Delay between restart prompts

Figure 6-39 displays the full list of available settings that can be configured for Windows Updates through Microsoft Intune.

FIGURE 6-39 Update policies for Microsoft Intune agent settings.

Manage update history

The update history for a computer can be managed in similar ways that you manage applying the updates. For example, on a computer that you are managing the updates locally, the same computer can view the history of installed updates. This is also true for computers that have updates that are managed by WSUS, Microsoft Intune, or Configuration Manager. Through the WSUS console, you can decline or remove updates that have been previously approved, and remove them from the targeted computers.

For computers that are being managed locally, the advanced settings of the Updates page enables you to view the update history. Figure 6-40 displays the update history for a Windows 10 client.

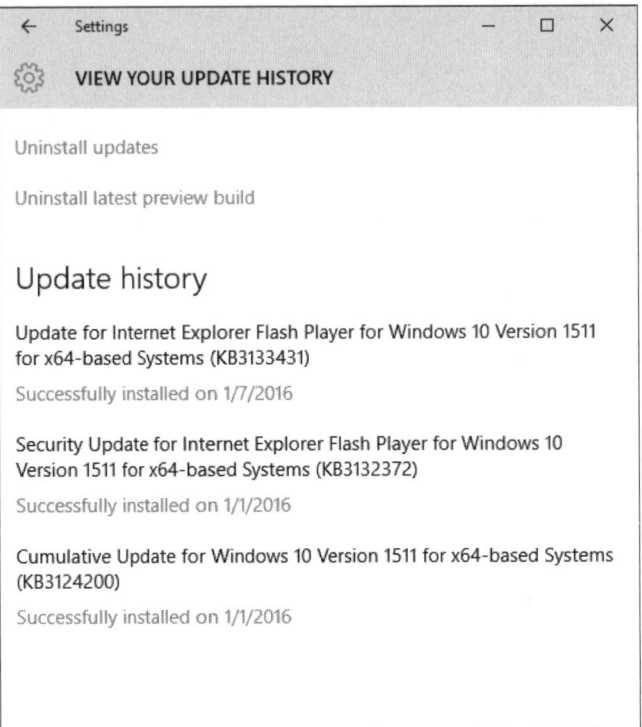

FIGURE 6-40 Update history for a Windows 10 client

On the View Your Update History page, you can also uninstall updates that have been installed. Additionally, if the client is a member of the Windows Insider program, you can also uninstall the latest preview build that was installed on the client. This is useful if an update or build has unexpected results with the applications or hardware that you are using.

Roll back updates

To uninstall an update that was applied locally, click the Uninstall Updates link on the View Your Update History page. The Installed Updates list appears, displaying the full list of updates that have been installed. Simply select the desired update to remove, and click Uninstall. Figure 6-41 shows the list of installed updates with a Microsoft Office 2013 update selected to be uninstalled.

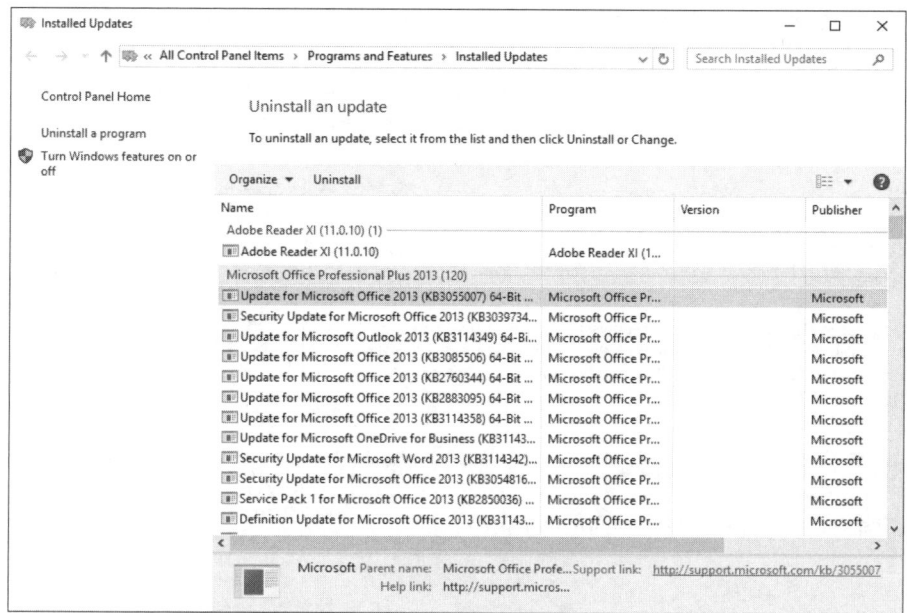

FIGURE 6-41 Uninstalling an update

After updates have been approved through WSUS, they are available to install on the clients. If you need to remove the update, simply decline it. Figure 6-42 displays the interface for declining an update.

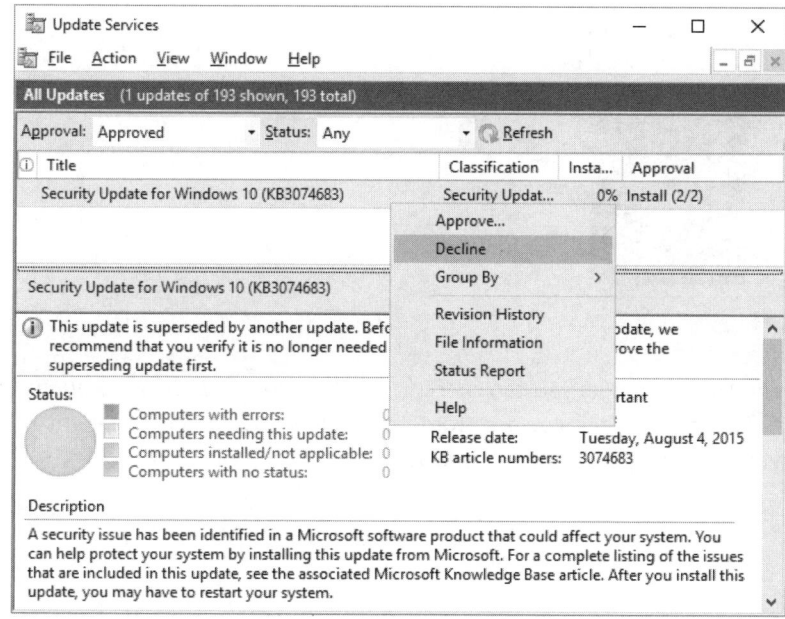

FIGURE 6-42 Declining an update on the WSUS console

When declining an update that has been previously approved, you are prompted to confirm that you want to remove the update. Since the update was previously approved, it is removed from view of the targeted clients, and any events or reports for the specified update are removed from the WSUS database. Figure 6-43 displays the confirmation prompt.

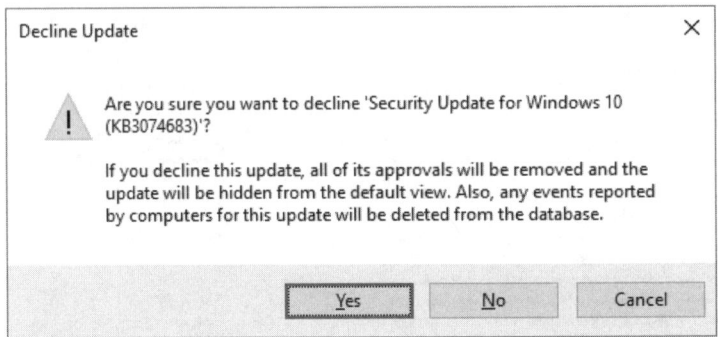

FIGURE 6-43 The WSUS decline update confirmation prompt

Design for Windows Store app updates

After Windows Store applications have been installed, they can also be updated from the Windows Store. To check for updates for apps, launch the Store app and click Downloads and Updates from the menu. Figure 6-44 displays the menu from the Store app.

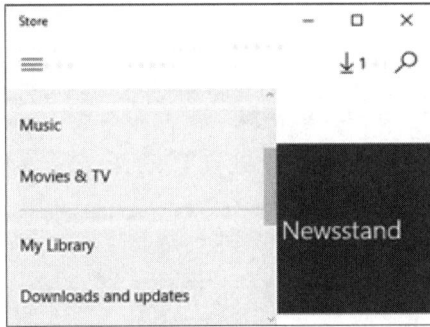

FIGURE 6-44 Windows Store app menu

On the Downloads and Updates page, you can check for updates for installed applications. If any applications have available updates, you are prompted to download and install them. Figure 6-45 displays an available update for the Windows Calculator on Windows 10.

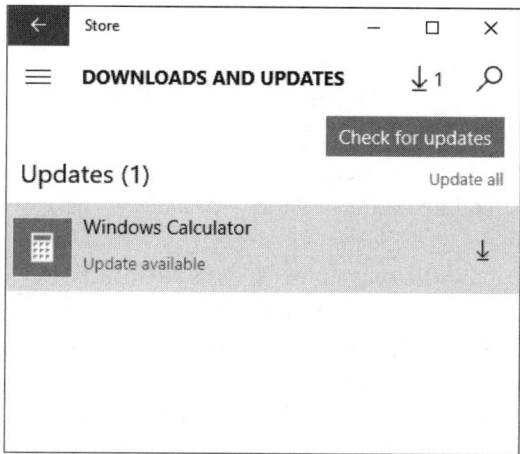

FIGURE 6-45 Available updates for a Store app

After updating the available applications, the update status changes to Completed. Figure 6-46 displays the updates page after the update has completed.

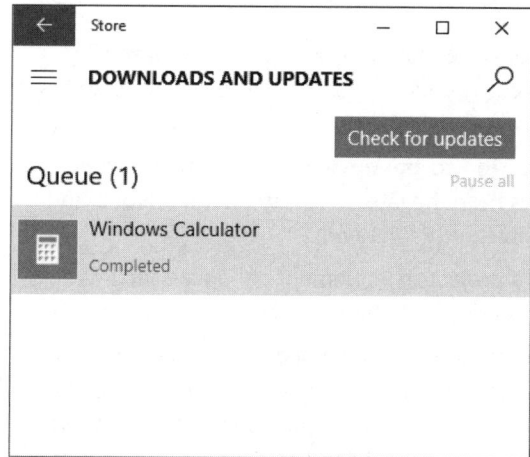

FIGURE 6-46 A completed update

Summary

- Windows Update can be managed by several services depending on the environment.
- WSUS provides enterprise management and deployment of Windows updates.
- Configuration Manager can be used with WSUS to deploy third-party updates.
- Microsoft Intune can be used to manage clients and devices, but not servers.
- Rolling back updates is a simple process to uninstall the update.
- Windows Store app updates are performed from within the store.

Thought experiment

In this thought experiment, demonstrate your skills and knowledge of the topics covered in this chapter. You can find answer to this thought experiment in the next section.

You are a systems administrator for Alpine Ski House, a luxury mountain sports provider of mountain lodging, recreational activities, and special events facilities and services. The current IT environment consists of a single-domain AD DS forest. All servers run Windows Server 2012 R2 and all client computers run Windows 10. All member servers are located in the Servers OU. All client computers are located in the Client Computers OU. The company operates 14 locations worldwide. Each location has a local server room. The main site, located in Jackson, Wyoming, hosts the company's primary datacenter. The company recently moved their disaster recovery site to Microsoft Azure. There are currently 2200 employees and 350 contractors. The following trouble spots have been identified:

- **Personal devices are being used often**   Employees and contractors have been bringing their personal devices to work and have been asking to connect them to the company networks. The result is a rapidly expanding supported device list and many security exceptions because the company does not have many written policies for personal device use.

- **File recovery is inconsistent**   Because of the multitude of devices that are being used, file storage has not been centralized. Therefore, when a user needs to restore a file, it is not always available in a backup.

- **Devices out of compliance are connecting to the network**   Users that have brought their own devices do not always have the latest security updates. Not being up-to-date makes these devices vulnerable on the network.

The company has recently decided to start a project to implement enhanced security company-wide. The following requirements have been identified:

1. All company devices must implement a solution so that company data can be recovered, regardless of the device that it was stored on. Data must also be recovered by individual versions, regardless of the device that the data is being stored on. If a user has more than one device, the data must also be synchronized between the devices.

2. The operating systems of user devices must have the ability to be restored in case of a hardware failure.

3. Devices must be able to meet compliance by having the latest Windows Updates.

And now let's examine some scenarios based upon these requirements:

1. You need to design a solution to protect company data on company client computers. What should you do?

    A. Install OneDrive on each client device. Configure all user data to be stored in the OneDrive libraries.

B. Install OneDrive for Business on each client device. Configure all user data to be stored in the OneDrive for Business libraries.

C. Implement File Recovery on each device.

D. Enable File History on each client device. Configure a flash drive to be used for the file recovery location.

2. You need to design a solution to provide a recovery method in case of hardware failure for the devices.

A. Enable restore points on each device. Create a scheduled task to create a new restore point every night.

B. Create a recovery drive based on the device. Instruct the user to store the drive in a safe place in case of an emergency.

C. Enable File History on the device. Configure File History with a flash drive. Instruct the user to store the drive in a safe place in case of an emergency.

D. Perform a driver rollback. On the affected devices, roll back the latest driver installation.

3. You need to design a solution to ensure that user devices are compliant with the Windows Update policy.

A. Manage all of the devices with a Microsoft Intune subscription. Create a policy to automatically update all of the managed devices.

B. Configure a local GPO to enabled client-side targeting on each of the devices. Point the client to the WSUS server for updates.

C. Configure a software distribution point within Configuration Manager. Manage the updates from the Configuration Manager console.

D. Modify the local Windows Update installation options. Enable automatic updates, but set the devices to defer upgrades.

Thought experiment answer

This section contains the solution to the thought experiment. Each letter answer explains why the answer choice is correct or incorrect.

1. The correct answer is B.

   A. **Incorrect.** OneDrive would provide cloud storage, but would not enable a version history with the files that are saved.

   B. **Correct.** OneDrive for Business provides the cloud storage that is necessary for the scenario. This enables users to restore individual versions of a file that has been synchronized to the cloud service.

C. **Incorrect.** File Recovery is only used for restoring versions of local files that have been identified through a restore point or File History. Because the scenario requires the data to be synchronized between devices, File Recovery won't accomplish the goal.

D. **Incorrect.** File History can be enabled to provide a backup copy of data on a local drive, or a network location. However, the data that is stored cannot be synchronized to other devices. It is still only a local backup option.

2. The correct answer is B.

A. **Incorrect.** Restore Points are useful for machines that still startup, but are not functioning as expecting. This can be because of an application update, or another third-party update.

B. **Correct.** Creating a recovery drive ensures that the operating system can be recovered in the event of a hardware failure. The recovery drive is useful to restore anything that could be lost during a migration or transition.

C. **Incorrect.** File History does not provide any redundancy on the device, and is only useful for creating a version history of files on a flash drive or network location.

D. **Incorrect.** A driver roll back does not work in this scenario because the hardware is not starting up. A driver rollback is useful if you've recently updated a driver, and that specific piece of hardware is not working as expected.

3. The correct answer is A.

A. **Correct.** A Microsoft Intune subscription enables you to create a policy to manage all of the devices that are being brought by the users.

B. **Incorrect.** A local GPO with client-side targeting would work if all of the devices were on the domain, and a WSUS server was available. However, with devices being brought by employees, this is not a feasible solution.

C. **Incorrect.** Configuring a software distribution point through Configuration Manager is useful only if all devices are on the domain.

D. **Incorrect.** Setting the local settings to update automatically does not ensure that the updates are received. We also do not want to defer the upgrades in the environment.

Index

A

B

C

About the authors

 **BRIAN SVIDERGOL**, lead author, builds Microsoft infrastructure and cloud solutions with Windows, Active Directory, Microsoft Azure, System Center, Office 365, and related technologies. He holds a bunch of Microsoft and industry certifications, including the Microsoft Certified Trainer (MCT) and Microsoft Certified Solutions Expert (MCSE) - Server Infrastructure. Brian authored Exam Ref 70-695 Deploying Windows Devices and Enterprise Apps, co-authored Virtualizing Desktops & Apps with Windows Server 2012 R2Inside Out, and co-authored the Active Directory Cookbook, 4th Edition.. He served as an MCT Ambassador at TechEd North America 2013 and at Microsoft Ignite 2015. Brian works as a subject matter expert (SME) on many Microsoft Official Curriculum courses and Microsoft certification exams. He has authored a variety of training content, blog posts, practice test questions, and has been a technical reviewer for over 25 books.

 **BOB CLEMENTS** is a systems administrator, writer, and tech enthusiast, specializing in enterprise device management and server infrastructure. With more than 10 years of experience in the IT industry, he has an in-depth knowledge of System Center Configuration Manager, Microsoft Deployment Toolkit (MDT), Windows Server, Active Directory, and virtualization.

 **CHARLES PLUTA** is a technical consultant and Microsoft Certified Trainer (MCT) that has authored several certification exams, lab guides, and learner guides for various technology vendors. As a technical consultant, Charles has assisted small, medium, and large organizations deploy and maintain their IT infrastructure. He is also a speaker, staff member, or trainer at several large industry conferences every year. Charles has a degree in Computer Networking, and holds over 15 industry certifications. He makes a point to leave the United States to travel to a different country once every year. When not working on training or traveling, he plays pool in Augusta, Georgia.

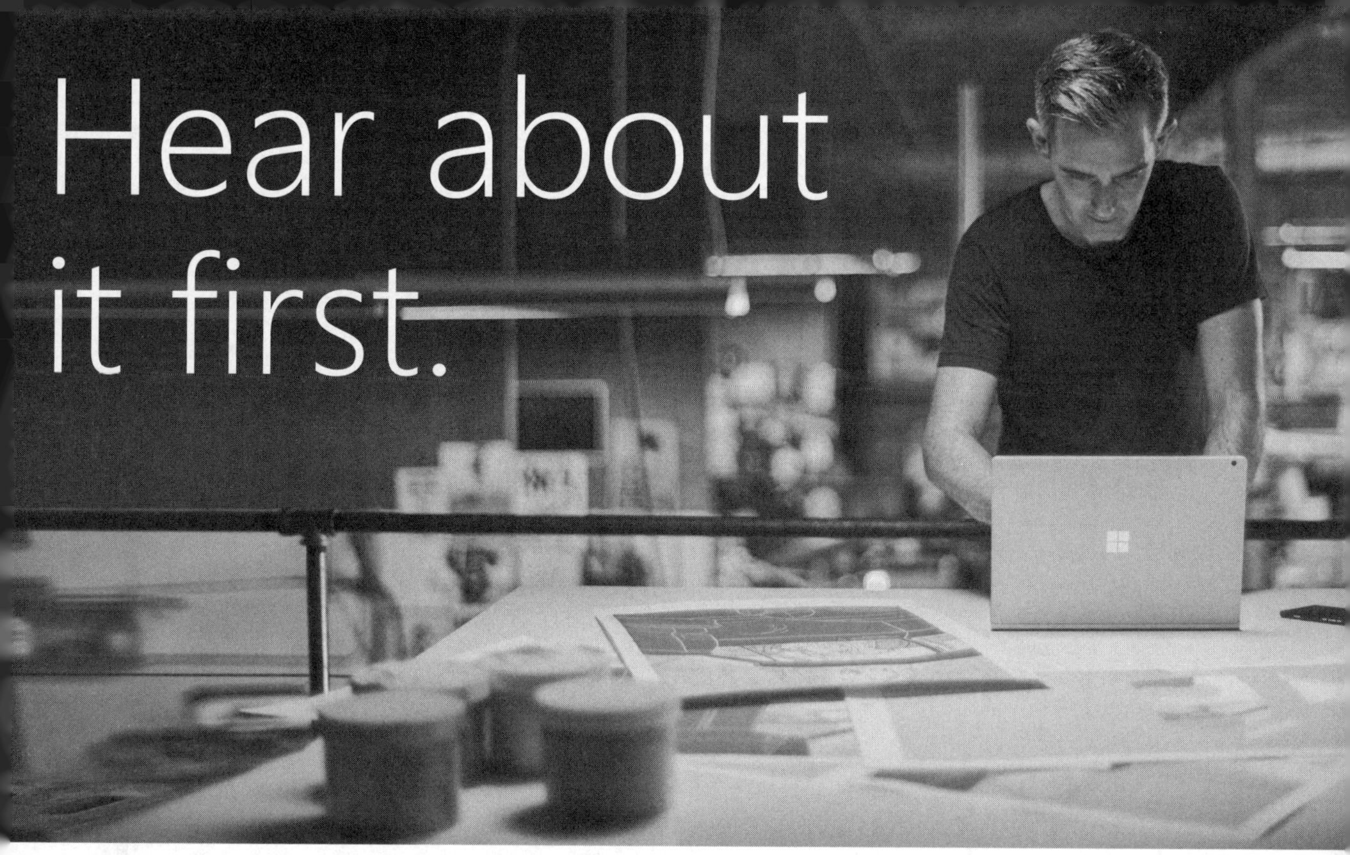

Hear about it first.

Get the latest news from Microsoft Press sent to your inbox.

- New and upcoming books

- Special offers

- Free eBooks

- How-to articles

Sign up today at MicrosoftPressStore.com/Newsletters

 Microsoft

Visit us today at

microsoftpressstore.com

- **Hundreds of titles available** – Books, eBooks, and online resources from industry experts

- **Free U.S. shipping**

- **eBooks in multiple formats** – Read on your computer, tablet, mobile device, or e-reader

- **Print & eBook Best Value Packs**

- **eBook Deal of the Week** – Save up to 60% on featured titles

- **Newsletter and special offers** – Be the first to hear about new releases, specials, and more

- **Register your book** – Get additional benefits

 Microsoft

From technical overviews to drilldowns on special topics, get *free* ebooks from Microsoft Press at:

www.microsoftvirtualacademy.com/ebooks

Download your free ebooks in PDF, EPUB, and/or Mobi for Kindle formats.

Look for other great resources at Microsoft Virtual Academy, where you can learn new skills and help advance your career with free Microsoft training delivered by experts.

Microsoft Press

Now that you've read the book...

Tell us what you think!

Was it useful?
Did it teach you what you wanted to learn?
Was there room for improvement?

Let us know at http://aka.ms/tellpress

Your feedback goes directly to the staff at Microsoft Press,
and we read every one of your responses. Thanks in advance!

 Microsoft